Levees and the National Flood Insurance Program

Improving Policies and Practices

Committee on Levees and the National Flood Insurance Program:

Improving Policies and Practices

Water Science and Technology Board

Division on Earth and Life Studies

NATIONAL RESEARCH COUNCIL
OF THE NATIONAL ACADEMIES

THE NATIONAL ACADEMIES PRESS
Washington, D.C.
www.nap.edu

THE NATIONAL ACADEMIES PRESS **500 Fifth Street, NW** **Washington, DC 20001**

NOTICE: The project that is the subject of this report was approved by the Governing Board of the National Research Council, whose members are drawn from the councils of the National Academy of Sciences, the National Academy of Engineering, and the Institute of Medicine. The members of the panel responsible for the report were chosen for their special competences and with regard for appropriate balance.

Support for this study was provided by the U.S. Federal Emergency Management Agency under grant contract number HSHQDC-11-D-00009 and task order HSFEHQ-11-J-0001. Any opinions, findings, conclusions, or recommendations expressed in this publication are those of the author(s) and do not necessarily reflect the views of the organizations or agencies that provided support for the project.

International Standard Book Number-13: 978-0-309-28290-1
International Standard Book Number-10: 0-309-28290-X
Library of Congress Control Number 2013940694

Cover image courtesy of the U.S. Army Corps of Engineers, Mississippi Valley Division.

Additional copies of this report are available from the National Academies Press, 500 Fifth Street, NW, Keck 360, Washington, DC 20001; (800) 624-6242 or (202) 334-3313 (Washington metropolitan area); http://www.nap.edu.

Printed in the United States of America.

COMMITTEE ON LEVEES AND THE NATIONAL FLOOD INSURANCE PROGRAM: IMPROVING POLICIES AND PRACTICES

GERALD E. GALLOWAY, *Chair*, University of Maryland, College Park
PATRICK L. BROCKETT, University of Texas, Austin
SUSAN L. CUTTER, University of South Carolina, Columbia
DAVID T. FORD, David Ford Consulting Engineers, Sacramento, California
CLIVE Q. GOODWIN, FM Global Insurance Company, Johnston, Rhode Island
KARIN M. JACOBY, Spica Consulting, LLC, Kansas City, Missouri
DAVID I. MAURSTAD, OST Inc., McLean, Virginia
MARTIN W. McCANN, Jack R. Benjamin and Associates, Inc, Menlo Park, California
ANDRE D. McDONALD, Fort Bend Flood Management Association, Sugar Land, Texas
EARTHEA A. NANCE, University of New Orleans, Louisiana
KENNETH W. POTTER, University of Wisconsin, Madison
J. DAVID ROGERS, Missouri University of Science and Technology, Rolla

NRC STAFF

LAURA J. HELSABECK, Senior Program Officer
MICHAEL J. STOEVER, Research Associate

Acknowledgment of Reviewers

This report has been reviewed in draft form by individuals chosen for their diverse perspectives and technical expertise, in accordance with procedures approved by the National Research Council's Report Review Committee. The purpose of this independent review is to provide candid and critical comments that will assist the institution in making its published report as sound as possible and to ensure that the report meets institutional standards for objectivity, evidence, and responsiveness to the study charge. The review comments and draft manuscript remain confidential to protect the integrity of the deliberative process. We wish to thank the following individuals for their review of this report:

GREGORY B. BAECHER, University of Maryland
RUDOLPH BONAPARTE, Geosyntec Consultants
MARK BROWNE, University of Wisconsin
MICHAEL K. BUCKLEY, Dewberry
TIMOTHY A. COHN, U.S. Geological Survey
DARRYL W. DAVIS, U.S. Army Corps of Engineers
JEROME B. GILBERT, Consulting Engineer
HOWARD KUNREUTHER, University of Pennsylvania
LARRY LARSON, Association of State Floodplain Managers
LEWIS E. LINK, University of Maryland
JEFFREY F. MOUNT, University of California
TIMOTHY TINKER, Booz Allen Hamilton

Although the reviewers listed above have provided many constructive comments and suggestions, they were not asked to endorse the conclusions or recommendations nor did they see the final draft of the report before its release. The review of this report was overseen by **Henry J. Vaux**, University of California, and **Michael C. Kavanaugh**, Geosyntec Consultants. Appointed by the National Research Council, they were responsible for making certain that an independent examination of this report was carried out in accordance with institutional procedures and that all review comments were carefully considered. Responsibility for the final content of this report rests entirely with the authoring committee and the institution.

Preface

Over the last decades, the United States has seen significant increases in disastrous flooding and rising losses in both human and property damage along rivers and coastlines from extreme storms and hurricane events. Hurricanes Katrina and Sandy will long be remembered for their human toll and economic costs—two of the most destructive natural disasters in U.S. history. These events not only had extensive personal and economic consequences for those directly and indirectly affected, but also threaten the nation's long-standing efforts to reduce the impacts of such events. Nations around the world face similar challenges.

Many of the damages that have resulted from flooding have involved either the failure of levees and related structures to withstand flood and hurricane stress, the overtopping of these levees once their design height was exceeded, or both. The majority of losses in New Orleans during Hurricane Katrina in 2005 stemmed from levee failures and overtopping. Many levees along the Missouri and Upper Mississippi rivers and their tributaries that were hit by historic floods in 2008 and 2011 experienced overtopping and failure. In 2007, the U.S. Army Corps of Engineers (USACE) reported that 122 levees under its oversight were at risk of failing, and later that year, the Congress passed the National Levee Safety Act. In 2009, a National Committee on Levee Safety, formed by the Act, stressed the threatening nature of this situation and indicated that a significant number of levees across the country are not well maintained and do not meet the standards required by the National Flood Insurance Program (NFIP), USACE levee programs, or accepted engineering practices. Also in 2009, the Infrastructure Report Card of the American Society of Civil Engineers assigned a grade of "D–" to the nation's levees and cited a maintenance backlog of nearly $50 billion.

The National Research Council (NRC) and its Water Science and Technology Board were asked by the Federal Emergency Management Agency (FEMA) to examine the manner in which levees are addressed in the NFIP and to provide advice as to what actions might be taken to improve program efficiency and effectiveness. The NRC established the *Committee on Levees and the National Flood Insurance Program: Improving Policies and Practices* to carry out this task. Committee members spent 16 months reviewing the program's history, examining numerous documents and reports, receiving briefings from relevant stakeholders including FEMA and other government agency officials. It held 12 virtual and in-person meetings in Washington, D.C., and in the Houston Texas, Sacramento, California, and St. Louis, Missouri areas to meet with those who supervise, operate, and maintain or are affected by levees.

For the most of the last three centuries, the nation has relied heavily on using structural means such as levees and dams to deal with flooding. Levees that are properly constructed, maintained, and upgraded when necessary have prevented inundation of major communities in the face of major flood events and have helped avert billions of dollars in damages. The performance of the levees on the Lower Mississippi River during the record 2011 floods provided strong testimony to the value of levees.

Beginning with studies in the 1940s and 1950s, including work from the late geographer Gilbert F. White, it was increasingly appreciated that a comprehensive approach to managing flooding that complemented levees and dams with nonstructural measures such as land-use planning, zoning, floodproofing, and flood insurance was needed. Over the last two decades of the 20th century, the comprehensive approach advocated by White began a transition to the concept of flood risk management that is founded on the understanding that no matter what actions are taken to reduce structural system risks, the residual risk of structural failure will always remain and that management strategies will take advantage of a broad portfolio of mitigation measures. This approach has been increasingly adopted around the globe and forms the basis of the European Union's 2007 Flood Directive. This report focuses on technical and programmatic aspects of the NFIP's treatment of levees and leveed areas that exist within the context of flood risk management and recognizing that NFIP-related levees represent only 5 to 10 percent of the nation's levees.

This report does not address standards for levels of flood protection. The one percent annual chance (100-year) flood originally was used by USACE and the Tennessee Valley Authority as one measure of explaining flood threat. This recurrence interval was later legislated by the Congress to designate areas that would be regulated in the NFIP. Over time, however, this standard has frequently been presumed also to be a flood safety standard for floodplain occupancy. Although an important topic, the committee concluded that it would be outside the charge to recommend a safety standard for levees in heavily urbanized area, but did note that many other expert reports have suggested that the one percent annual chance flood standard is inadequate for large urban areas. For example, a 2009 report from a committee of the NRC recommended that the protection system for the City of New Orleans be designed for a hurricane storm surge event with an expected recurrence interval of 400 to 1,000 years. The committee also discussed the natural and beneficial functions of floodplains and the impacts that activities in the floodplain have on these functions, but determined that this topic also was outside the charge.

In this report, the committee has been asked to not only address an overall subject, "Levees and the NFIP," but also to examine separate, but somewhat interdependent, topic areas. As a result, the report provides both a continuous discussion of the levee issues in general and individual chapters focused on the different topic areas. Although the recommendations and conclusions offered in this report are focused on levees in the NFIP and the land behind them, they may also offer insights for the NFIP as a whole. To illustrate, the majority of losses from Hurricane Sandy in late 2012 were not levee related, as there are few levees or similar structures on the Delaware, New Jersey, and New York coastline. However, Sandy resulted in significant claims under the NFIP, and recovery efforts will involve a thorough examination of possible new risks and how to appropriately balance a range of possible structural and nonstructural alternatives for addressing these flood and storm risks. Many of the conclusions and recommendations offered by this report have been drawn and offered in the past, and the committee has identified these earlier reports (Appendix G). Unfortunately, some of these historical lessons have yet to be fully appreciated and respected.

From the beginning of our effort, Doug Bellomo and Roy Wright, senior officials at FEMA, made themselves available to the committee and for that, the committee is most appreciative. Their openness accelerated the committee's ability to quickly gain an understanding of the key issues that they faced. The committee also extends its appreciation to the numerous individuals who provided highly informative presentations and information regarding their collective experience with levees and the NFIP (Appendix D). The committee extends thanks to FEMA staff for its transparency and generosity of time in answering the hundreds of questions posed by the committee. The work of Siamak Esfandiary, the FEMA liaison to the committee was especially helpful and ensured that the information needed by the committee was made available. David Bascom, Bill Blanton, Kelly Bronowicz, Paul Huang, Andy Neal, Dave Stearrett, and Jeff Woodward of FEMA and Ray Alexander, Bryan Baker, Eric Halpin, and Mike Jordan of USACE gave considerable assistance to the committee. The committee appreciates the challenges that both FEMA and stakeholders face and current and past efforts made to overcome these challenges. The committee offers this report in the hope that it will assist FEMA in executing its most important mission.

Finally, the committee owes much to the untiring efforts of our NRC Study Director, Dr. Laura J. Helsabeck, whose intellectual acumen, organizational skills, writing abilities, collaborative temperament, and exceptionally long hours brought the report together and kept us focused on our mission. She was ably assisted by our research associate, Michael Stoever, who deftly handled research and logistics activities for the committee over its term. Dr. Jeff Jacobs, the Director of the Water Science and Technology Board, provided much appreciated oversight of and support to the committee.

Gerald E. Galloway, *Chair*
Committee on Levees and the
National Flood Insurance Program:
Improving Policies and Practices

Contents

APPENDIXES

Acronyms and Initialisms

AMPO	Association of Metropolitan Planning Organizations
ASFPM	Association of State Floodplain Managers
BFE	base flood elevation
CA DWR	California Department of Water Resources
CEA	California Earthquake Authority
CLD	California Levee Database
CRS	Community Rating System
DELV	damage by elevation
DFIRM	Digital Flood Insurance Rate Map
DRMS	Delta Risk Management Study
FBFM	Flood Boundary and Floodway Map
FEMA	Federal Emergency Management Agency
FHBM	Flood Hazard Boundary Map
FIA	Federal Insurance Administration
FIS	Flood Insurance Study
FIMA	Federal Insurance and Mitigation Administration
FIRM	Flood Insurance Risk Map
FMA	flood mitigation assistance
GFIP	group flood insurance policy
GIS	geographic information system
HMA	hazard mitigation assistance
HMGP	Hazard Mitigation Grant Program
HUD	U.S. Department of Housing and Urban Development

IFMRC	Interagency Floodplain Management Review Committee
LAMP	Levee Analysis and Mapping Procedure
LFC	levee fragility curve
Lidar	light detection and ranging
MLI	Mid-Term Levee Inventory
MPO	metropolitan planning organization
MPR	mandatory purchase requirement
NAFSMA	National Association of Flood & Stormwater Management Agencies
NCLS	National Committee on Levee Safety
NFIP	National Flood Insurance Program
NLD	National Levee Database
NRC	National Research Council
PAL	Provisionally Accredited Levee
PDM	Pre-Disaster Mitigation
PE	professional engineer
PELV	probability of elevation
PRA	probabilistic risk assessment
PRP	preferred risk policy
RFC	repetitive flood claim
SFHA	Special Flood Hazard Area
SPF	standard project flood
SPFC	State Plan of Flood Control
SRL	severe repetitive loss
TVA	Tennessee Valley Authority
USACE	U.S. Army Corps of Engineers

Summary

The Federal Emergency Management Agency's (FEMA) Federal Insurance and Mitigation Administration manages the National Flood Insurance Program (NFIP), which is a cornerstone in the U.S. strategy to assist communities to prepare for, mitigate against, and recover from flood disasters. The NFIP was established by the Congress with passage of the National Flood Insurance Act in 1968, to help reduce future flood damages through NFIP community floodplain regulation that would control development in flood hazard areas, provide insurance for a premium to property owners, and reduce federal expenditures for disaster assistance. The flood insurance is available only to owners of insurable property located in communities that participate in the NFIP.[1] Currently, the program has 5,555,915 policies[2] in 21,881 communities[3] across the United States.

The NFIP defines the one percent annual chance flood (100-year or base flood) floodplain as a Special Flood Hazard Area (SFHA).[4] The SFHA is delineated on FEMA's Flood Insurance Rate Maps (FIRMs) using topographic, meteorologic, hydrologic, and hydraulic information. Property owners with a federally backed mortgage within the SFHAs are required to purchase and retain flood insurance, called the mandatory purchase requirement (MPR).

Levees and floodwalls, hereafter referred to as levees, have been part of flood management in the United States since the late 1700s because they are relatively easy to build and a reasonable infrastructure investment. A levee is a "man-made structure, usually an earthen embankment, designed and constructed in accordance with sound engineering practices to contain, control, or divert the flow of water so as to provide protection from temporary flooding."[5] A levee system is a "flood protection system which consists of a levee, or levees, and associated structures, such as closure and drainage devices, which are constructed and operated in accordance with sound engineering practices."[6] Under NFIP regulations, homes and commercial buildings located in the SFHA within a participating community may be exempted from the MPR and land-use regulations when located behind a levee system that has been recognized by FEMA as providing protection against the one percent annual chance flood

[1] The NFIP primarily insures residential properties (single- and multifamily homes) but does insure a smaller number of commercial properties.

[2] As of December 2012.

[3] As of June 2012.

[4] One percent annual chance flood, base flood, and the 100-year flood are terms commonly used to describe a hydrologic event that has a 100-year (average) recurrence interval. That is, a flood that has in any year a 1 in 100 chance of being equaled or exceeded.

[5] See Code of Federal Regulations, Title 44, Section 59.1 (44 CFR §59.1).

[6] See 44 CFR §59.1.

event, that is, "accredited." Certification is the technical evidence provided by a levee owner to FEMA demonstrating that the levee system meets the requirements to reduce risk from at least the one percent annual chance flood.[7]

Recognizing the need for improving the NFIP's treatment of levees, FEMA officials approached the National Research Council's (NRC) Water Science and Technology Board and requested this study. The NRC responded by forming the ad hoc Committee on Levee and the National Flood Insurance Program: Improving Policies and Practices, charged to examine current FEMA treatment of levees within the NFIP and provide advice on how those levee-related policies and activities could be improved. The study addressed four broad areas—risk analysis, flood insurance, risk reduction, and risk communication—regarding how levees are considered in the NFIP. Specific issues within these areas include current risk analysis and mapping procedures behind accredited and nonaccredited levees, flood insurance pricing and the MPR, mitigation options to reduce risk for communities with levees, flood risk communication efforts, and the concept of shared responsibility. For the full statement of task, see Chapter 1, Box 1-4. The report's principal conclusions and recommendations are highlighted in bold in this Summary.

TREATMENT OF LEVEES WITHIN THE NATIONAL FLOOD INSURANCE PROGRAM

After the establishment of the NFIP, efforts to identify the nation's flood hazard began. Because many flood protection works, mostly levees, had, in the past, successfully passed what had been considered a one percent annual chance flood, both local communities and contractors raised the issue of excluding areas protected by a one percent annual chance or greater structure from the SFHA. As a result, areas behind levees were not labeled on early flood maps as being within the SFHA where levees had provided or were thought to have provided protection from the one percent annual chance flood or had been constructed by U.S. Army Corps of Engineers (USACE) to withstand the one percent or higher annual chance flood.

As the program evolved, levees were required to be certified and accredited in order for the areas behind them to be mapped out of the SFHA. When a levee in a community provides less than 1 percent protection, it is not shown on a FIRM as providing protection and the flood risk is assessed as if the levee does not exist. Not recognizing levees that are not accredited on FIRMs is referred to in this report as a "without levee" analysis. An all-or-nothing approach, this either gives communities credit for a one percent annual chance levee (through exclusion from the SFHA and exemption from the MPR, etc.) or does not; that is, a levee that is not accredited is not recognized on the FIRM.

In 2003, with the support of communities and organizations to improve the quality of flood mapping, FEMA moved forward with a program to modernize maps by converting to a digital format and, when possible, providing a new engineering analysis. During "Map Mod," FEMA officials became concerned that the levees encountered in this process might not meet the one percent annual chance standard despite being labeled as such. Affected communities, faced with the potential movement of the previously exempt areas into the floodplain and the mandatory requirement to purchase flood insurance at rates that might be the same as areas without protection, argued that some credit should be given to the presence of the deficient levees (i.e., some protection less than the one percent annual chance flood). Members of the Congress urged FEMA to discontinue use of the without-levee analysis in updating FIRMs. In agreement, FEMA is currently developing a new approach, the Levee Analysis and Mapping Procedure (LAMP), to replace the without-levee approach.

MOVING TO FLOOD RISK MANAGEMENT AND A MODERN RISK ANALYSIS

Flood risk management represents efforts to continuously carry out analyses, assessments, and related mitigation implementation activities to reduce flood risk. It focuses on assessment of flood hazards, the consequences resulting from flooding, the significance of risks identified, and the development of risk management strategies to deal with flood risk, which is gaining momentum across the global floodplain management community. Effective flood risk management requires use of the best available science in its execution. A flood risk analysis is that

[7] Certification may be accomplished by a registered professional engineer or a federal agency with levee responsibilities.

analytical part of flood risk management that provides information about or quantifies consequences and probabilities of a flood event.

The current NFIP flood hazard analysis is a partial risk-based analysis used with respect to performance of levee systems, where many parts of the analysis are deterministic in fashion. A levee system that is not accredited is not considered in the analysis used to quantify flood risk, even though it provides some (potentially considerable) protection against floods. This encourages communities to construct levee systems that protect only to the one percent annual exceedance flood, enabling new development in areas with significant, but unquantified exposure to catastrophic flood risk. Thus, protection against the one percent annual chance flood event is the de facto design standard for most levees seeking accreditation in the United States, with limited attention given to the consequences ("residual risk") should a levee fail or be overtopped. Furthermore, levee systems that only marginally meet certification standards are vulnerable to loss of accreditation status. If not properly maintained, the performance of levee systems degrades over time due to erosion, rodent damage, subsidence, and other factors. Further, the frequency and magnitude of flood hazards can increase over time due to natural and anthropogenic causes. Loss of accreditation status can be very disruptive to the affected communities in terms of safety and insurance cost.

Thus, the current NFIP approach to flood hazard analyses leads to an incomplete description of the flood hazard in many areas. This is not unknown to FEMA or relevant stakeholders, including policy makers. A more modern approach to flood risk analysis addresses problematic aspects of the without-levee analysis.[8] The elements of a modern risk-based approach include:

- flood hazard analysis or an estimate of the frequency and magnitude of flooding;
- levee and other system component fragility analysis, which estimates structure performance as a function of flood levels;
- systems analysis, which includes all features that impact flooding along a levee system;
- levee breach and inundation assessment; and
- consequence analysis that estimates the damage to structures that are inundated in a flood event.

All elements of this analysis are subject to uncertainty; the evaluation of these uncertainties is also integral to a modern risk analysis.

The modern risk-based analysis offers several advantages. It would provide an in-depth technical evaluation of flood hazards and directly account for the performance of levee flood protection systems whether accredited within the NFIP or located within an NFIP community. It would account for all features of the flood protection system (levees, gates, other structures) that significantly affect flood risk, as long as they meet minimum design, operation, and maintenance standards. Hence credit could be given for establishing flood insurance rates for flood protection systems that do not protect to the one percent annual chance standard, but at the same time provide some level of protection. In other words, the modern risk-based approach acknowledges that risk across a floodplain varies across the landscape. It has the impact of softening, but not eliminating, the one percent annual chance decision line.

Also, a risk-based approach would estimate flood risk for lands protected by levee systems that do protect to the one percent annual chance standard or higher. In these circumstances, the risk analysis results would inform communities of the flood protection system limitations and potential vulnerabilities, the actual flood risk being faced, and inform flood insurance ratings. Risk analysis results would also differentiate between protection provided by one percent annual chance levees and 0.2 percent levees. It would be able to recognize difference between low areas behind levees and higher ground. It would also provide guidance to the communities for development of new regulations that would encourage floodplain management. **The NFIP should move to a modern risk analysis that makes use of modern methods and computational mapping capacity to produce state-of-the-art risk estimates for all areas that are vulnerable to flooding.**

[8] In this report the term "modern" is used to indicate characterization and analysis of risk using the best science available to the risk management field.

Levee Analysis and Mapping Procedure

The LAMP approach is based on sound algorithms and is a reasonable first cut at dealing with the potential contributions of levees that do not meet the standards for protection against the one percent annual chance flood. LAMP represents a response to the issues raised by members of Congress that, given funding and personnel constraints, will likely take several years to fully implement. However, it represents a diversion from the need to move the NFIP to a modern risk-based analysis. Because it would be highly unlikely for the NFIP to move forward concurrently with implementation of LAMP and development of a modern risk-based approach, a move to LAMP would simply postpone the inevitable need to move the NFIP to the risk-based foundation that other agencies and nations have already endorsed. **FEMA should move directly to a modern risk-based analysis for dealing with areas behind levees and not implement the Levee Analysis and Mapping Procedure (LAMP).** Should a decision be made not to implement LAMP, dealing with the current situation is still necessary; interim steps are proposed in Chapter 4 of the report.

INSURING PROPERTIES BEHIND LEVEES

Flood insurance is an essential part of the NFIP used to address catastrophic loss potential from floods. It provides a method for prefunding (usually over time) the adverse consequences of potential flood losses. FEMA evaluates the flood risk of policyholders and determines rates for the "full-risk" class through a balance of factors including the extent and type of flood hazard, the base elevation of the insured structure and the structure type,[9] the contents location (first floor, second flood, etc.), and whether or not the community participates in the Community Rating System (CRS)[10] that provides discounts for communities that actively manage their flood risk.

The notion of actuarial soundness in the context of the NFIP is somewhat subtle and elusive because of the differences between a government-run program with mandated discounts, no profit requirement, no significant interest income, and reliable, significant borrowing power and the private sector from which the notions of actuarial soundness arose and are routinely applied. The major structural drawback to the fiscal soundness of the NFIP is *not* actuarial unsoundness due to inappropriate or inadequate rate-setting methodology or inadequacies, but rather the fiscal unsoundness caused by not being able to apply actuarial principles uniformly across all classes. The NFIP mandated "discounted risk" class was anticipated to eventually decrease substantially over time, and this has indeed been the case.[11] Still, there are a substantial number of discounted policies in the NFIP. The difference between the full-risk premium for a discounted policy and the discounted premium for the policy is revenue lost to the NFIP (this lost revenue is not compensated for in any other part of the premium structure). FEMA is legally unable to raise premiums in a manner sufficient to allow the NFIP to be financially sound or to build a contingency reserve fund sufficient to pay for a catastrophic future loss. Without this ability, the NFIP is destined by law not by actuarial methodology inadequacy to run short of money in the long run, with occasional dramatic shortfalls. Furthermore, repetitive loss properties continue to drain the fiscal soundness of the NFIP. **The NFIP is constructed using an actuarially sound formulaic approach for the full-risk classes of policies, but is financially unsound in the aggregate because of constraints (i.e., legislative mandates) that go beyond actuarial considerations.** The Biggert-Waters Flood Insurance Reform Act directs adjustment of fiscal practices to move the NFIP to a more fiscally sound approach.

Risk-based insurance pricing is key to efficient and equitable financial transfer of the risk. With a modern

---

[9] Condominium, single- or multifamily home, residential or commercial use, number of floors, with or without a basement, ventilated crawlspace, etc.

[10] FEMA's Community Rating System (CRS) is a voluntary incentive program that encourages floodplain management activities exceeding the minimum NFIP requirements.

[11] The discounted risk class applies when (1) a structure was built before the flood insurance maps were available, that is, a "pre-FIRM" structure or built before December 31, 1974; (2) a structure was built in a V zone before 1981 and before maps that consider wave height were adopted in setting flood insurance rates; (3) a structure is in an AR or A99 zone for levees in the course of reconstruction or construction (and so their current actuarial risk does not correspond to their current risk charge); or (4) the policyholder participates in a group flood insurance policy. Currently, approximately 20 percent of NFIP policies are discounted policies.

risk-based analysis, a gradation of probabilities that can differ from property to property even within the same geographic area or flood zone can be applied. **Rate setting for properties behind levees, accredited or not accredited, should be improved by using a modern risk analysis method employing advances in hydrology, meteorology, and geotechnical engineering to more precisely calculate the probability of water inundation levels and the associated damage estimates throughout the area behind the levee in a graduated fashion.** This will replace the current approach and more appropriately reflect the risk.

The Mandatory Purchase Requirement

When the NFIP was established, property owners could purchase insurance voluntarily. In 1973, given the limited voluntary purchase that had taken place to that date, the mandatory purchase requirement (MPR) was implemented to ensure that the exposure of the federal government through mortgages on properties within the SFHA was covered by insurance, thus decreasing the financial burden on the federal government for uninsured losses. Although the MPR applies to properties within the SFHA that hold a federally backed mortgage, apart from those behind accredited levees, the NFIP encourages all those living in areas subject to flooding to purchase insurance.

The effectiveness of the MPR can be explored by examining the percentage of those living within the SFHA who have insurance (i.e., market penetration) and the percentage of those living within the SFHA with a federally backed mortgage who have insurance (i.e., MPR compliance). Review of studies indicate that national market penetration in the SFHA is between 30 and 50 percent and national MPR compliance is between 50 and 78 percent. The more recent and more detailed information indicates that penetration and compliance are likely closer to the lower values in both ranges.

The development of market penetration and compliance estimates are confounded by several factors including varied structure type across assessment (single-family homes versus all residential structures), whether or not the structure owner carries a mortgage, and whether or not the structure is insured with a federal lender. Regardless, national market penetration and compliance studies call into question the efficacy of the MPR. **The current policy of mandatory flood insurance purchase appears to be ineffective in achieving widespread purchase of NFIP flood insurance policies.** To date, relying on federal supervisory agencies to oversee and lenders to require purchase has not achieved widespread compliance with the MPR. Extending the MPR to areas behind levees where there are large numbers of structures, incomplete determination of flood risk, and varied evidence supporting its effectiveness is imprudent. **At this time, there is no sound reason to institute mandatory purchase of flood insurance in areas behind accredited levees.**

A perception exists that if the NFIP simply sold more policies, the program's fiscal issues (e.g., debt to the U.S. Treasury) would be solved. However, simply adding new policies does not automatically improve the program's fiscal soundness. Adding new polices could lower the program's overall risk *if* the policies increased the program's diversification of risk acquired through those policies. However, adding new policies very well may increase the NFIP's overall risk through concentration of policies in, for example, high-risk areas such as the SFHA. Furthermore, adding new policies would also increase the program's exposure to risk accumulation, or the additive risk of multiple events occurring at the same time. If NFIP policies continue to be concentrated in the same high-risk areas around the country, the diversification of risk and the fiscal soundness of the program will remain relatively unchanged.

Upon implementation of a modern risk analysis, insurance rates will more accurately reflect flood risk behind a levee. If the insurance rate goes up, it indicates that the property is at more peril than previously understood to be; if the insurance rate goes down, it indicates the property is at less peril than previously understood to be. Thus, FEMA's moving to a modern risk-based analysis and pricing premiums accordingly will have an impact on NFIP policy purchase. (The Biggert-Waters Flood Insurance Reform Act directs full, risk-based premiums and elimination of certain discounts, also promoting a change to the NFIP that affects policy purchase.) More accurate pricing and more information created by a modern risk analysis has the potential to prompt additional policy purchase. Property owners would be more favorably inclined to buy flood insurance if individual risk is well known and understood and insurance rates are priced to match the probability of flooding and financial impact of flooding events. This includes areas behind levees, both accredited and not accredited, in the SFHA at large, and also non-

SFHA areas. It is likely that if a lender has more information through a risk-based analysis, the lender is more aware that requiring flood insurance purchase is in line with its fiduciary responsibility. However, this premise is challenged by nuances in risk communication principles, specifically, that awareness does not necessarily lead to actions to mitigate risk.

As the use of risk-based pricing becomes widespread, the thinking behind the necessity of the MPR in the SFHA may change, along with the decision of waiving the MPR behind an accredited levee. **A modern risk-based analysis has the potential to impact the purchase of flood insurance, diversify the NFIP's exposure to flood risk, and generate a fiscally sounder program. Once the risk-based approach has been put in place and matures, FEMA should review and study the necessity of the mandatory purchase requirement behind levees and throughout the SFHA.**

IMPLEMENTING FLOOD RISK MANAGEMENT STRATEGIES

There is some level of risk to all locations within the floodplain. The magnitude of that risk is a function of the flood hazard, the characteristics of a particular location (e.g., elevation, proximity to the river or coast, and susceptibility to fast-moving flows and surges), measures that have been taken to mitigate the potential impact of flooding, the vulnerability of people and property, and the consequences that result from a particular flood event. The initial risk to an area is represented by the area's characteristics without consideration of mitigation and risk transfer measures and the vulnerability of the population. Mitigation and risk transfer measures reduce the overall risk to some degree, but it is impossible to completely eliminate risk, resulting in residual risk.

A flood risk management strategy identifies and implements measures that reduce the overall risk. Available measures include structural (levees, dams, pumps, etc.), nonstructural (elevation, relocation, etc.), and risk transfer methods, that is, insurance. In developing a flood risk management strategy, those responsible judge the costs and benefits of each measure taken and its overall impact in reducing the risk. Levees represent one method of reducing the impacts of flooding on a community or a region.

There are varied circumstances that dictate flood risk behind levees. These circumstances can include anything from proximity to the coast, local climate, population growth, existing infrastructure, to community preparedness. Some risk scenarios are associated with long time horizons, such as flooding associated with climate change, and might be less perceptible to the at-risk community compared with, for example, a design deficiency in a levee discovered during the Map Mod process. The importance of climate change and population growth is further highlighted when considering the future impact on the size of the nation's floodplain, both geographically and in terms of population, and the corresponding number of insurance policies needed in the floodplain.[12] Complicated scenarios arise as communities grapple with unique situations. A mix of flood management mitigation and risk transfer measures tailored to the risk of a particular community is the most effective flood risk reduction strategy. **There is a clear need for a comprehensive, tailored approach to flood risk management behind levees that (1) is designed and implemented at the local level; (2) involves federal and state agencies, communities, and households; (3) takes into account possible future conditions; and (4) relies on an effective portfolio of structural measures, nonstructural measures, and insurance to reduce the risk to those behind levees.**

To reduce the flood risk to those behind levees, FEMA should encourage communities to develop and implement multimeasure risk management strategies for areas behind accredited levees. This strategy would contemplate the full range of foreseeable flood scenarios involving levee overtopping or failure and future conditions. The development of new risk management analysis tools will facilitate the identification of risks and suggest feasible approaches to dealing with the risks to those behind levees.

Development and maintenance of supporting policies affecting the portfolio of mitigation measures are both critical and effective floodplain management strategies. In the case of levees, the principal regulatory document that establishes the procedures for inclusion of levees in the program is Title 44 of the *Code of Federal Regulations*, which contains several parts pertaining to levees in the NFIP. There is a need for revisions to this document that address the handling of embankments (i.e., railroads or existing roads) in a levee system, provide a unique

[12] Hydrologic nonstationarity associated with environmental change makes these uncertainties even more complex.

designation for areas behind accredited levees on FIRMs, require actions to verify to FEMA that an accredited levee maintains integrity, and expand requirements for maintenance to ensure the carrying capacity of the river channel.

A modern risk analysis and effective flood management strategy options require knowledge of the location, condition, and ownership of levees. Currently, the majority of the knowledge of the extent of levees in the United States is at the federal level. The USACE's National Levee Database (NLD) is the biggest single resource for levee data in the United States and currently includes more than 2,500 levees and associated structures (also known as levee systems) totaling more than 14,500 miles in length, catalogued in digital format. The FEMA Mid-Term Levee Inventory (MLI) database, designed to complement the NLD, contains data on 5,100 miles of NFIP-accredited levees and an additional 24,700 miles of other levees identified by FEMA.

There is currently no method to determine whether a levee in the NLD is also in the FEMA MLI except through a one-to-one comparison of certain levee characteristics. Because some data elements in the two inventories are different, USACE and FEMA have been attempting to manually look for overlaps, but have yet to complete the effort. A preliminary analysis by both agencies indicates that approximately 3,400 miles of levees, or 66 percent, may be operating under oversight of both agencies. There is a reasonable understanding of the number of federally owned and operated levees in the country, although opportunities for improvement exist. Efforts to inventory levees at the state level vary widely, with the majority of the states not having levee inventories. The definition of what constitutes a levee also varies widely in inventory efforts, ranging from structures built under federal and state supervision to other structures, such as embankments, highways, railroads, irrigation water and supply canals, that work to reduce flooding. A comprehensive inventory of the nation's levees, along with a consistent definition of a levee, is needed to ensure public safety and develop appropriate floodplain management strategies. The states have an important role in inventorying the number of levees within their respective borders. This information can then be combined with a federal inventory to create an accurate assessment and inventory of the nation's levees.

RISK COMMUNICATION

Development of appropriate strategies for risk communication is part of the overarching formulation of flood risk management strategies. Challenges associated with the public perception of flood risk need to be considered when developing risk communication strategies, including:

- The meaning of the 100-year floodplain and, to a lesser degree, the one percent annual chance flood is often misunderstood.
- Awareness of risk does not always lead to actions to reduce or mitigate risk.
- People assume that flood control structures are "safe," with no need to take additional actions to mitigate risk.
- Experience with floods matters and influences the actions that residents and communities take to mitigate the impacts.
- People will seek advice from friends, family, and neighbors on the determination of risk and actions to reduce flood risk.
- Personalizing risk on the basis of the experience and feelings is more likely to motivate behavioral action than simply providing technical information about risks and consequences.

Communicating the concept of residual risk behind a levee presents an additional challenge. Furthermore, the variety of levee-related stakeholders, including several federal agencies with different roles and responsibilities who are sources and recipients of risk information, adds additional complexity. So, rather than one entity confronted with the challenge of communicating risk to the public, levee-related flood risk communication efforts face the challenge of presenting unified, effective communication among all the governmental agencies, as well as between agencies and the public.

Historically, there has been little information about hazards associated with levees available to local communities, other than Flood Hazard Boundary Maps and FIRMs that reflect only the flood hazard area (the area flooded by the one percent annual chance event). However, this is changing with continued efforts to communicate flood risk behind levees, largely with nonregulatory products.[13] To be effective, risk communication efforts need to be anchored in several principles:

- delivered at the local level;
- tailored to individual households, communities, and other stakeholders;
- delivered from a credible and trusted source;
- long term;
- have consistent, correct, and nonconflicting content;
- encourage or motivate some behavior;
- account for the values of target audiences or communities;
- employ multimodal networks; and
- provide repeat messaging.

FEMA and others involved in risk communication at all levels should incorporate contemporary risk communication principles in the development of flood risk communication strategies and implementation efforts.

It appears that risk communication efforts surrounding NFIP levees have an advantageous window of attention during three situations: (1) after a flood disaster, (2) through the mapping programs, and (3) when a community participates in FEMA's CRS program. It also appears that during these three situations, in particular situations 1 and 2, considerable confusion and controversy or considerable success in communicating risk can occur. **FEMA should communicate flood risk through a collaborative approach that works with and provides strong support to local communities.**

SHARED RESPONSIBILITY IN MOVING TO FLOOD RISK MANAGEMENT

Several federal agencies, state and local governments, and other stakeholders play important roles in managing the nation's flood risk. Federal roles are dictated by their legislated authorities and presidential guidance. As such, coordination challenges exist. Collaborative and coordinated efforts that share responsibility for levee-related flood risk across the federal, state, and local levels are not as robust as they could be. Increased coordination and cooperation among and between governmental agencies, state and local entities, and the public at large would result in increased effectiveness of and improved efficiency within the NFIP, and would improve the nation's approach to floodplain management and treatment of levees. Promising developments do exist, but more progress is necessary. A coordinated, national approach that shares responsibility across federal, state, and local entities is necessary to appropriately manage the nation's flood risk behind levees.

FEMA and USACE are challenged by different approaches to risk analysis. The technical and programmatic differences between FEMA and USACE as they relate to the evaluation of flood hazards and levee systems have been and continue to be an issue that compromises interagency communication and causes confusion and frustration in the local communities that work with the agencies and in the public that is being served. **FEMA and USACE should jointly develop a common, risk-based approach to levee assessment in a timely manner and apply this approach to all levees assessed by the two agencies. This includes a joint methodology, procedure, and where feasible, the sharing of models and other risk analysis tools.**

The utility of levee-related flood risk communication developed and communicated at the federal level is challenged by uncoordinated messaging. **One federal message using consistent terminology, transparent data, and open discussion and decisions about the determination of flood risk is critical to inform the affected communities who, in turn, communicate and manage risk at the local level. FEMA should assume a leadership role in providing direction for research, development, and release of flood risk communication products and maps.**

[13] The committee interprets regulatory flood risk communication products as FIRMs and nonregulatory products as all other FEMA products that communicate leveerelated flood risk.

MOVING FORWARD

This report provides advice on the current treatment of levees within the NFIP. Key to this advice is the transition of FEMA to a flood risk management approach and the use of a modern, risk-based analysis of flood risk. This transition will, no doubt, be challenging. New or revised regulations, procedures, and training systems will need to be established, and clear, consistent messaging to those affected by the new approaches is critical. A modern risk analysis will bring about changes in insurance ratings and modify current perceptions of risk. Communicating flood risk in general and flood risk behind levees will require close collaboration among all levels of government and within governmental agencies. There are currently legal challenges and more will most likely arise. The appropriate balance between public and private investment will continually be considered. Despite these challenges, the advice in this report is provided to move FEMA's NFIP toward floodplain management anchored in a modern risk-based analysis, a move that is increasingly recognized as imperative in the floodplain management community.

1

Introduction

The Federal Emergency Management Agency's (FEMA) Federal Insurance and Mitigation Administration[1] manages the National Flood Insurance Program (NFIP), which is a cornerstone in the nation's strategy to assist communities in preparing for and recovering from flood disasters. The NFIP was established by Congress with passage of the National Flood Insurance Act in 1968, because despite the nation's efforts to address flood hazards through construction of civil engineering works such as dams, levees, and floodwalls and seawalls, flood losses were rising and unwise development within floodplains was continuing. The Act was passed to reduce future flood damages through implementation of community floodplain ordinances that would control development in flood hazard areas, provide flood insurance for a premium to property owners, and, as a result, reduce federal expenditures for disaster assistance. The flood insurance is available only to owners of insurable property located in communities that participate in the NFIP.

Today, most of the nation's communities with significant flood hazards have joined the NFIP. Currently, the program has approximately 5,555,915 policies[2] in 21,881 communities[3] across the United States (Siamak Esfandiary, FEMA, personal communication, February 2013). Although community participation is voluntary, communities must agree to a floodplain management ordinance that requires that new buildings be built at or above the level of the one percent annual chance flood.[4] In turn, FEMA will issue Flood Insurance Rate Maps (FIRMs) and make flood insurance available throughout the community (FEMA, n.d.).

A key feature of the NFIP is that structures in areas within a community that are determined to be located within the one percent annual chance flood (100-year or base flood) floodplain are referred to as being within the Special Flood Hazard Area (SFHA) (Box 1-1). These areas are subject to mandatory flood insurance purchase requirements.

The SFHA is delineated on FEMA's FIRMs using topographic, meteorologic, hydrologic, and hydraulic information. It is difficult to estimate the degree of exposure to flood risk in the United States. FEMA currently estimates that 6.5 percent of U.S. housing units, or 5.8 percent of the population, are located in SFHAs, which equates to about 7.3 percent of the nation's area (Table 1-1).

Levees and floodwalls, hereafter referred to as levees, have been part of flood management in the United States since the late 1700s because they are relatively easy to build and a reasonable infrastructure investment. A levee

---

[1] See http://www.fema.gov/what-mitigation/federal-insurance-and-mitigation-administration-fima#0.
[2] As of December 2012.
[3] As of June 2012.
[4] The requirements related to this ordinance are found in Title 44 of the *Code of Federal Regulations*, including Parts 59 and 60.

BOX 1-1
One Percent Annual Chance Flood

One percent annual chance flood, base flood, and the 100-year flood, are terms commonly used to describe a hydrologic event that has a 100-year (average) recurrence interval, that is, a flood that has in any year a 1 in 100 chance of being equaled or exceeded. The National Flood Insurance Program (NFIP) refers to properties in the one percent annual chance floodplain as those in the Special Flood Hazard Area. An illustration showing the impact of a one percent annual chance flood on a natural floodplain with various levee heights is shown in Figure 1-1. Over the lifetime of a 30-year mortgage, this equates to at least a 26 percent chance that the property will be flooded if the property is in the one percent annual chance floodplain. Probability theory is used to derive the value 26 percent, where each of the 30 years is accounted for as having a one percent annual chance flood (USGS, 2010.)

Similarly, hydrologic events of any size are described using the same nomenclature, for example, the"50-year" or "500-year" describing events that have a 1 in 50 chance or 1 in 500 chance of occurring (2 percent annual chance flood and 0.2 percent annual chance flood, respectively). Levees are also classified in the same way; that is, a 100-year levee is of adequate height to withstand the one percent annual chance flood.

There is a degree of interchangeability in general use of these terms. However, the use of certain recurrence interval terms, such as the "100-year flood," can cause confusion because it is interpreted as a flood that occurs once every 100 years (USGS, 2010).

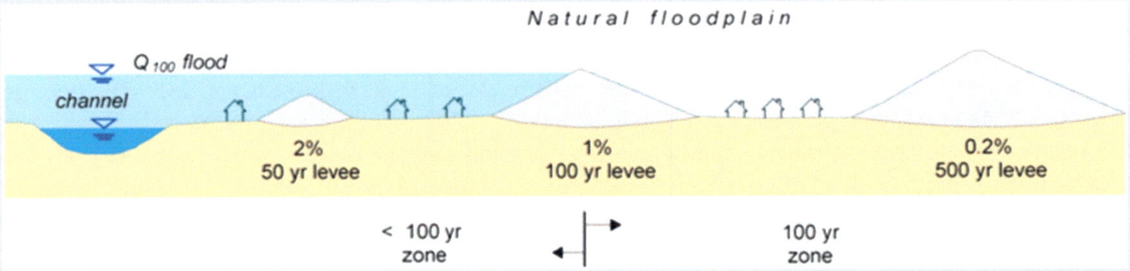

FIGURE 1-1 Illustration of the one percent annual chance flood (Q100) and the impact on a natural floodplain with various levee heights.

is a "man-made structure, usually an earthen embankment, designed, and constructed in accordance with sound engineering practices to contain, control, or divert the flow of water so as to provide protection from temporary flooding."[5] Levees are generally designed to control water up to a given water elevation and, unlike dams, levees do not typically have spillways to reduce structural damages when water levels exceed the design criteria and overtop the structure (Box 1-2). A levee system is a "flood protection system which consists of a levee, or levees, and associated structures, such as closure and drainage devices, which are constructed and operated in accordance with sound engineering practices."[6] Under NFIP regulations, homes and commercial buildings located in the SFHA

[5] See Code of Federal Regulations, Title 44, Section 59.1 (44 CFR §59.1).

[6] See 44 CFR §59.1. An expanded definition of a levee system, adapted from the National Committee on Dam and Levee Safety, is as follows: A levee system comprises one or more components that collectively provide flood risk reduction to a defined area. Breach or malfunction of one component within a system constitutes breach or malfunction of the entire system. The levee system is inclusive of all components that are interconnected and necessary to ensure exclusion of floodwaters from the associated leveed area. The leveed area may also be considered as the associated separable floodplain or separable consequence area. Structures and features include levee and floodwall sections, closure structures, pumping stations, culverts, interior drainage works, and other structures and features, such as highway and railroad embankments that function as components of the system whether or not intentionally constructed as part of the system.

TABLE 1-1 National Special Flood Hazard Area (SFHA) Population, Housing Units, and Square Miles Distribution

	Total in SFHA				
	Coastal	Riverine	Total	Total Outside SFHA	Total National
Population within SFHA (million)	6.9	11.3	18.2	294.2	
National percentage population in SFHA	2.2	3.6	5.8	94.2	312.4
Housing units within SFHA (million)	3.8	4.9	8.6	124.7	
National percentage housing units within SFHA	2.8	3.7	6.5	93.5	133.3
Area within SFHA (square miles)	42,677	222,621	265,299	3,375,949	
National percentage area in SFHA	1.2	6.1	7.3	92.7	3,641,247

SOURCE: RAMPP (2012).

BOX 1-2
Use of the Word "Protection"

In the literature, the word "protection" is often used to describe a levee that "protects" against a one percent annual chance flood. For example, from 44 CFR §65.10 when describing levee design criteria, "For levees to be recognized by FEMA, evidence that adequate design and operation and maintenance systems are in place to provide reasonable assurance that protection from the base flood exists must be provided." This type of language leads to the inaccurate conclusion that levees are "safe" and protect, without fail, to the one percent annual chance flood. However, all levees, even accredited levees, can fail or be overtopped—that is, achieving zero risk is not a possibility. Thus, in trying to make areas safer, in reality, the potential for catastrophic losses is increased (Burby, 2006).

The use of the word protection is unavoidable in discussion of levee-related flood risk. However, it is important to keep in mind that although levees do offer "protection" from floodwaters, they only separate the area behind a levee from floodwaters until the point where the levee fails or is overtopped.

within a participating community may be exempted from the mandatory purchase requirement and land-use regulations when located behind a levee system that has been recognized by FEMA as providing protection against the one percent annual chance flood event, that is, accredited.

For a levee system to gain accreditation status, a community or other party must submit an accreditation package to FEMA (Figure 1-2). This package includes certification by a registered Professional Engineer (P.E.) or federal agency with responsibility for levee design that the levee system has been adequately designed and constructed to provide protection against the base flood according to the structural requirements of the FEMA criteria as found in Title 44, Section 65.10 of the *Code of Federal Regulations* (44 CFR §65.10; see Appendix A of this report for the full text of this regulation). Accreditation also requires an operation plan and a maintenance plan for the levee system, the details of which are also found in 44 CFR §65.10. FEMA's Procedure Memorandum 63, *Guidance for Reviewing Levee Accreditation Submittals*, discusses the process of review for compliance with 44 CFR §65.10 and accreditation (FEMA, 2010).

Levees can provide some flood control benefits that range from minor to substantial. For example, during record-setting floods in 2011, the levee system on the Lower Mississippi River was credited with preventing over $110 billion dollars in damages to the lands behind them (MRC, 2011). At the same time, no levee can provide absolute protection against all floods and failures of levee systems, such as those that occurred during Hurricane Katrina, which can cause billions of dollars worth of damages and threaten lives. All levees are subject to struc-

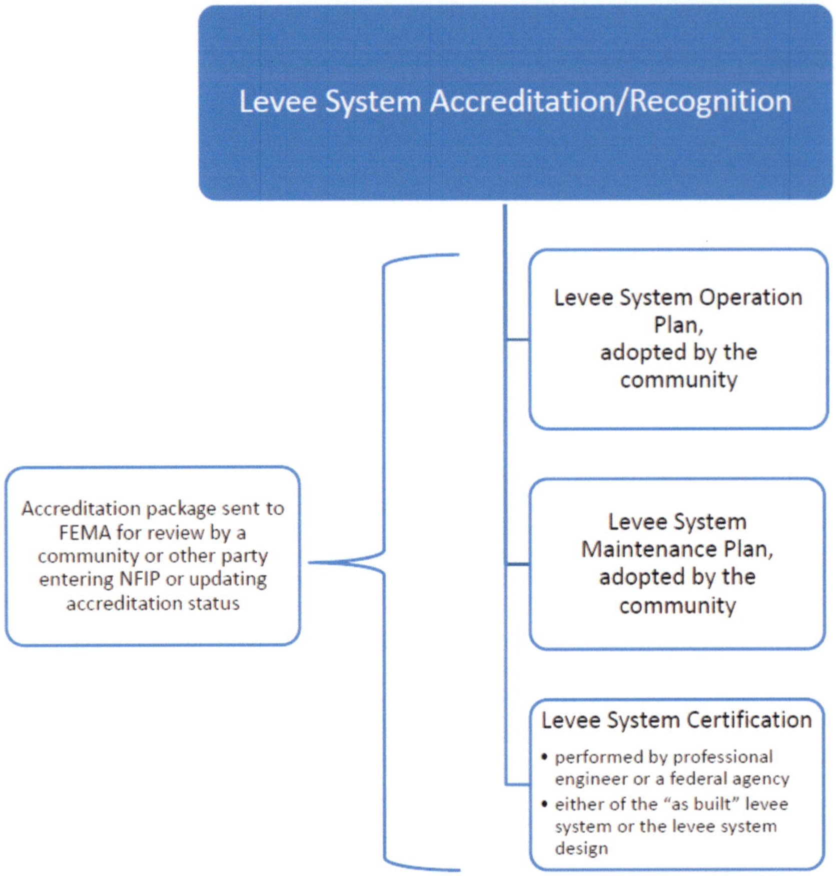

FIGURE 1-2 FEMA's levee system accreditation/recognition process.
NOTES: Recognition is used in parallel with accreditation in 44 CFR §65.10; design criteria for levee systems to be recognized by FEMA are provided in 44 CFR §65.10(b)(1)-(7).

tural failure. Thus, there are always risks to property, infrastructure, and inhabitants located behind levees, that is, residual risk (Box 1-3).[7]

The concept of residual risk behind levees confounds and complicates many public policy and investment decisions. For example, a levee may provide benefits in the form of flood damages prevented, but such benefits must be balanced against the costs of building and maintaining a levee and the potential costs of a levee being overtopped or failing. If a levee is capable of controlling the water flow of the one percent annual chance flood event, there is still a 26 percent probability that the levee will be overtopped at least once in the next 30 years. There is also some probability that the levee will not maintain its structural integrity as the flood waters rise against it, or that because of poor maintenance or operation it will not function as designed. Also, initial levee construction may have been faulty.

Moreover, residual risk generally changes over time with changes in land-use patterns, development, and hydrologic variability. If development behind a levee substantially increases, the consequences of a levee failure or overtopping will also substantially increase. If development upstream of the levee site increases, the rainfall

[7] Overtopping is the condition that occurs when floodwaters exceed the levee height. Levee failure is when the design component of a levee is reached. Overtopping and failure occur in a variety of scenarios related to the performance of the levee, including breach before or after overtopping, overtopping without a breach, and malfunction of system components (e.g., gates, pumping plants, or floodwalls). These scenarios result in different degrees of inundation, with failure causing the most severe flooding.

BOX 1-3
Risk, Residual Risk, and Uncertainty

Many definitions of risk (i.e. disaster risk) exist in the literature (Peduzzi et al., 2009; IPCC, 2012; NRC, 2012). A broad definition of risk is adopted in this report when dealing with physical hazards and their consequences. Risk is the potential for adverse effects from the occurrence of a particular hazardous event, which is derived from the probable combination of physical hazards (physical characteristics), the exposure and vulnerabilities of people and property subject to danger or damage from the hazard, and the consequences (impact or damage) caused by the hazard. In the context of this report, the physical hazard discussed is flooding.

Note, however, that in insurance nomenclature, risk is the term used to designate the loss consequence of the realization of the uncertain peril, and it may be financial or nonfinancial in nature (e.g., reputational loss). The cause of loss is called a peril, and a condition or situation that increases either the likelihood or severity of the peril occurring is called a hazard. This latter definition is used in Chapter 5 in discussing insurance rating.

Similarly, many definitions of residual risk exist in the literature (USACE, 2006; UNISDR, 2009). A broad definition of residual risk is adopted in this report. Residual risk is the risk that remains after considering the mitigating effects of structural, nonstructural, and other risk reduction measures. Residual risk is always present behind a levee, because no levee is fail-safe.

Determination of risk requires analyses that are dependent on knowledge of the factors that influence the risk. There is uncertainty connected with the definition of these factors. A 2010 National Research Council committee examining risk for the Department of Homeland Security found that

> uncertainty is always present in our ability to predict what might occur in the future, and is present as well in our ability to reconstruct and understand what has happened in the past. This uncertainty arises from missing or incomplete observations and data; imperfect understanding of the physical and behavioral processes that determine the response of natural and built environments and the people within them; and our inability to synthesize data and knowledge into working models able to provide predictions where and when we need them. (NRC, 2010)

The specific nature of the elements of uncertainty are examined in detail in Chapter 3.

runoff into the river and the level of the flow increase. An additional challenge to federal, state, and local officials is that the concept of residual risk behind levees causes confusion for residents living or working behind levees. Because areas behind the 100-year levee are not shown on FIRMs as subject to the one percent annual chance flood and residents are not subject to the MPR, many residents assume that they are protected against all flooding, not just the one percent annual chance flood.

Stationarity is the idea that hydrologic systems fluctuate in an unchanging envelope of variability. Water management systems have been designed on the basis of this assumption throughout the world (Milly et al., 2008). However, factors such as climate change, land use, and demographic change impact the water cycle and are causing the Earth's hydrologic system to depart from the historical behaviors (NRC, 2011). Changes in hydrology could dramatically change exposure to flood risk. Furthermore, nonstationarity means that the elevation of the one percent annual chance flood will change over time. For programs based on the one percent annual chance flood, this has the potential to cause conflict and confusion.

THE CURRENT STUDY

In 1982, the NRC issued a FEMA-sponsored report regarding levee policies for use in administering floodplain management, insurance, and risk mapping aspects of the NFIP (NRC, 1982). The report's recommendations were taken into account in 1986 when requirements for levee system accreditation were established (44 CFR §65.10; see Appendix A). The regulations focus on requirements that must be met if a community or levee owner requests that a levee system be recognized on the NFIP FIRMs.

The most recent mapping techniques for accredited and not accredited levees were developed and published in 2002 (FEMA, 2002). During its Map Modernization Program or "Map Mod,"[8] funded for fiscal years 2003 to 2008, FEMA determined that analysis of the role of levees in flood risk reduction would be an important part of the effort and noted that the status of the nation's levees was not well understood and the condition of many levees had not been assessed since their original inclusion in the NFIP, if at all. FEMA also learned that a large number of accredited levees might no longer meet the requirement of 44 CFR §65.10 because of either structural deficiencies or insufficient information on which to base continued compliance with regulatory requirements. Deaccrediting a number of large levees that were previously accredited and making distinctions between a levee versus road or railroad embankments that have served as part of the levee system are topics with significant social and political ramifications and are top priorities for FEMA.

Recognizing the need for improving the risk assessment and mapping techniques for areas behind levees, improving regulations governing levees in NFIP, raising public awareness of the levee risk, and enhancing institutional relationships, FEMA officials approached the NRC Water Science and Technology Board and requested the present study. The NRC responded by forming the ad hoc Committee on Levee and the National Flood Insurance Program: Improving Policies and Practices (Appendix B). The committee's charge was to examine current FEMA treatment of levees within the NFIP and provide advice on how those levee-related policies and activities could be improved. The study addressed risk analysis, flood insurance, risk reduction, and risk communication regarding how levees are considered in the NFIP (Box 1-4).

This report offers conclusions and recommendations to improve the science and engineering behind the NFIP's treatment of levees. Often, nonscientific issues are integral to solving challenging questions such as those captured in the statement of task for this report, for example, available resources; social and cultural values; and a variety of federal, state, and local regulations and laws. Defining and adhering to the boundary between science and engineering and policy is challenging for scientists, decision makers, and stakeholder groups. There is a role for science and engineering to inform policy decisions, a role that is explored in this report.

The NFIP offers insurance to all property owners in communities that participate in the program regardless of where the property is located. This includes properties behind levees and those not behind levees but within the SFHA. This also includes geographic areas that are vulnerable to flooding that might extend beyond the SFHA. This report provides guidance on FEMA's treatment of levees in the NFIP, considering risk analysis, risk reduction, insurance rates, and risk communication. Many of the issues covered in this report are not limited to managing property and life risk behind levees—FEMA grapples with these issues throughout the SFHA and often beyond. Thus, some of the content of this report can be extended throughout both the SFHA and the surrounding floodplain despite the fact that it was written within the framework of levees. The report explicitly states when observations relate to the SFHA, rather than just the area behind levees. The report also distinguishes between the SFHA and the floodplain at large, when appropriate.

Despite relevance to the report, providing specific advice on some topics was determined to be outside of the scope of the statement of task. For example, the report does not provide conclusions or recommendations related to the appropriateness of the current standard for the NFIP (i.e., the base flood elevation or the one percent annual chance flood event). However, the report does provide perspective on the current view of this standard in the scientific literature and identifies efforts to utilize alternative standards. The report also does not specifically address fiscal issues associated with construction, maintenance, and upgrades of levees within the NFIP, other than to identify where such issues may arise. The report does not address the subjects of fiscal responsibility for disaster

[8] Map Mod was a 5-year program to modernize FEMA flood maps by converting from paper to digital format and updating out-of-date maps, when possible. Map Mod is discussed further in Chapter 2.

BOX 1-4
Statement of Task

An ad hoc committee will examine current FEMA treatment of levees within the National Flood Insurance Program, and provide advice on how those levee-related policies and activities could be improved. The study will address the following topics regarding how levees are considered in the NFIP: (1) risk analysis, (2) flood insurance, (3) risk reduction, and (4) risk communication. More specifically, within this framework the committee will consider the following key issues and topics:

(1) RISK ANALYSIS

- Current risk analysis and mapping procedures and guidelines of levees implemented to determine risk premium rate zones for flood insurance purposes.
- Existing Special Flood Hazard Areas (SFHA) and their corresponding risk premium rates for areas behind accredited and non-accredited levees.
- Existing requirements for levee accreditation under 44 CFR §65.10.

(2) FLOOD INSURANCE

- Flood insurance pricing options for areas behind levees.
- Direct annualized flood loss estimates for residential and commercial structures behind levees.
- Waiving mandatory flood insurance purchase requirements for areas behind accredited levees.

(3) RISK REDUCTION

- Floodplain management, building standards, and land use practices employed behind levees across the nation.
- Engineered overtopping and breaches as a risk reduction or mitigation measure for levee systems.
- Existing and proposed levee-related grants and personal assistance policies. How can communities maximize benefits and reduce risks by using these?
- Mitigation options for communities with levees to help offset risks as investments grow in them and in areas behind levees

(4) RISK COMMUNICATION

- Existing FEMA levee outreach activities, programs, and material.
- The concept of "shared responsibility" for flood risk management and how it might be promoted.
- Incentives for communities to participate in mitigation activities that reduce levee-related risks.
- Non-regulatory products that can complement regulatory products in terms of risk communication.

The study and report will examine the efficacy of these practices and policies, make observations regarding levee policies, and analyze options for improving FEMA's treatment of levees within the National Flood Insurance Program. All conclusions and recommendations will be grounded in physical and social science and engineering concepts and evidence.

assistance or the relationship between federal disaster assistance and flood insurance, flood damage mitigation, or "affordability" of levees for a community.

In the course of deliberations and information-gathering efforts, it became clear that use and interpretation of critical levee-related terminology varies across relevant stakeholders, in the scientific literature, and among committee members from different backgrounds. The committee adopted definitions for each relevant term, which are defined upon first usage and used consistently throughout most of the report; where terminology is not consistent, the reasoning is discussed upon first usage. Appendix C contains a glossary of terms, and a list of acronyms is provided in the front matter of the report.

The Committee on Levees and the National Flood Insurance Program: Improving Policies and Practices met five times during the 20-month study period. In its information-gathering activities, the committee spoke to a variety of federal, state, and local flood experts and practitioners. Numerous persons contributed to the development of this report (Appendix D). The following report is a result of this work and contains consensus advice from the committee to help FEMA in pursuing its responsibilities related to levees in the context of the NFIP. The report is also designed to be of use to FEMA contractors, government agencies at all levels, various other entities, academics, and the public in understanding and dealing with risks from flooding in levee-protected areas.

The report is organized in eight chapters. Chapter 2 provides a history of and background information on the treatment of levees within the NFIP. Chapter 3 and 4 discuss the NFIP's approach to risk analysis, both currently and the committee's vision for the future. Chapter 5 explores the program's current approach to setting flood insurance rates and provides advice for improving the approach in the context of the committee's vision for risk analysis in the future. Flood risk management strategies, including mitigation measures, are explored in Chapter 6. Communicating flood risk behind levees is explored in Chapter 7. Finally, the shared responsibility at the local, state, and federal levels in managing flood risk and implementing the NFIP is discussed in Chapter 8. The report's conclusions and recommendations are highlighted in bold throughout the report.

REFERENCES

Burby, R. J. 2006. Hurricane Katrina and the paradoxes of government disaster policy: Bringing about wise governmental decisions for hazardous areas. Annals of the American Academy of Political and Social Science 604: 171-191.

FEMA (Federal Emergency Management Agency). 2002. Guidance for mapping areas protected by levee systems. Appendix H in Guidelines and Specifications for Flood Hazard Mapping Partners. Available online at http://www.fema.gov/library/viewRecord.do?id=2206. Accessed August 7, 2012.

FEMA. 2010. Guidance for Reviewing Levee Accreditation Submittals. Procedure Memorandum No. 63. Available online at http://www.fema.gov/library/viewRecord.do?id=4331. Accessed January 14, 2013.

FEMA. n.d. Managing Floodplain Development Through the National Flood Insurance Program. Home Study Course. Available online at http://www.fema.gov/sites/default/files/orig/fema_pdfs/pdf/floodplain/is_9_complete.pdf. Accessed December 7, 2012.

IPCC (Intergovernmental Panel on Climate Change). 2012. Managing the Risks of Extreme Events and Disasters to Advance Climate Change Adaptation. A Special Report of Working Groups I and II, C. B. Field, V. Barros, T. F. Stocker, D. Qin, D. J. Dokken, K. L. Ebi, M. D. Mastrandrea, K. J. Mach, G.-K. Plattner, S. K. Allen, M. Tignor, and P. M. Midgley, eds. Cambridge, UK, and New York: Cambridge University Press.

Milly, P. C. D., J. Betancourt, M. Falkenmark, R. M. Hirsch, Z. W. Kudzewicz, D. P. Lettenmaier, and R. J. Stouffer. 2008. Stationarity is dead: Whither water management? Science 319: 573-574.

MRC (Mississippi River Commission). 2011. 2011 MR&T Flood Report. Available online at http://www.mvd.usace.army.mil/mrc/pdf/MRC_2011_Flood_Report.pdf. Accessed November 14, 2012.

NRC (National Research Council). 1982. A Levee Policy for the National Flood Insurance Program. Washington, DC: National Academy Press.

NRC. 2010. Review of the Department of Homeland Security's Approach to Risk Analysis. Washington, DC: The National Academies Press.

NRC. 2011. Global Change and Extreme Hydrology: Testing Conventional Wisdom. Washington, DC: The National Academies Press.

NRC. 2012. Disaster Resilience: A National Imperative. Washington, DC: The National Academies Press.

Peduzzi, P., H. Dao, C. Herold, and F. Mouton. 2009. Assessing global exposure and vulnerability to hazards: The Disaster Risk Index. Natural Hazards Earth System Science 9: 1149-1159.

RAMPP (Risk Assessment, Mapping, and Planning Partners). 2012. Analysis and Update of National Flood Hazard Layer Demographics and NFIP Policy and Claims Data. Project Report for Task Order #HSFEHQ-11-J-002. Washington, DC: FEMA.

USACE (U.S. Army Corps of Engineers). 2006. Risk Analysis for Flood Damage Reduction Studies. Regulation No. 1105-2-101. Available online at http://www.corpsnedmanuals.us/FloodDamageReduction/FDRID042_ER11052101Sumry.asp. Accessed November 27, 2012.

USGS (U.S. Geological Survey). 2010. 100-Year Flood—Its All About Chance. Haven't We Already Had One This Century? Available online at http://pubs.usgs.gov/gip/106/pdf/100-year-flood-handout-042610.pdf. Accessed March 12, 2013.

UNISDR (United Nations International Strategy for Disaster Reduction). 2009. UNISDR Terminology on Disaster Risk Reduction. Available online at http://www.unisdr.org/files/7817_UNISDRTerminologyEnglish.pdf. Accessed November 20, 2012.

2

Treatment of Levees Within the National Flood Insurance Program

Along with the establishment of the National Flood Insurance Program (NFIP), the passage of the National Flood Insurance Act of 1968 established a requirement for the U.S. Department of Housing and Urban Development (HUD) to "identify and publish information with respect to all floodplain areas, including coastal areas located in the United States, which ha[ve] special flood hazards, within five years following August 1, 1968. . . ."[1] This direction led to the establishment within HUD and subsequently within the Federal Emergency Management Agency (FEMA) of programs to map flood-prone areas in communities wishing to participate in the NFIP.

As the NFIP and mapping began, there was no established standard for what constituted a special flood hazard area, although the one and two percent annual chance floodplains had been identified as benchmark levels of flood control in some efforts of the U.S. Army Corps of Engineers (USACE) and the Tennessee Valley Authority in connection with their floodplain management activities. In 1969, HUD asked the University of Chicago to convene a group of experts to recommend what might be a logical definition of the special flood hazard area. The group recommended that the one percent annual chance flood be considered *as an initial standard for the NFIP.* After some consideration within federal agencies, in 1971, HUD issued a rule establishing the one percent annual chance flood as the minimum standard for the regulation required by the NFIP, that is, the Special Flood Hazard Area (SFHA).

The Federal Insurance Administration (FIA), which had been established as an element of HUD, began mapping the nation's flood risk through contractors and federal agencies. Flood Hazard Boundary Maps (FHBMs), delineating the SFHA, and Flood Insurance Rate Maps (FIRMs), delineating the SFHA and risk premium zones, were established. Also, land was classified according to flood risk on FIRMs and FHBMs as high-, moderate-, or low-risk zones (Appendix E). The SFHA is considered high-risk, and this land is depicted as either Zone A or V (A, AE, A1-A30, AH, AO, AR, A99, V, VE, and V1-V30). Moderate- to low-risk areas, those outside the SFHA that are still subject to some level of flooding, are depicted as Zones B, C, or X. Those with undetermined flood risk can be classified as Zone D.

The mapping entities encountered existing flood protection works that were considered to have reduced the flood threat to the communities located behind them. Because many of these structures, mostly levees, had, in the past, successfully passed what had been considered a one percent annual chance flood, both local communities and the contractors raised the issue of excluding areas with a one percent annual chance or greater structure from

[1] The National Flood Insurance Act of 1968, §1360, 42 U.S.C. §4101.

BOX 2-1
"Without-Levee" Analysis

The practice of not recognizing nonaccredited levee systems on FIRMs, and recognizing areas behind these structures as part of the SFHA, is referred to in this report as FEMA's "without-levee" analysis. An all-or-nothing approach, it either gives communities credit for a one percent annual chance levee system, exclusion from the SFHA, and exemption from the mandatory purchase requirement or it does not; that is, a levee that is not accredited is not depicted on the FIRM as having any flood reduction impact. A levee system that is not accredited is not considered in flood hazard analyses to determine the reduction of risk behind the levee system, even though some risk reduction at a level below that offered by an accredited levee system might be provided.

the SFHA. Following 1973 congressional hearings that raised similar points, FIA concurred in this approach and authorized the removal of areas behind levees from the SFHA where the levees provided 100-year protection (Richard W. Krimm, personal communication, 2012; Box 2-1). It is not apparent that any regulations or rules were developed to formalize this practice.

Because the focus was on mapping the unprotected areas, little attention was given to the physical status of many of the levees. In many cases, there were no specific inspections or analyses of the levees. Their acceptance was based primarily on having been constructed by the federal government or having previously withstood a significant flood. In a 1974 amendment to the 1968 National Flood Insurance Act, the Congress affirmed the one percent annual chance flood as the minimum standard for the program. The amendment, known as the Brooks Amendment, was made a part of the Housing and Community Development Act of 1974[2] and also included provisions for allowing certain structures to be accredited if work was under way to achieve recognition.

While the one percent annual chance standard was being used in mapping areas subject to regulation under the Act, discussion continued between HUD and USACE over the appropriateness of the standard. In 1977, USACE indicated to HUD that setting the design of levees in urban areas at the one percent annual chance level could be imprudent because that is not a high degree of protection (IFMRC, 1994). At the same time, senior engineering officials within HUD recommended to FIA management that it strongly consider a federal policy that would recognize only levees designed to provide protection against the standard project flood (SPF)[3] (Crompton, 1977). USACE recommended use of a larger design flood such as the SPF for new levees (Wilson, 1978). USACE headquarters also issued guidance to its field elements indicating "on the assumption that exceedance of the design flow would cause a catastrophe, the standard project flood (SPF) is the desirable minimum level of protection that should be recommended for high levees, high floodwalls and high velocity channels in urban areas" (Wilson, 1978).

In 1977, Executive Order 11988 was issued, requiring federal agencies to "avoid to the extent possible the long and short term adverse impacts associated with the occupancy and modification of floodplains and to avoid direct or indirect support of floodplain developments wherever there is a practicable alternative. . . ." It required each agency to "provide leadership and take action to reduce the risk of flood loss, to minimize the impacts of floods on human safety, health, and welfare, and to restore and preserve the natural and beneficial values served by floodplains in carrying out its responsibilities . . . ."

In 1979, responsibility for the NFIP was transferred from HUD to the newly established FEMA, and FIA became a part of FEMA. In 1980, the FIA administrator recommended to the FEMA director "consideration of a standard in excess of the 100-year flood, such as the Standard Project Flood, for local protection works to be recognized by FEMA." The FIA administrator also noted that USACE had recommended this to FIA (Jimenez, 1980). FIA's concerns were focused on the following:

[2] Public Law 93-383.

[3] The SPF is a derived discharge estimate that represents a flood that can be expected from the combination of the most severe meteorologic and hydrologic conditions that are considered reasonably characteristic of a region (USACE, 1965).

- "It is estimated that levee overtopping or failure [was] involved in approximately 1/3rd of all flood disasters;
- "The 100-year flood is generally found to be a low design standard for structures protecting densely populated areas;
- "Only a fraction all earthen levees built at the computed hundred year flood elevation can be expected to provide protection to the true 1 percent elevation;[4]
- "The degree of protection to be expected from a 100 year levee is less than that obtained by elevating individual buildings to hundred year flood elevation;
- "Crediting a levee with protection against the hundred year flood would remove essentially all NFIP floodplain management requirements;
- "Results of USACE nonfederal dam inspections suggest that a large percentage of private or locally built levees as well as dams are or can be expected to be poorly designed and maintained" (NRC, 1982).

FEMA was also concerned that "the use of a 100 year standard was encouraging construction of levees to the 100 year design level for the sole purpose of removing an area from the special flood hazard designation" (Jimenez, 1980).

MOVING AHEAD WITH LEVEES IN THE NFIP

As FEMA worked to integrate levees into the NFIP, the White House, the Congress, and FEMA directed that several studies be conducted to provide insights into the role of levees in the NFIP. Recommendations from these studies are found in Appendix F, and summaries of the most relevant are found in subsequent paragraphs in this chapter.

The 1982 National Research Council Study

As a result of concerns within FEMA and USACE in the late 1970s, FEMA issued an interim levee policy in 1981 and asked the National Research Council (NRC) to examine the levee issue and recommend provisions for a "comprehensive levee policy for use in administering floodplain management, insurance, and risk mapping aspects of the National Flood Insurance Program" (NRC, 1982). In 1982, the NRC committee submitted its report to FEMA. The report contained recommendations covering engineering criteria for levee recognition, levee inspection and evaluation, requirements of levee owners, requirements of local communities, liability of local governments and levee districts, floodplain mapping approaches, and treatment of levees in the insurance aspects of the NFIP.

The NRC report concluded that because the basic objective of the NFIP was to mitigate flood damages and that the basic program prohibits construction below the 100-year level, FEMA should require construction of all new levees to at least the one percent annual chance flood. Existing levees that provided protection against 25-year or larger floods should be recognized for the purposes of reducing insurance rates. However, all levees that are given credit for reducing flood risks should meet standard minimum engineering criteria with respect to geometric parameters, freeboard,[5] soils and foundation, interior drainage,[6] closure devices, and rights-of-way. Recognizing the residual risk behind levees, the committee recommended that FEMA require, "purchase of flood insurance in all areas where the ground is lower than the unconfined 100 year flood level except where protected by levee built to contain the 500 year flood."

---

[4] The accuracy of the one percent annual chance flood elevation on a FIRM is affected by a variety of uncertainties in mapping data, for example, the choice of contour interval for terrain. Thus, the "true" one percent annual chance flood elevation is rarely, if ever, known (NRC, 2007).

[5] As defined in 44 CFR §59.1, freeboard is: "a factor of safety usually expressed in feet above a flood level for purposes of floodplain management. Freeboard tends to compensate for the many unknown factors that could contribute to flood heights greater than the height calculated for a selected size flood and floodway conditions, such as wave action, bridge openings, and the hydrological effect of urbanization of the watershed."

[6] The drainage that is required to evacuate rainwater from areas behind levees.

The committee also recommended that each levee be individually evaluated, hydrologically and structurally, and that FEMA should inventory all levees previously credited as providing protection from the one percent annual chance flood, set priorities, and schedule communities for restudy to reevaluate the levees.

As a follow-up to the NRC report and other internal actions, in 1983 FEMA initiated a system to maintain an inventory of levees, by community name, that were accredited as providing 100-year protection on NFIP maps. In 1985, FEMA issued regulations prescribing the criteria for levee recognition by the NFIP but did not reevaluate existing levees, alter the required level of protection, or move on other nonengineering recommendations (Galloway et al., 2006). The regulations were codified as Title 44, Section 65.10, of the *Code of Federal Regulations* (44 CFR §65.10; see Appendix A for the text of this section) and established that to be recognized as removing an area from floodplain regulation and mandatory purchase requirements in the NFIP, a levee must provide protection at the 100-year or higher level plus a specified level of freeboard; must meet engineering design standards similar to those used by the USACE; and must be properly maintained and operated under the control of a government entity. The regulations also require that a professional engineer certify the data submitted to support a given levee system's compliance with the structural requirements set forth in the regulation and must submit certified as-built plans of the levee. Where appropriate, a federal agency with responsibility for levee design may certify that the levee has been adequately designed and constructed to provide protection against the base flood. No provisions were made to reduce the rates for those behind levees providing less than one percent annual chance flood protection or to conduct periodic inspections, and 44 CFR §65.10 has remained unchanged since 1985.

Interagency Floodplain Management Review Committee

From April to October of 1993, the Mississippi and Missouri rivers and their tributaries inundated the surrounding floodplain, causing one of the costliest and most devastating floods in U.S. history. The flood of 1993 brought new attention to the condition of levees both within and outside of the NFIP. In 1994, a White House–based committee studying the 1993 flood reported that activities in floodplains of the United States, even with levee protection, remain at risk, finding that during the Mississippi-Missouri river floods:

> - Many locally constructed levees breached and/or overtopped. Frequently, these events resulted in considerable damage to the land behind the levees through scour and deposition [of material from the river channel and the levee itself];
> - Levees can cause problems in some critical reaches by backing water up on other levees or lowlands;
> - The current flood damage reduction system in the upper Mississippi River Basin represents a loose aggregation of Federal, local, and individual levees and reservoirs;
> - Many levees are poorly sited and [as a result] will fail again in the future. (IFMRC, 1994)

The White House study committee also indicated that the SPF flood should be the target flood for population centers and critical facilities protected by levees and that the government should require those behind levees who provide protection against less than the SPF discharge to purchase actuarially based insurance. It also found that federal agencies should ensure proper siting, construction, and maintenance of nonfederal levees. No action was taken by either Congress or the administration on these recommendations.

MAP MODERINIZATION AND LEVEE DECERTIFICATION

As a result of pressure from communities and professional organizations to improve the quality of flood mapping, the President proposed and the Congress, in 2003, authorized FEMA to undertake a 5-year billion-dollar program to modernize its maps by converting existing paper maps to a digital format and in the process, and where possible, provide new engineering analysis. As work began in 2004 on map modernization or "Map Mod," FEMA representatives became concerned that many of the levees encountered in the mapping process might not, under review, meet the standards of 44 CFR §65.10 and, as a result formed a committee to examine the issue. In April 2005, a small internal Interim Levee Coordination Committee reported that treatment of levees in Map Mod across FEMA regions was inconsistent and that guidance should be issued immediately to ensure that appropriate information about levees was obtained and considered in the map modernization process. It also recommended that a levee coordination committee be established to address this issue (FEMA, 2005).

On August 22, 2005, a week before Hurricane Katrina (Box 2-2), FEMA issued Procedure Memorandum 34, Interim Guidance for Studies Including Levees, which established that as Map Mod was implemented, currently credited levees should be checked against the 44 CFR §65.10 criteria (see Appendix G). When combined with a requirement to review "grandfathered" levees, this action sought to ensure that the accreditation status for all levees in the program was validated. The memorandum further indicated that when the levee owner was not able to produce data or was unable to otherwise substantiate that the levee met criteria, the area protected by the levee would be mapped as being in the floodplain and the requirements for mandatory purchase of insurance and regulation would go into effect when the revised map became effective.

In 2005, the Interagency Levee Policy Review Committee was established by FEMA to develop further guidance on treatment of levees in the NFIP. Membership on the committee included representatives of 10 federal agencies.[7] The committee reported in 2006 that:

> • Many levees shown on current NFIP maps as providing protection from the 100-year flood have never been evaluated against the criteria of 44 CFR 65.10 . . . . For levees that are found not to meet 44 CFR 65.10, the areas behind the levees will be identified as SFHAs. The impacts on the community of such an action will be significant.
>
> • The protection afforded by many certified levees decays over time due to various factors, such as changed hydrology or channel characteristics, erosion or deterioration of the embankment due to rodents or tree growth, general lack of proper maintenance, subsidence, or partial levee failures. However, there is no mechanism in place to regularly verify continuing accreditation of levees against 44 CFR 65.10. . . .
>
> • FEMA does not certify levees, but rather relies on certification provided by Federal agencies or State, local, or other parties seeking accreditation on NFIP maps. . . . Frequently, communities do not have the requisite resources or technical expertise to properly evaluate the levees against the 44 CFR 65.10 requirements and will have to hire engineering firms to do this work. . . .
>
> • As Procedure Memorandum 34 is implemented, many levees ultimately will not meet the criteria of 44 CFR 65.10 and will have to be shown as "failed" on FIRMs. The performance of detailed levee failure analysis as required by . . . Guidelines and Specifications will be costly, particularly for countywide [digital flood insurance rate map (DFIRM)] conversion projects and in counties and communities with large numbers of levees. . . .
>
> • The regulatory requirements of 44 CFR 65.10 lack adequate specificity for minimum operation and maintenance standards. . . .
>
> • Risks in areas behind certified levees are not adequately communicated to map users. . . .
>
> • Current requirements treat levees in a binary manner—either they meet 44 CFR 65.10 or they do not. Therefore, residual risk is greatly under-represented, and differences in risk that may exist to those behind certified levees are disregarded. In addition, no distinction between graceful versus chaotic capacity exceedance exists in the way levees are mapped, and there is no identification of the most highly hazardous areas immediately behind/along levees in the event of levee failure. (ILPRC, 2009, pp. 22-23)

The Interagency Levee Policy Review Committee urged FEMA and, as appropriate, other federal agencies to take the following actions immediately:

> • When communities are not immediately able to document whether currently certified levees can be recertified and the levees are not known to be at risk, FEMA should notify the communities of the need for a more detailed examination of the levees and provide a specific schedule for that activity. FEMA should require these communities to notify all people in areas behind the levees of the reexamination and identify for them the residual risks and the lack of information about the levees.
>
> • FEMA should define, as a matter of policy, a new flood insurance zone (Zone XL) for areas behind levees that meet the requirements for inclusion in the NFIP. Zone XL would include those areas behind the levee that would be subject to inundation by the 100-year flood if there were no levee. The area between the 100-year and 500-year lines should be shown as a shaded Zone X.

[7] Representatives from six nongovernmental organizations and four states observed committee meetings.

BOX 2-2
Hurricane Katrina

Hurricane Katrina was one of the deadliest and costliest hurricanes in U.S. history. Forming over the Bahamas, Katrina crossed Florida as a Category 1 storm and strengthened in the Gulf of Mexico. It made landfall in southeastern Louisiana on August 29, 2005, impacting Louisiana as well as Mississippi, Alabama, Georgia, and the Florida panhandle to various degrees. The hurricane's storm surge that pushed ashore from the Gulf of Mexico was over 29 feet, the highest ever recorded in the United States. In total, over 1,700 people were killed and hundreds of thousands were displaced (New York Times, 2012).

The levee system protecting New Orleans, Louisiana, built by USACE and operated and maintained by the local levee districts, failed in the aftermath of Katrina, and a significant portion of the city and surrounding areas were flooded (Figure 2-1). The NFIP experienced an unprecedented number of claims, 212,235, in the 2005 calendar year and over $17 billion dollars in loss dollars paid.[a] The levee failure cast a spotlight on USACE, prompting litigation, congressional investigation, and a variety of investigations regarding the levee failure.

The impacts of Katrina have been long-lasting economically, environmentally, and from the perspective of human and societal welfare of the city and surrounding area. Levee-related policy was reformed; for example, the National Levee Safety Program Act (2007), which directs USACE to carry out activities to enhance the safety of U.S. levees, was amended, and the USACE Levee Design Manual (USACE, 2000) was revised to incorporate risk-based concepts into levee design.

FIGURE 2-1 Aerial photograph of flooding as a result of breached levees in New Orleans in the wake of Hurricane Katrina.

[a]See http://www.fema.gov/policy-claim-statistics-flood-insurance/policyclaim-statistics-flood-insurance/policy-claim-13-17.

SOURCE: Taken by Marty Bahamonde on August 29, 2005, available online at http://www.fema.gov/photolibrary/photoCollectionView.do?id=81, accessed December 6, 2012.

- FEMA should initiate the revision of 44 CFR 65.10 and other appropriate FEMA guidance and specification documents, to address the technical changes recommended [in] this report.
- As a matter of policy, FEMA should require levee sponsors to conduct annual inspections of their levee systems; biennially submit to FEMA the results of the annual inspections, levee system operation and maintenance records, and an assessment of the levee systems during any flood events that occurred within the reporting period; and every 10 years submit a report that recertifies the engineering and geotechnical conditions of the levee system.
- FEMA and USACE, recognizing the important State and local roles, should develop incentives and support mechanisms to ensure that State and local agencies effectively carry out their responsibilities with respect to levee operation and maintenance.
- FEMA should exclude embankments, such as roads and railroads, from the accreditation of new levee systems unless those embankments are engineered specifically to provide damage/risk reduction from flooding and meet the engineering criteria required of levees. (ILPRC, 2006, pp. 61-62)

The Interagency Levee Policy Review Committee, recognizing the need for additional study and regulation development, urged FEMA to initiate action on the following longer term recommendations:

- . . . [R]evise 44 CFR 65.10 to phase out, over the next 10 years, the freeboard-based approach and to substitute the risk analysis methodology, with a 90-percent assurance of passing the 100-year flood, to determine the required levee heights. . . .
- . . . [T]ake immediate steps to examine the feasibility of not recognizing, for NFIP purposes, levees that protect highly urbanized areas unless they provide protection from events greater than 100-year floods (e.g., 500-year floods).
- . . . [A]ddress the challenge of the residual risk to structures behind levees by seeking legislative change to require property owners to purchase some level of flood insurance for structures behind levees (in the new Zone XL).
- FEMA, working closely with other Federal agencies, States, and communities, should examine how best to deal with climate change, sea-level rise, linked levee systems, and future development as major drivers of future conditions [hydraulics and hydrology (H&H)], with a goal of determining a BFE based on possible, but not fully quantifiable, future conditions.
- FEMA, in coordination with other Federal agencies, States, and communities, should initiate studies to determine how levees could be identified under a scientifically based levee risk-classification system (e.g., high, medium, or low). This system would be based on factors that include the potential depth of flooding in the event of failure or overtopping; the type and density of development in protected areas behind the levee; steps taken to ensure that levee failure does not occur when the levee capacity is exceeded (overtopping); warning times; and the number and types of egress so that people who may be inundated may move out of harm's way. This levee risk classification would be designated on the FIRM and published in the FIS report. (ILPRC, 2006, p. 62)

(For additional information regarding the advice of the Interagency Levee Policy Review Committee, see Appendix F.) In 2009, FEMA completed the initial Map Modernization Program, digitizing FIRMs covering areas containing 92 percent of the U.S. population (NRC, 2009).

Report of the National Committee on Levee Safety

Title IX of the 2007 Water Resources Development Act (Public Law 110-114), the National Levee Safety Act, directed the establishment of a federal, state/tribal, local/regional, and private-sector committee on levee safety to "develop recommendations for a national levee safety program, including a strategic plan for implementation of the program." In January, 2009, the committee submitted its report to the Administration (NCLS, 2009).

The report provided a framework for developing a long-term approach to levee safety management, recommending establishment of a National Levee Safety Commission. It also recommended the development of a levee hazard potential classification system, tolerable risk guidelines,[8] and national levee safety standards; peer review

[8] Tolerable risk guidelines "categorize the nature of risks in ways that can assist in assessing their acceptability or non-acceptability, and in prioritizing actions for reducing risks" (USACE, 2010).

of the FEMA levee certification process; development of a national levee risk communication campaign; establishment of a national levee rehabilitation, improvement, and flood mitigation fund; delegation of levee safety responsibilities to the states; and mandating purchase of risk-based flood insurance in leveed areas. Other recommendations can be found in Appendix F.

LEVEE DEACCREDITATION

Issuance of Procedure Memorandum 34 (FEMA, 2005) and the potential of many levees under study at the time no longer being accredited caused a strong pushback from levee owners and affected communities. Recognizing the need to give those responsible for providing certification information the opportunity to gather that information, in September 2006, FEMA issued Procedure Memorandum 43, *Guidelines for Identifying for Provisionally Accredited Levees* (PAL), which provided a 2-year period for affected levee owners to complete those actions necessary to retain their accreditation (Appendix G). The Memorandum, which was reissued the following March, provides more detailed guidelines that allow

> mapping partners to issue preliminary and effective versions of DFIRMs while the levee owners or communities are compiling the full documentation required to show compliance with 44 CFR Section 65.10. . . . If the levee qualifies for the PAL designation, FEMA will provide the community 90 days to sign and return an agreement indicating that the full documentation for 44 CFR Section 65.10 will be provided within 24 months of the signed agreement.

The Memorandum also provided guidelines for levees whose maintenance was deficient, allowing time for these deficiencies to be remediated (FEMA, 2006).

In April 2009, FEMA issued Procedure Memorandum 53, *Guidance for Notification and Mapping of Expiring Provisionally Accredited Levee Designations* (see Appendix G). This Memorandum provided instructions to FEMA field units concerning the steps that were to be taken if PAL-designated levees had not completed the required activity within the 24-month period. In these cases, the mapping partners were instructed to map the area as being within the SFHA or as "without levee," thus removing areas previously exempted from compliance with mandatory insurance and regulatory actions from this status (FEMA, 2009).

Affected communities, faced with the potential movement of the previously exempt areas into the floodplain and the necessity to purchase flood insurance at rates that might be the same as those without any protection argued that some credit physically and fiscally should be given to the presence of the deficient levees in any analysis of without-levee conditions. In February 2011, 27 U.S. senators sent a letter to FEMA, urging the Administrator:

> "to use your current authorities to discontinue the Federal Emergency Management Agency's (FEMA) use of "without levees" analysis to determine new Flood Insurance Rate Maps (FIRMs) in cases where a final determination has not been made and an affected community objects to such analysis in favor of more precise methods of flood modeling . "Without levee" modeling methods assume that a levee or flood control structure that exists in physical reality does not exist for the purposes of modeling, reducing the precision or flood maps and eroding public confidence in the mapping process itself." (U.S. Senate, 2011)

At the same time, 49 members of the U.S. House of Representatives sent a similar letter to Administrator Fugate.

In response, FEMA Administrator Fugate directed FEMA staff to replace the without-levee modeling approach with a suite of methodologies that would be technically sound, credible, and cost-effective, and would better meet the needs of citizens while providing more precise results that better reflect the impact of levees on flood risk. This procedure would not replace the need for levee owners or the associated communities to remain engaged in flood risk management activities or change the existing requirements for them to provide levee certification information as outlined in 44 CFR §65.10 (Fugate, 2011a). The development and current status of this new approach, the Levee Analysis and Mapping Procedure, is discussed in Chapter 4.

In June 2011, testifying before the U.S. Senate Committee on Banking, Housing, and Urban Affairs, FEMA Administrator Fugate pointed out that

> while FEMA is committed to working closely with communities to develop the most accurate flood maps possible, the current "in or out" nature of the SFHAs (one is either in an SFHA or not) has left the program with a perceived credibility problem, as there is no gradation of risk identified within a flood zone. (Fugate, 2011b)

The Biggert-Waters Flood Insurance Reform and Modernization Act

The Biggert-Waters Flood Insurance Reform and Modernization Act of 2012 (heretofore known as the Biggert-Waters Act) was signed into law on June 29, 2012, extending the NFIP until September 30, 2017.[9] Reforms and modernization directives in the Biggert-Waters Act call for a variety of changes that apply to the NFIP, many of which are relevant to this report. These changes are discussed throughout the chapters, when relevant, and summarized here.

Section 100207 of the Biggert-Waters Act directs full, risk-based premiums or a "premium adjustment to reflect the current risk of flood" in area participating in the NFIP:

> any property located in an area that is participating in the national flood insurance program shall have the risk premium rate charged for flood insurance on such property adjusted to accurately reflect the current risk of flood to such property, subject to any other provision of this Act. Any increase in the risk premium rate charged for flood insurance on any property that is covered by a flood insurance policy on the effective date of such an update that is a result of such updating shall be phased in over a 5-year period, at the rate of 20 percent for each year following such effective date.

Related to this report, the FEMA administrator is also directed to contract with the National Academy of Sciences to study exploring methods for understanding graduated risk behind levees and the associated land development, insurance, and risk communication dimensions.[10]

The FEMA administrator is required to form a Technical Mapping Advisory Council to deal with map modernization issues.[11] This includes recommendations on how to "ensure that flood insurance rate maps incorporate the best available climate science to assess flood risks" and that the impact of sea level rise is considered. The Biggert-Waters Act directs FEMA to review, update, and maintain NFIP rate maps with respect to

(i) all populated areas and areas of possible population growth located within the 100-year floodplain;
(ii) all populated areas and areas of possible population growth located within the 500-year floodplain;
(iii) areas of residual risk, including areas that are protected by levees, dams, and other flood control structures;
(iv) areas that could be inundated as a result of the failure of a levee, dam, or other flood control structure; and
(v) the level of protection provided by flood control structures.[12]

The Biggert-Waters Act also requires FEMA to use the most accurate topography and elevation data available and "any relevant information . . . relating to the best available science regarding future changes in sea levels, precipitation, and intensity of hurricanes" and coastal inundation, land subsidence and coastal erosion when updating maps.[13]

Several requirements in the Biggert-Waters Act call for changes that adjust or study fiscal practices of the program. For example, the Act calls for phasing out both subsidies and grandfathering for some properties,[14] establishment of a reserve fund,[15] requiring the FEMA administrator to develop a plan for repaying the debt

---

[9] Moving Ahead for Progress in the 21st Century Act, H.R. 4348, Title II, Subtitle A, Flood Insurance Reform and Modernization, 112th Cong. (2012).

[10] Section 100231.

[11] Section 100215.

[12] Section 100216(b)(1)(A).

[13] Section 100216(b)(3).

[14] For example, subsidies for nonprimary residences, severe repetitive loss properties, and businesses.

[15] Section 100212.

largely incurred by Hurricane Katrina,[16] assess opportunities associated with privatization and reinsurance,[17] and imposition of minimum deductibles.[18]

The Biggert-Waters Act directs the study of the impact, effectiveness, and feasibility of amending the NFIP to include building codes as part of floodplain management criteria developed under NFIP.[19]

The administrator is called to contract with the National Academy of Sciences to study full-risk–based premiums and means-tested federal assistance in comparison with the current system of flood insurance rates and federally funded disaster relief:

> shall conduct . . . an economic analysis of the costs and benefits to the Federal Government of a flood insurance program with full risk-based premiums, combined with means-tested Federal assistance to aid individuals who cannot afford coverage, through an insurance voucher program. The analysis shall compare the costs of a program of risk-based rates and means-tested assistance to the current system of subsidized flood insurance rates and federally funded disaster relief for people without coverage.[20]

With respect to levee communication, the Biggert-Waters Act directs the administrator to enhance communication efforts. This includes outreach to states and local communities about map changes and associated insurance rate changes and education of property owners about continued flood risk even when no longer subject to the mandatory purchase requirement.[21] Interagency collaboration is also promoted. The Biggert-Waters Act directs the FEMA administrator and the secretary of the Army, acting through the chief of engineers, in cooperation with the National Committee on Levee Safety, to establish a Flood Protection Structure Accreditation Task Force to develop a process to better align the information and data collected by or for USACE's Inspection of Completed Works (ICW) Program with the flood protection structure accreditation requirements so that information and data collected for either purpose can be used interchangeably and information and data collected by or for USACE under the ICW Program are sufficient to satisfy the FEMA flood protection structure accreditation requirements.[22]

The Past and Present

Over the past 40 years, the treatment of levees within the NFIP has been a continuous topic of discussion and often, of controversy. Many studies have examined what should be done to ensure that when levees are brought into the NFIP, their presence does not increase the risk faced by those who are located behind them. For fiscal, administrative, and political reasons, few of the recommendations from these studies have been fully adopted. The chapters that follow examine once again many of these recommendations, and offer new recommendations in the interest of improving the NFIP and floodplain management in the United States.

REFERENCES

Crompton, F. M. 1977. Memorandum to Richard Krimm on HUD Policy on Levees, Engineering Division Recommendations. Washington, DC: HUD.

FEMA (Federal Emergency Management Agency). 2005. Interim Guidance for Studies Including Levees. Procedure Memorandum 34. Washington, DC: FEMA.

FEMA. 2006. Guidelines for Identifying Provisionally Accredited Levees. Procedure Memorandum 43. Washington, DC: FEMA.

FEMA. 2009. Guidance for Notification and Mapping of Expiring Provisionally Accredited Levee Designations. Procedure Memorandum No. 53. Washington, DC: FEMA.

---

[16] Section 100213.
[17] Section 100232.
[18] Section 100201.
[19] Section 100235.
[20] Section 100236(b).
[21] Section 100243.
[22] Section 100226.

Fugate, C. 2011a. Letter to Senator Thad Cochran regarding discontinuation of without-levees analysis. Washington, DC: FEMA. Available online at http://www.pnwa.net/new/Articles/Fugate%20response%20to%20Senator%20Cochran.pdf. Accessed March 7, 2013.

Fugate, C. 2011b. Testimony on the National Flood Insurance Program Reform, June 9, Before the Senate Committee on Banking, Housing, and Urban Affairs. Available online at http://www.banking.senate.gov/public/index.cfm?FuseAction=Files.View&FileStore_id=c6f08bf5-5daa-4406-b461-1781159ec9c1.

Galloway, G. E., G. B. Baecher, D. Plasencia, K. G. Coulton, J. Louthain, M. Bagha, and A. R. Levy. 2006. Assessing the Adequacy of the National Flood Insurance Program's 1 Percent Flood Standard. Washington, DC: American Institutes for Research. Available online at http://www.fema.gov/library/viewRecord.do?id=2595.

IFMRC (Interagency Floodplain Management Review Committee). 1994. Sharing the Challenge: Floodplain Management into the 21st Century. Washington, DC: IFMRC.

ILPRC (Interagency Levee Policy Review Committee). 2006. The National Levee Challenge: Levees and the FEMA Flood Map Modernization Initiative. Available online at http://www.fema.gov/library/viewRecord.do?id=2677. Accessed August 8, 2012.

Jimenez, G. M. 1980. Memorandum to John W. Macy, Director of FEMA; Subject: Levees, February 1. Washington, DC: Federal Insurance Administration.

NCLS (National Committee on Levee Safety). 2009. Recommendations for a National Levee Safety Program. Washington, DC: NCLS.

NRC (National Research Council). 1982. A Levee Policy for the National Flood Insurance Program. Washington, DC: National Academy Press.

NRC. 2007. Elevation Data for Floodplain Mapping. Washington, DC: The National Academies Press.

NRC. 2009. Mapping the Zone: Improving Flood Map Accuracy. Washington, DC: The National Academies Press.

New York Times. 2012. Hurricane Katrina. Available online at http://topics.nytimes.com/top/reference/timestopics/subjects/h/hurricane_katrina/index.html. Accessed December 10, 2012.

USACE (U.S. Army Corps of Engineers). 1965. Standard Project Flood Determinations, EM 110-2-14-11. Washington, DC: USACE.

USACE. 2000. Design and Construction of Levees. EM 1110-2-1913. Washington, DC: USACE.

USACE. 2010. Proceedings of the Workshop Exploration of Tolerable Risk Guidelines for the USACE Levee Safety Program. Institute for Water Resources Report 10-R-9. Washington, DC: USACE.

U.S. Senate. 2011. Letter to Administrator Fugate. Available online at http://www.fema.gov/library/viewRecord.do?id=4984. Accessed March 15, 2013.

Wilson, D. 1978. Letter to Richard Krimm. Washington, DC: U.S. Army Corps of Engineers.

3

Moving to Flood Risk Management and a Modern Flood Risk Analysis

Society has been attempting to deal with flooding for several millennia. Initially, those at risk of flooding moved out of the way of rising waters. If that became impractical, leaders would turn to methods that would control floods by keeping them away from the population through use of embankments, channels, and elevating structures. This emphasis on flood control began in the third millennia B.C. and continued through to the middle of the 20th century when it was recognized that structures alone could not overcome the flood challenges (Sayers et al., in press). In the mid-20th century, other tools such as floodproofing, building codes, land-use management, and early warning systems joined flood control structures in flood damage reduction efforts. However, as population and development grew, flood losses continued to rise and the need for the establishment of priorities for the use of scarce resources became apparent (Figure 3-1).

Following in the path of several centuries of experience in the insurance and financial industries, flood management professionals across the globe began to examine risk management methodologies that could be used to assess the flood hazards that had to be faced and the consequences that would result from flooding, and the relative significance of the risks that were identified. By the end of the 20th century, the concept of flood risk management had been widely accepted in Europe and was beginning to take hold in the United States. This evolution of approaches to dealing with floods is described in Figure 3-2.

Flood risk is a function of the characteristics of the hazard, the exposure to the hazard, the vulnerability[1] of that which is exposed, and the consequences that could occur should the hazard reach the exposed, and vulnerable elements of the potentially flooded area. Probabilities are a part of all elements of risk—the probability of the event taking place, the probability that the exposed area will be flooded, the probability that the flood mitigation systems in place will successfully reduce vulnerability, the probability that the people and property in the target area will remain in the area and be subject to flooding, and the probability that they have insurance, etc. Flood risk management represents comprehensive efforts, to *continuously carry out analyses, assessments, and related mitigation implementation activities to reduce flood risk* (FLOODsite, 2004; Sayers et al., 2012).

Movement to flood risk management requires that those responsible enter into an iterative process that seeks to identify the hazards faced by the community and assess the exposure, vulnerability, and potential consequences of these hazards should a flood event occur. Provided with this information, the community then develops and

[1] Vulnerability is "the potential for harm to the community and relates to physical assets (building design and strength), social capital (community structure, trust, and family networks), and political access (ability to get government help and affect policies and decisions). Vulnerability also refers to how sensitive a population may be to a hazard or to disruptions caused by the hazard" (NRC, 2012b).

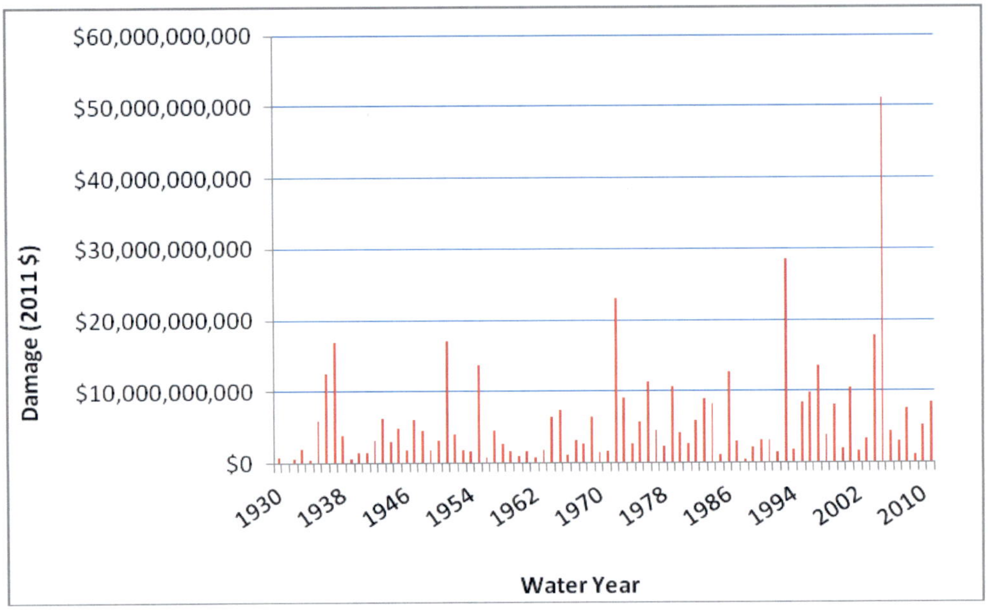

FIGURE 3-1 Estimated Total Flood Damages in the United States, 1934-2000, billions of dollars. Data adjusted to inflation by using the Construction Cost Index (CCI) from the McGraw-Hill Construction Engineering News-Record. To adjust each year's number, the 2011 CCI was divided by the CCI of the year in question and then multiplied by the raw damage amount for that year.
SOURCE: NOAA (2012).

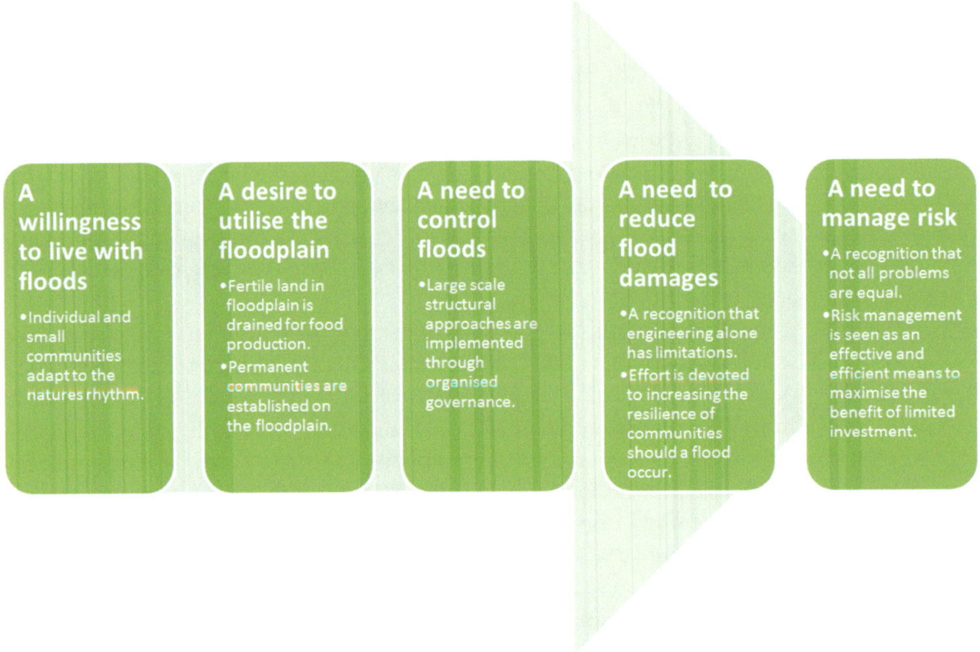

FIGURE 3-2 The evolution of flood management practice through history. It can be anticipated that in the years ahead, advances in technology will provide capabilities that will permit more effective and efficient capabilities to identify and deal with risk. At the same time, increased communications capabilities will better prepare the population at large to understand and participate in the development and use of risk strategies.
SOURCE: Sayers et al. (in press).

implements a strategy for coping with the risk. The strategy and implementation remain under continuous review, and when circumstances dictate, adjustments are made to the strategy. Risk communication is a continuous effort throughout the process (Figure 3-3). The effectiveness of the flood risk management process is linked to sound data and modern computational tools.

A risk analysis is a detailed examination:

> performed to understand the nature of adverse consequences from a particular event to human life, property, or the environmental; an analytical process that provides information about or quantifies probabilities and consequences of an unwanted event. Oftentimes broad definitions of risk analysis include examination of risk communication, risk perception, and risk management alternatives. (NRC, 2010)

A subset of a risk analysis, a flood risk analysis is an analytical process that provides information about or quantifies probabilities and consequences of a flood event. A flood risk analysis is a subset of the broader risk-based approach to flood risk management described throughout this report (Figure 3-3).

The use of risk-based methods to evaluate the performance of civil infrastructure systems such as nuclear power plants and dams (URS/JBA, 2008; ASME/ANS, 2009) and to serve as the basis for building codes (DOE, 2002), evaluation of terror threats (NRC, 2010) is now well established. Similarly, in the context of the insurance industry, risk analysis methods are used routinely to conduct portfolio assessments (AIR, 2012; RMS, 2012). The advancement of risk analysis methods to evaluate major infrastructure systems and extensive portfolios parallels the improvement in technology; software, hardware, and key datasets (e.g., Lidar, building inventories). As a consequence, software tools and hardware to support risk-based analyses have advanced to the point that evaluation of large portfolios has been performed (URS/JBA, 2008; IPET, 2009; AIR, 2012; RMS, 2012).

FIGURE 3-3 The flood risk management cycle illustrates the multistep flood risk management process.
SOURCE: Sayers et al. (in press), NRC (2012a), Moser et al. (n.d).

The use of risk analysis methods to evaluate flood risks and specifically to consider the performance of levees has been ongoing and improving for over a decade (USACE, 1996; URS/JBA, 2008; IPET, 2009). In the 1990s, USACE developed a probabilistic approach and software tools to evaluate flood damages (USACE, 1996). This methodology was used as part of planning studies to evaluate flood loss reduction benefits. The National Research Council (NRC) reviewed the USACE flood loss reduction methodology and reported on its findings in 2000 (NRC, 2000). The NRC found the USACE techniques a significant step forward and commended the agency for embracing risk analysis techniques. The NRC also noted that the USACE approach was technically sound, but aspects of the methodology as it related to the evaluation of uncertainties required improvement.

Following Hurricane Katrina, USACE established the Interagency Performance Evaluation Task Force (IPET) to determine the causes of levee and floodwall failures in New Orleans. Early work by the IPET, which involved over 300 professionals from government agencies, academe, and the private sector, used risk-based techniques in its analysis and the IPET recommended that USACE begin a move to risk-based approaches, to include geotechnical analyses in its work. Shortly thereafter, USACE announced that it would become a risk-based organization and that efforts would be made to accelerate the use of risk-based techniques not only in studies of future work but in analysis of existing structures (USACE, 2008).

Title 44, Section 65.10, of the *Code of Federal Regulations* (44 CFR §65.10) refers to the "risk study" that is performed when a community is seeking recognition of a levee system. Although the regulation does not specifically spell out what is meant by a risk study, in practice, it is a flood hazard analysis. The flood hazard analysis currently used within the NFIP to assess flood risk establishes a legislated level of protection (the one percent annual chance flood), using standard hydrologic methods that compute the expected elevation of that flood at a given location and assumes that any flood protection structures at that location designed to pass the one percent annual chance flood will in fact do so. To account for uncertainty, structures are required to be built to the one percent annual chance flood elevation plus a specified number of feet of freeboard.

To elaborate, under current NFIP policies a partial risk-based analysis is used with respect to performance of levee systems, where many parts of the analysis (e.g., the geotechnical component) are deterministic in fashion. A levee system is only recognized for its benefits if it is "accredited," meaning that the levee system has been determined to meet minimum design, operation, and maintenance standards that are consistent with a level of protection associated with the ability to pass the one percent annual chance flood, as specified in section 65.10 . In other words, the levee system is only considered if it performs perfectly up to standard; if not, it is ignored. Thus, if a levee system is accredited, its potential failure is included in the flood hazard analysis and it is simply considered as part of the river hydraulics. A levee system that does not meet accreditation standards is not considered in the analysis used to quantify flood risk, even though it provides some (potentially considerable) protection against flooding.

The current approach to flood risk analysis does not address certain components that are critical to a modern flood risk analysis. These include the uncertainties in the hydrology, the probabilities that a protection structure might fail at less than the design elevation, the consequences that will result from the actual flooding, and the probabilities of the success of actions such as, for example, evacuation of the elderly and disabled. Capturing these factors in the NFIP's approach to flood risk analysis is the source for the committee's advice that the program move to a modern risk-based analysis.

To elaborate on these issues, communities construct levee systems to protect up to the one percent annual chance flood, enabling new development in areas with significant, but un-quantified exposure to catastrophic flood risk. Thus, protection against the one percent annual chance flood event is the de facto standard for most levees in the United States, with limited regard to residual risk—that is, the consequence of capacity exceedence.[2] Furthermore, levee systems that only marginally meet certification standards are vulnerable to decertification. If not properly maintained, the performance of levee systems degrades over time due to erosion, rodent damage, and subsidence. Further, the frequency and magnitude of flood hazards can increase over time due to natural and anthropogenic causes.

[2] The State of California recently legislated that levees protecting urban areas should provide protection against the 0.5 percent annual chance flood (200-year) (California Government Code § 65007(l)).

Under current NFIP policies, lands behind accredited levee systems are not designated as being in a Special Flood Hazard Area (SFHA) and are not required to manage flood risk, even though the risk can be significant. Floods greater than the one percent annual chance flood can and do occur. Accredited levees can fail and have failed during smaller floods, causing catastrophic damage to structures and endangering residents. For example, in February 1986 the Linda-Olivehurst levee, a certified levee located on the Lower Yuba River in California, failed (Figure 3-4). Economic and life-safety risks associated with accredited levee systems are not accounted for in the current NFIP policy, which generally means that flood risks have not been effectively evaluated and, as a result, not effectively managed.

FIGURE 3-4 Images of the certified Linda-Olivehurst levee failure in 1986. The failure occurred after the flood had crested 8.6 feet below the levee crest. (*Top*) The breach was 170 feet wide and (*bottom*) entire neighborhoods were flooded in the surrounding community. The economic loss was $1.5 billion and lives were lost.
SOURCE: Rogers (2012).

The summation of these factors is that, under current FEMA policies and flood hazard analyses, the flood risk for many lands is not described completely, for a variety of reasons. This is not unknown to FEMA or relevant stakeholders including policy makers. In its 2006 report, the Interagency Levee Policy Review Committee recommended that within 5 years FEMA should, in the evaluation of levees, begin to use the risk-based hydrologic techniques being employed by USACE (ILPRC, 2006). In 2007, USACE, in coordination with FEMA, established the National Flood Risk Management Program, emphasizing the intention of the USACE to move to risk-based approaches to flood management. In 2009, with the creation of the Risk MAP program, FEMA identified risk-based analysis as a critical element in efforts to accurately identify the risk to those who occupy flood-prone areas.[3] As indicated in Chapter 2, the Biggert-Waters Act placed emphasis on risk analysis in NFIP activities and directed the formation of the USACE-FEMA Flood Protection Structure Accreditation Task Force to better coordinate the data development of the two agencies in levee accreditation.[4]

ADVANTAGES OF A MODERN RISK-BASED APPROACH TO FLOOD RISK ANALYSIS AND INSURANCE

In this report, a modern flood risk analysis is defined as a risk analysis using the best available science and analytical methods to evaluate flood risk. This is more precisely defined as an analysis where the likelihood of adverse consequence (loss, personal injury) is quantitatively evaluated, taking into account

1. the likelihood of flood events occurring (the randomness of future events);
2. the likelihood that the capacity of levee and other flood protection systems may be overwhelmed by the flood (be overtopped) or fail (the randomness of structure performance for given levels of loading), leading to flooding of protected areas; and,
3. given flooding has occurred, an assessment of the consequences that occur.

Each element of the analysis is subject to uncertainty; neither the chance that different-size flood events can occur nor their magnitude can be estimated exactly—they are uncertain. These uncertainties and the related implications to the estimate of risk are quantitatively evaluated in a modern risk-based analysis.

There are a number of clear and specific attributes of a flood risk analysis that are not inherent to the definition, but at the same time, define what it means to perform such an analysis. For instance, the analysis considers the occurrence and consequences for a full range[5] of flood events that can occur, impact a community, and affect the assessment of insurance rates; it models and analyzes the hydrologic system, including levees and flood protection systems as systems, physically and logically integrated features that impact the outcome of future events; and it quantitatively evaluates the sources of uncertainty that affect the results and the estimation of future losses.

The ILPRC (2006) and many others (Appendix F) recommend that the NFIP should move to implement a more complete approach to the risk analysis that is the cornerstone of its mapping and insurance program. The committee concurs with these recommendations. A modern risk analysis would take advantage of new computational and mapping capacity to produce state-of-the art risk estimates for all areas that are vulnerable to flooding hazards. The analysis would provide a foundation for the NFIP to directly evaluate flood hazards and the performance of levee and other flood control systems and assess the consequences of flooding using sound engineering assessment and scientific and probabilistic methods. Specifically, the modern risk analysis would

---

[3] Although Risk MAP is focused on new and better analysis to support existing criteria and methodologies of the NFIP, this focus does not prevent the program from considering risk in the broader sense. The committee considers this program an important step in FEMA's attempt to move toward risk-based analysis.

[4] Section 100221, Interagency Coordination Study.

[5] The term "full range" is used in general terms to reflect that the likelihood of small as well as large flood events are considered in the analysis, not just a single event. The range that is considered should be broad enough to include the frequency and consequences of events that are meaningful to floodplain management and determination of insurance rates.

- provide direct engineering- and reliability-based assessment of the performance of flood protection systems in the evaluation of flood hazards in protected and unprotected areas;[6]
- account for the performance of risk reduction that is attributable to nonstructural measures;
- provide a means to establish an insurance program that is based on a continuum of flood risks as opposed to the current approach that is partially deterministic;
- offer a clear and informed basis for risk communication;
- provide a reliability-based means for the assessment of levees and flood protection systems and their integrity;
- improve the long-term stability of flood hazard and risk estimates that are a basis for insurance pricing, presuming that uncertainty and changing conditions are accounted for properly throughout the analyses;
- require that levees and appurtenant structures and subsystems be evaluated and modeled as part of flood protection "systems" whose reliability has a significant impact on the flood risks that communities face;
- afford the opportunity for communities and others to understand their flood risks at the community level and to assess the economic benefit of alternative floodplain management options; and
- provide risk information at a community as well as an individual property scale, which would offer the opportunity for communities to more clearly understand the value of community participation in the NFIP.

Such an analysis would provide an in-depth technical evaluation of flood hazards and directly account for the performance of levee flood protection systems, whether accredited within the NFIP or located within an NFIP community (i.e., levee systems that protect against the one percent annual chance flood or not). It would account for all elements of the flood protection system (levees, gates, other structures, etc.) that significantly affect flood risk, as long as they meet minimum design, operation, and maintenance standards. Hence credit could be given when establishing flood insurance rates for flood protection systems that do not protect to the one percent annual chance standard (or any new standard), but at the same time provide some level of protection. Risk analysis results would also differentiate between protection provided by levees that protect to the one percent annual chance flood and higher, for example, 0.2 percent levees. It would be able to recognize the difference in risk between low areas behind levees and higher ground.

In addition to the foregoing, a risk-based analysis offers other advantages. A modern risk-based analysis will account for the uncertainties that affect the quantitative representation of the flood hazard and the estimated performance of flood protection systems. The current analysis for flood risk estimation neglects significant sources of uncertainty in the hydrologic and hydraulic modeling and it does not account for any aspect concerning the performance of flood protection systems other than the binary aspect (meets or fails to meet criteria). And it does not account for uncertainties in the consequences of flooding. The proposed modern flood risk analysis directly accounts for these uncertainties, building on techniques that have been developed and applied in the evaluation of risks associated with other natural phenomena (e.g., earthquakes) and that have been used in the flood and levee areas (USACE, 2006; URS/JBA, 2008; IPET, 2009).

In a broader context, utilizing a modern risk analysis will provide a sound foundation for transitioning to risk-informed floodplain management. For example, it will provide flood hazard and risk information (at the individual property and at the community level) to communities in areas protected by levees and will identify their risk. The present system has implied to communities that live behind accredited levees that they are "safe." This implication and the lack of specific risk information can be addressed with risk analysis products that are designed, with the support of risk communication experts, to better inform communities and decision makers about flood risks. Also, a modern risk analysis would provide guidance to the communities for development of new strategies that would encourage wise floodplain management.

In addition to advancing the technical basis for flood risk analysis, the approach described here is, in many ways consistent with the methods being developed and implemented by USACE (Box 3-1). However,

[6] The terms "reliability" and "performance" are used differently across the flood risk community. For the purposes of this report, the geotechnical behavior of levees is referred to as performance. Reliability is used to refer to the likelihood under a given condition that the levee or levee system will perform in a way for which it was designed (IPET, 2009).

BOX 3-1
USACE Risk-Related Guidance

USACE began to develop an approach to using modern risk concepts in the evaluation of flood management projects in the early 1990s. In 1996, the agency published its first formal guidance document on risk, *Risk-Based Analysis for Flood Damage Reduction Studies* (USACE, 1996). This manual describes how the risk concepts would be used in USACE flood damage reduction studies in the conduct of hydrologic, hydraulic, geotechnical, and other analyses. The manual became the standard for conduct of hydrologic and hydraulic analyses for planning studies but was not adopted by the geotechnical and other communities within USACE. In 2000, USACE issued *Design and Construction of Levees* to promulgate the basic principles that should be used in design and construction of levees (USACE, 2000). That manual noted the use of a risk-based approach and hydrologic and hydraulic design, but indicated that geotechnical would continue to use a deterministic methodology.

In 2006, in the wake of Hurricane Katrina, USACE issued *Risk Analysis for Flood Damage Reduction Studies*, to update the 1996 manual on risk-based analysis, noting that

> The ultimate goal is a comprehensive approach in which the values of all key variables, parameters, and components of flood damage reduction studies are subject to probabilistic analysis. Not all variables are critical to project justification in every instance. In progressing toward the ultimate goal, the risk analysis and study effort should concentrate on the uncertainties of the variables having a significant impact on study conclusions. (USACE, 2006).

Structural and geotechnical analyses were to be included in overall risk-based analyses.

In 2006, in the wake of levee failures during Hurricane Katrina, USACE directed the conduct of an evaluation of the performance of the New Orleans hurricane protection system. The results of this evaluation, which was carried out by the Interagency Performance Evaluation Task Force, highlighted the need for the use of modern risk analyses in the design and construction of levees and related flood damage reduction structures. This evaluation led to the establishment of the USACE Levee Safety Program with the mission "to assess the integrity and viability of levees and recommend courses of action to make sure that levee systems do not present unacceptable risks to the public, property, and environment." In response to congressional and administration direction, it also initiated a levee inventory project that has been transformed into the National Levee Database (NLD) (see Chapter 6).

Faced with the need to evaluate many levees that were part of both the USACE and NFIP programs, in 2010, USACE issued the *USACE Process for the National Flood Insurance Program (NFIP) Levee System Evaluation*. This document provided guidance to field activities on the evaluation of NFIP levees to determine if USACE would certify that the levees met the standards of 44 CFR §65.10. The document indicated that it was "USACE policy to apply a probability and uncertainty analysis framework to NFIP levee system evaluations." It noted however that, "Probability of exceedance and uncertainty-based methodologies are under development and emerging for structural and geotechnical engineering elements but are not yet sufficiently mature for direct application in NFIP levee system" (USACE, 2010).

In 2008, USACE established a Risk Management Center to develop policies, methods, tools and systems to enhance its risk management activities overall, including as this applied to levees and dams. As part of its tool and policy development activity, the Risk Management Center is developing processes by which levee safety can be analyzed in the field and levee safety classifications determined for use in levee management.

USACE is currently using risk analysis in its programs and continues to develop, along multiple paths, its approach to use of these concepts in levee planning, design, and evaluation. No single approach has been promulgated, but all are founded on the use of risk-based methodologies. Although USACE's goal is to ultimately develop an integrated approach to risk analysis across its flood risk reduction portfolio, the agency has not yet issued fully coordinated policy guidance and technical instructions that would define the approach and how it will be implemented.

some differences may exist in the USACE approach and the analysis proposed here (e.g., in the evaluation of uncertainties). Yet, by adopting a risk-based approach, the NFIP and USACE can develop a common technical basis for evaluating flood protection systems and defining their future working relationship as it relates to levees in the NFIP.

Thus, the proposal to adopt a modern risk analysis is considerably more than implementing a different technical approach to the determination of insurance rates, but rather a program-wide adoption of and engagement in risk-informed management principles and practices. As such, the implementation of risk analysis methods will be an integrated program activity that extends beyond the NFIP actuaries and risk analysts.

SUMMARY DESCRIPTION OF A MODERN RISK-BASED APPROACH

The elements of a modern, flood risk analysis are schematically illustrated in Figure 3-5. The elements of the analysis are summarized here and more fully described in Appendix H. They include:

- Flood Hazard Analysis. The hazard analysis estimates the frequency of occurrence and the magnitude of flooding (flood elevations, flow velocities, forces) that may impact a region. A key component of the hazard analysis is an assessment of the sources of epistemic uncertainty and the effect that these uncertainties have on the estimate of the frequency of occurrence and the magnitude of future flood events.
- Levee and Other Hydraulic Structure and Component Fragility Analysis. The purpose of the fragility analysis for structures and components that make up a flood protection system is to estimate their performance as a function of the flood levels (elevations, forces, etc.) that may occur. The elements of the flood protection system may include structures (levees, floodwalls, gates, etc.), mechanical or electrical components (pumps, emergency power systems, etc.) and operators (personnel responsible for gate closures, pump system operation, etc.). The hazard level that causes failure of a structure or component is not exactly known; therefore, the conditional prob-

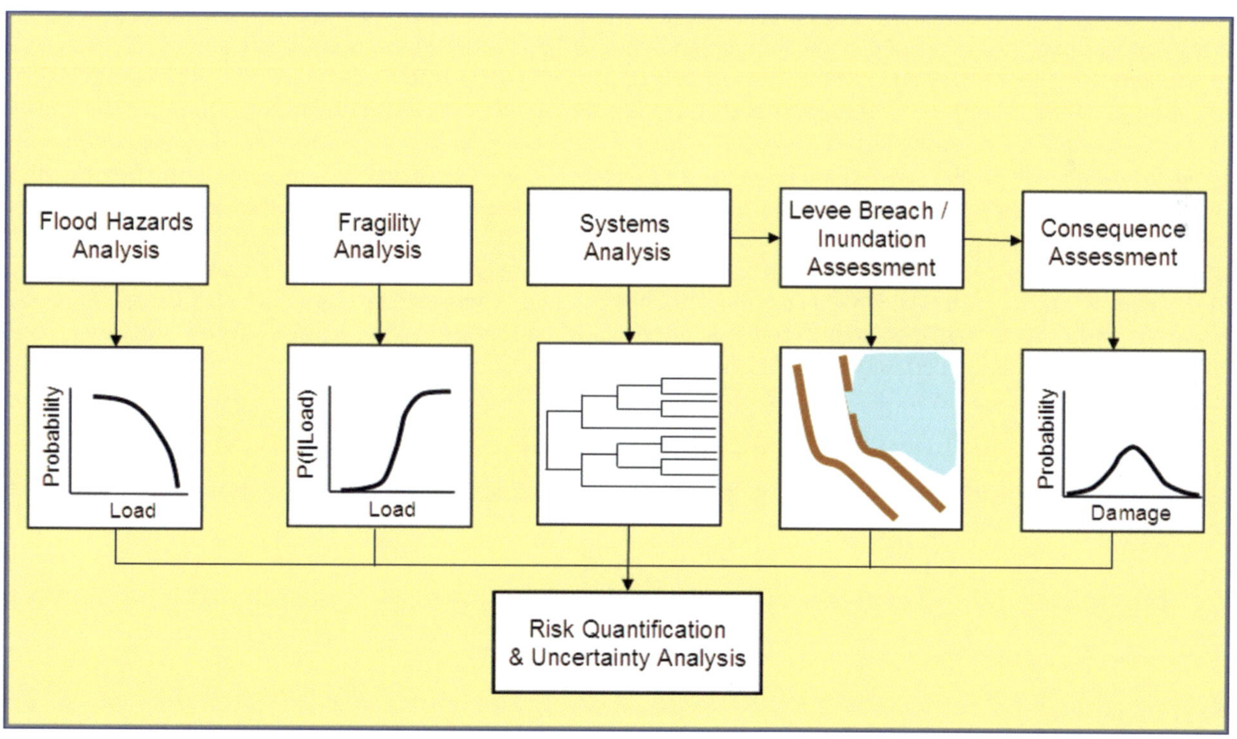

FIGURE 3-5 Elements of the modern flood risk analysis.
SOURCE: Modified from IPET (2009).

ability of failure or damage (the fragility) is uncertain. A fragility curve shows the variation of the conditional probability of structure or component failure, given a level of flooding. At low levels of flooding, the conditional probability is zero. As flood levels increase, the conditional probability of failure increases, until a point is reached where failure is certain. There are a number of uncertainties in estimating the fragility of a structure or component, thus making the assessment of conditional probability of failure uncertain.

- Systems Analysis. The potential for flooding in a region, and in particular, protected areas, depends on the performance (response) of the natural and man-made system during a flood event. This includes all features that affect flooding along the river system, for example, the hydrologic impact of upstream structures and non-structural measures. A system model defines the relationship between the hazard and the combination of events (levee performance including failure, nonfailure, and overtopping) that can lead to different flooding outcomes. A system model describes the relationships between different elements of the flood protection system and their performance and the interaction with the flood event itself and the likelihood that protected areas may experience flooding. These relationships may be physical/causal, functional (systems operations), or probabilistic (correlation of different factors).
- Inundation Analysis Including Levee Breach and Inundation. For a flood event that is modeled, the inundation that may occur in a region is a function of the flooding level, the performance of the structures and components that provide protection to a region, the interrelationship of structures and their performance, and the hydraulic assessment of inundation that may occur. In the event of levee or floodwall failures, an assessment of the breaching that may occur is considered. This includes the size and timing of the breach, as well as its location. When a breach occurs, properties that are located in the resulting inundation are likely to experience greater damage than if the same area had been flooded in a without-levee situation. There are a number of reasons for this. First, if there are structures in the area adjacent to the breach, these structures will be impacted by high-velocity flows. In addition, these structures may be completely destroyed as a result of the scour that takes place as the breach forms. Similarly, structures located beyond the scour zone will also experience high-velocity flows that otherwise might not have occurred in a without-levee case.
- Consequence Assessment. The purpose of the consequence assessment is to estimate the damage to structures that are inundated in a flood event. At this time, the routine consequence analysis estimates direct damages to structures. Indirect losses and such aspects as loss of life and social cultural impacts are in the earlier stages of evolution (IPET, 2009). The assessment of damage is often based on models derived from empirical loss data. However, as part of the consequence analysis, it is important to consider the range and type of damages that may occur for events where there are limited loss data. For instance, if a levee breaches, structures in the flooded area will be inundated to a certain level. Some structures may experience greater damage if they are located close to the breach scour zone.
- Risk and Uncertainty Quantification. This part of the risk analysis probabilistically combines the elements of the analysis and calculates the frequency of occurrence of flood events and the consequences (damage to structures). As part of the quantification, the epistemic uncertainties, discussed below, are also propagated through the analysis to estimate the uncertainty in the risk results.

A MODERN RISK-BASED ANALYSIS: THE FOUNDATIONAL BASIS

As a starting point for the development of an NFIP flood risk analysis, a conceptual framework for a flood risk analysis can be defined as

$$R = f(H,V,C) \tag{3-1}$$

where risk (R) is a function of

H = flood hazard,
V = vulnerability, and
C = consequence (e.g., economic impact, public safety).

This basic conceptualization of flood risk analysis has been adopted by the Department of Homeland Security and is consistent with Box 1-3 (NRC, 2010). In this context, risk is a function of the flood hazard to which a community is exposed, the vulnerability of flood protection systems, and the consequences associated with flooding and system failures and the damage to a community, including economic impact and life safety. Implicit in this conceptual framework is the probability that consequences may occur.

This conceptual framework is extended to the assessment of the probability (also expressed as a frequency of occurrence) of adverse consequences that could occur as a result of a flood hazard. In other words, this extension incorporates the concept of uncertainty into the framework by an expanded, quantitative definition of risk (Kaplan and Garrick, 1981; ASME/ANS, 2009; IPET 2009):

$$R = f(H,V,C,\nu,\rho) \qquad (3\text{-}2)$$

where

ν = frequency of occurrence or exceedance and is a measure of the aleatory uncertainty (the randomness of events),

ρ = probability as a measure of the confidence to which an estimate of ν is the true value, or the epistemic uncertainty in the estimate of ν.

Equation 3-2 is consistent with the definition adopted for the entire report in the sense that the estimate of risk takes into account the occurrence of the flood hazard, the exposure and vulnerabilities of people and property to the hazard, the consequences caused by the hazard, and the uncertainty about all these parameters. But most importantly, it establishes the quantitative representation of risk.

As mathematically depicted above, a flood risk analysis is an evaluation that is made on the basis of available information, in which all aspects of the analysis are subject to uncertainties: the flood hazard analysis, the assessment of levee system performance, the assessment of flooding/inundation, and the estimate of consequences. In its recent review of the DHS approach to risk analysis, the National Research Council (NRC, 2010) noted that the evaluation of uncertainties is integral to a risk analysis. In building a risk analysis capability and risk-informed culture within the NFIP, how uncertainty is defined and incorporated is critical. Experience in other fields suggests that there is considerable benefit to defining these fundamentals so that they become a core element of the methods and tools that are developed (SSHAC, 1997; USNRC, 2011).

There are different types of uncertainty to consider in a flood risk analysis. The first type is attributed to the inherent randomness of events in nature—aleatory uncertainty—and is, in principle, irreducible. Examples of this type of uncertainty are the outcome of a roll of the dice, the occurrence (the time and place) and magnitude of a flood event, the spatial variation of levee properties, and the performance of a levee during a flood. In the context of engineering or statistical models, aleatory uncertainty can also correspond to unique (often small-scale) details that are not explained by a "model." For a given model, one cannot reduce the aleatory uncertainty by collection of additional information. One may be able, however, to better quantify the aleatory uncertainty by using additional data.

The second type of uncertainty is attributed to lack of understanding (e.g., knowledge) about physical processes or a system that must be modeled. This source of uncertainty is referred to as epistemic (knowledge-based) uncertainty. There are a number of sources of epistemic uncertainty, including limited data to estimate model parameters, incomplete understanding of physical processes or systems, the limited ability of models (engineering or statistical) to predict events or outcomes of interest (modeling uncertainty), and the potential that multiple, alternative, technically defensible models or modeling approaches exist that give different results that affect the final assessment of risk. In principle, epistemic uncertainties can be reduced with improved knowledge and/or the collection of additional information.

The process of identifying and evaluating sources of epistemic uncertainties can vary, depending on the subject, the state of scientific or engineering understanding, observational and modeling experience, etc. For example, in a field or topical area where considerable observational experience exists and statistical models are used to develop predictive tools, the analysis of epistemic uncertainties may be an integral and in-depth part of the state-of-practice. In other fields, direct observational evidence may be limited and predictive models are based on theoretical understanding (i.e., analytical models), estimates of the model parameters, the analysts' experience, comparisons of model predictions with observations, etc. In areas where direct observation of events/parameters of interest is limited, competing models and/or scientific interpretations may exist, it is often necessary to elicit input from experts to evaluate and quantify epistemic uncertainties (Morgan and Henrion, 1990; SSHAC, 1997; USNRC, 2011).

One result from a flood risk analysis is the likelihood that a levee system will fail or a measure of the randomness of the flood hazard and the future performance of the levee system, that is, aleatory uncertainty (Figure 3-6, *top*). There is also epistemic uncertainty about the true value of the levee system failure, denoted as $\nu_{\text{Levee System Failure}}$ (ν in equation 3-2). As a result, there is a distribution of the estimate (Figure 3-6, *bottom*), which is an appropriate description of the true value of likelihood that the levee system would fail. This distribution can be viewed as a means to estimate confidence intervals on the estimate of the frequency of levee system failure.

With respect to levees, the current NFIP approach addresses epistemic uncertainty in the assessment of the frequency of flooding, that is, the flood hazard analysis, through requirements for levee freeboard. However, the

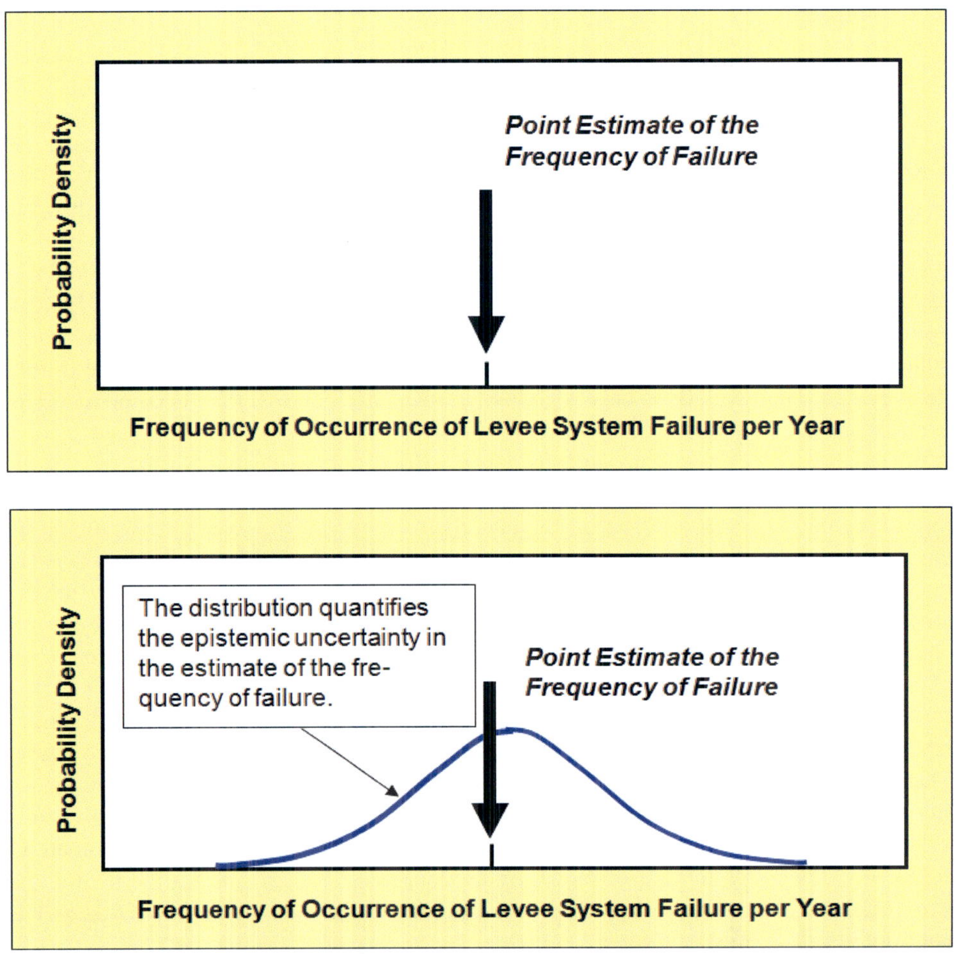

FIGURE 3-6 Estimate of the frequency of levee system failure as a single or best estimate, aleatory uncertainty (top), and an estimate that includes the quantification of both the aleatory and epistemic uncertainty (bottom) of one result from a flood risk analysis; the frequency of occurrence per year that a levee system would fail. The bottom figure can be used to make confidence statements about the estimate of the frequency of levee failure.

requirement for levee freeboard has been an ad hoc way of recognizing there are epistemic uncertainty in the assessment of the flood hazard (the one percent flood level). In other words, adding freeboard only implicitly accounts for uncertainty in a manner that does not make best use of the information and technology currently available that takes a "one size fits all" approach.[7] Thus, in some communities, more freeboard is needed and in others, less freeboard may be required. A more thorough assessment of the uncertainties in the flood hazard assessment and in levee performance would derive a situation-specific estimate of necessary freeboard.

An NFIP flood hazard analysis estimates future flood damages to insured properties for rate-setting purposes by taking into account the frequency of flooding, the performance of a levee system (in a binary fashion), and the damages that might occur to insured residences. The result of the analysis will be a distribution on the economic consequences, or the probability of certain values of loss (Figure 3-7, *top*). The result can be used to estimate expected annual flood losses, which is a measure of the flood risks that is used to determine flood insurance rates.

However, because of epistemic uncertainties in each of the elements of the risk analysis—in the NFIP flood hazard analysis, the evaluation of levee system performance, and the assessment of consequences—there is epistemic uncertainty in the distribution of economic consequences. A more modern flood risk analysis captures how epistemic uncertainty manifests in the estimate of the expected losses (Figure 3-7, *bottom*). This allows consideration of a range of values related to the confidence of a certain dollar loss (rather than a dollar value, Figure 3-7, *top*), which can be explicitly considered in the determination of insurance rates.

A Modern Risk Analysis: The IPET Risk and Reliability Analysis

As part of the IPET study commissioned by USACE, a risk analysis was performed that included both the aleatory and epistemic uncertainties in the hurricane analysis (i.e., storm surge and wave height analysis) and the levee fragility analysis (IPET, 2009). The IPET analysis is a recent, cutting-edge example of a risk-based approach to flood analysis utilizing the two different types of uncertainty; thus it is used as an example to further illustrate a modern risk analysis.

The concepts of aleatory and epistemic uncertainty are not always intuitive and thus can be difficult to visualize. In the IPET example, aleatory uncertainty is characterized in terms of a single, best estimate of hurricane storm surge, which is expressed in terms of the annual frequency of exceedance of peak surge elevations (Figure 3-8*A*, *top*). Similarly, in the levee fragility analysis, the aleatory uncertainty in levee performance is quantified in terms of the estimate of the conditional probability of levee failure as a function of this peak elevation (Figure 3-8*A*, *middle*). The results of the hurricane and levee fragility analyses can be combined to determine a point estimate of the annual frequency of levee system failure (Figure 3-8*A*, *bottom*). This part of the IPET analysis is consistent with the current NFIP flood hazard analysis.

The sources of epistemic uncertainty in the IPET example were based on the knowledge available in assessing the occurrence of hurricanes and sources of uncertainty in the hurricane modeling. There were a number of sources of epistemic uncertainty that were considered in the IPET hurricane analysis:

- uncertainty in the storm surge model ability to predict the storm surge and wave height;
- limits in available data including the estimated annual frequency of occurrence of hurricane events in the Gulf of Mexico, the frequency of hurricane central pressure deficits, and accuracy of bathymetry and near-shore and onshore hydraulic properties; and
- limits in knowledge of atmospheric processes, specifically the hurricane central pressure deficit and the radius to maximum winds.

The estimate of levee fragility in the IPET analysis evaluated two primary failure modes: stability failure of

[7] In section 65.10, the minimum standard for freeboard in NFIP communities is described: "(1) Freeboard. (i) Riverine levees must provide a minimum freeboard of three feet above the water-surface level of the base flood. An additional one foot above the minimum is required within 100 feet in either side of structures (such as bridges) riverward of the levee or wherever the flow is constricted. An additional one-half foot above the minimum at the upstream end of the levee, tapering to not less than the minimum at the downstream end of the levee, is also required."

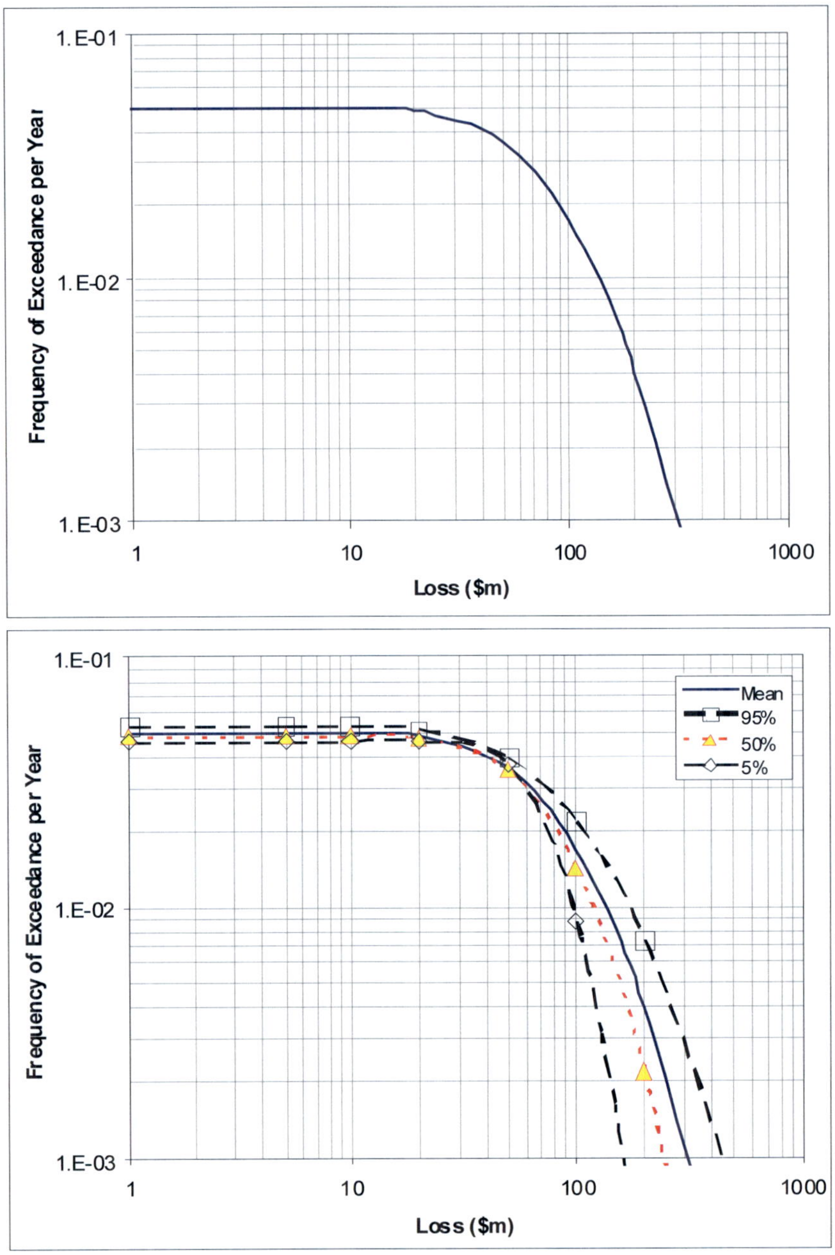

FIGURE 3-7 (*Top*) F/N curve illustrating the estimate of the frequency distribution on economic consequences without epistemic uncertainty . This shows the frequency that the annual maximum damage exceeds a specific value; this curve is the underpinning of the NFIP Hazus Multi Hazard (Hazus-MH) calculation of Average Annualized Losses due to flooding. (*Bottom*) Estimate of economic consequences that includes the quantification of the epistemic uncertainty.

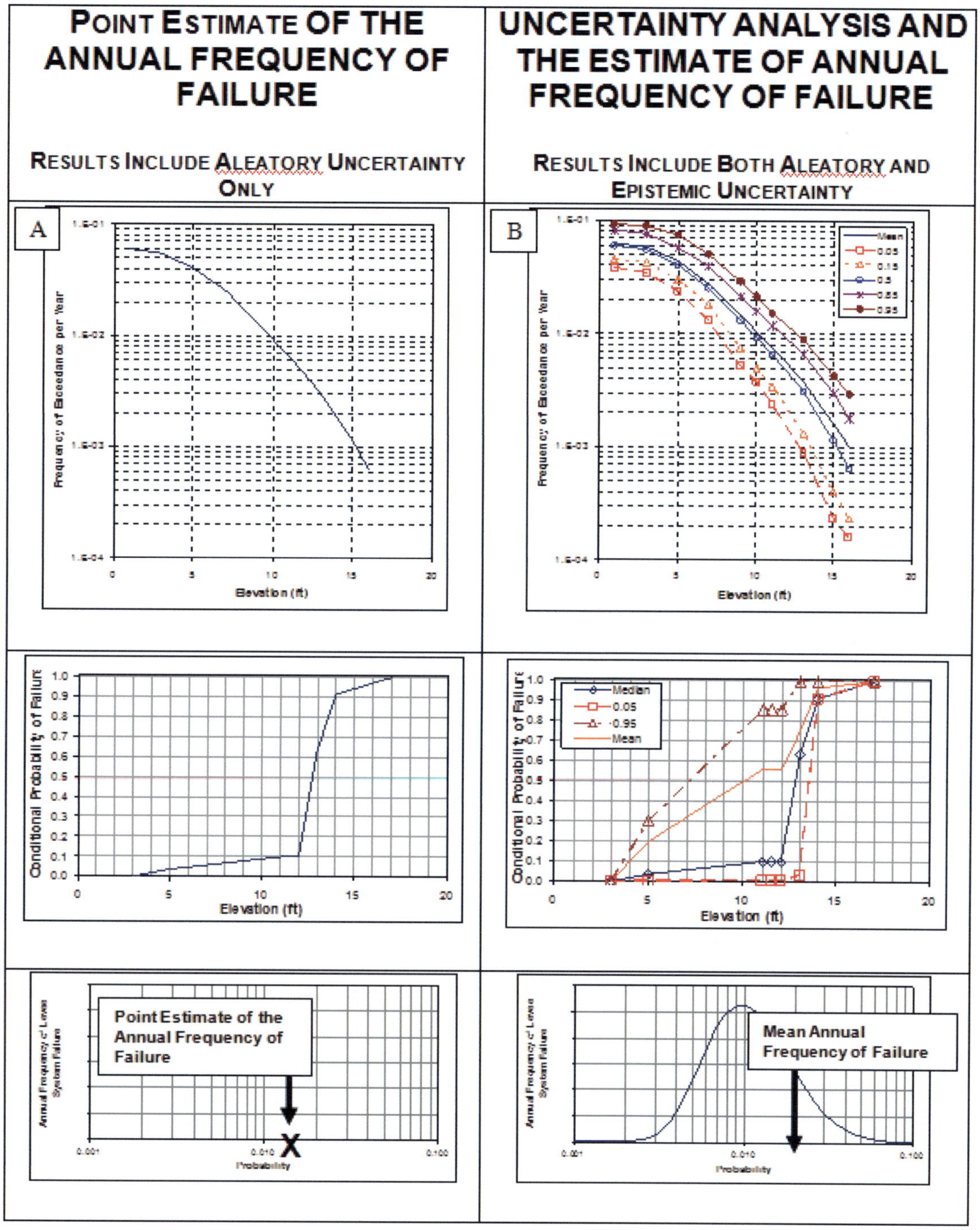

FIGURE 3-8 IPET description of the flood hazard (*top*); performance of the levee, that is, levee fragility curve (*middle*); and potential for flooding (*bottom*). Incorporation of aleatory uncertainty only is shown on the left. Both aleatory and epistemic uncertainty are shown on the right, giving a more complete estimate of the likelihood of inundation.

the levee embankment (for flood levels below the levee crest) and failure due to overtopping. The assessment of the fragility for these failure modes is subject to epistemic uncertainties; that is, an estimate of the conditional probability of failure given a peak storm surge and wave height elevation cannot be estimated exactly (Figure 3-8*B*, *middle*). For instance, epistemic uncertainties that affect the levee stability analysis include:

- uncertainty in the geotechnical stability model (model uncertainty),
- limits in available data including the estimate of soil properties and overtopping resistance of levees,
- limits in the knowledge of the levee feature.[8]

Given the epistemic uncertainty in the estimate of the frequency of hurricane surge and wave heights and in the estimate of levee fragility, there is a corresponding uncertainty in the IPET estimate of the frequency of level system failure. That is, rather than making a single point estimate of the frequency of failure, a distribution of estimates is made that quantitatively reflects the epistemic uncertainty in both the hurricane hazard and in the levee system fragility analyses. This distribution of levee failure frequency estimates is shown in Figure 3-8*B*, *bottom*, as a probability density function on the estimate of the frequency of levee system failure. From this result, the mean frequency of failure can be determined. In addition, the analyst can make "confidence" statements about the uncertainty in the levee system frequency failure.

RISK ANALYSIS RESULTS, PRODUCTS, AND USES

The NFIP is a unique government program that has a specific mandate and, at the same time, provides information to a wide range of users, establishes a standard of practice for flood hazard analysis, and floodplain mapping, and, by default, establishes what has been seen as a public safety standard for communities behind levees. The implementation of a more modern risk analysis as a part of the NFIP, as proposed and described earlier in this report, yields enhanced information about flood hazards, performance of flood protection measures, and consequences. With that information, FEMA can further support efforts to reduce flood losses in the United States (see Figure 3-9).

Table 3-1 shows a sample of the risk analysis products, uses, and users. Although uses are limited to FEMA's main missions (risk communication, insurance, and mitigation; column 3), this information is useful to the public or to public agencies meeting other objectives. For example, a local or state agency seeking to reduce flood risk could use information on levee performance to make investment decisions for strengthening levees or implementing a property buy-out program in areas of greatest potential damage.

STEPS TO DEVELOPMENT AND IMPLEMENTATION OF A RISK-BASED APPROACH

Realigning the NFIP to undertake a modern risk analysis, use products of such an analysis, and account for uncertainty in the inputs is a shift from "business as usual." Current NFIP analyses focus on hazard—primarily on the 1 percent and 0.2 percent annual chance exceedance events. Current analyses limit consideration of consequence of flooding to insurance rate setting, without geographic specificity that is possible as proposed here. Levee performance is considered in current practice in a standards-based manner, with accredited levees considered as surely protecting against inundation and unaccredited levees as surely not protecting. What is proposed herein expands the risk assessment to consider explicitly the entire range of flood events, integrates the hazard assessment with economic consequence estimation throughout the analysis and decision making, and accounts in a more robust manner for levee performance throughout the range of hazards. **The NFIP should move to a modern risk analysis that makes use of modern methods and computational mapping capacity to produce state-of-the-art risk estimates for all areas that are vulnerable to flooding.**

The analysis of flood risks is multidisciplinary (hydrologists, geotechnical and structural engineers, operations personnel, etc.). As procedures for the implementation of a modern NFIP risk analysis are developed, these pro-

[8] Estimate of the correlation length of levee properties. Correlation length refers to the length along a levee over which properties of the levee are estimated to be homogeneous and therefore its performance during the hurricane will be uniform.

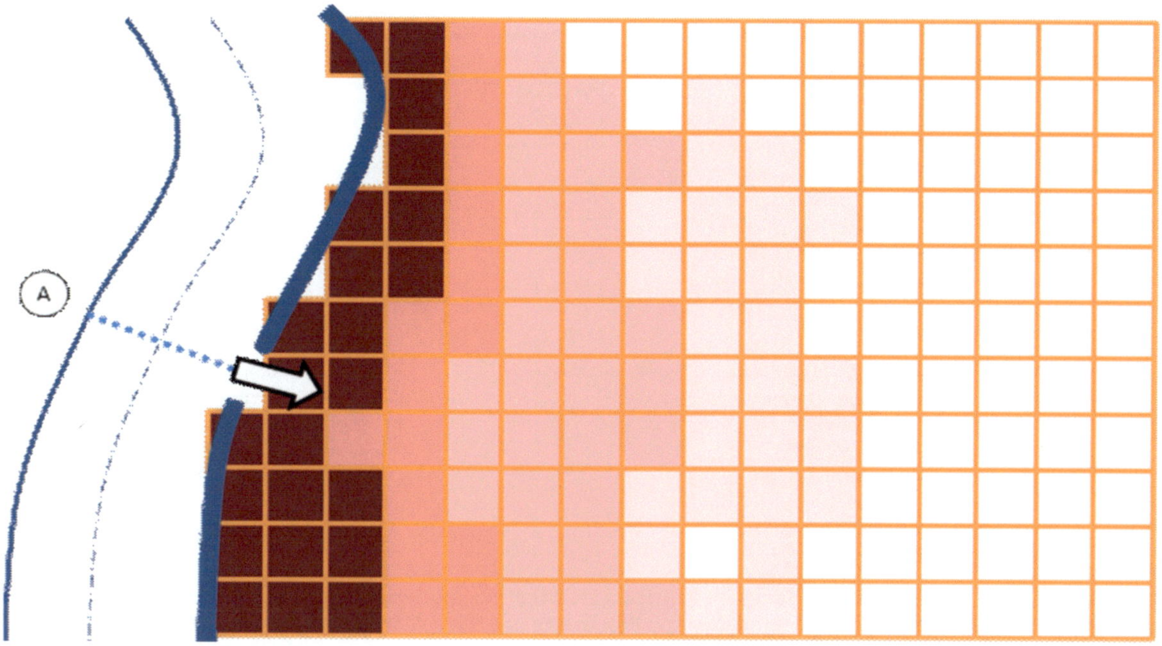

FIGURE 3-9 With results of risk analysis as proposed, annual exceedence probability for individual elements of floodplains can be mapped, shading those elements with probability > regulatory threshold. Darker elements have a greater probability of inundation; lighter elements have less. The A transect illustrates a cross section of the channel at which the flow moves from the channel onto the floodplain. This illustrates the shading with a fixed-grid two-dimensional floodplain flow model, coupled with a one-dimensional channel flood model. Thus, the probability of inundation for any location in the floodplain behind a levee could be computed and mapped, considering the natural hazard and the performance of the levee. This would enable quick inspection to determine areas of greater hazard.

cedures need to clearly define the responsibilities of the professionals involved in the study, the interdisciplinary coordination that is required, the evaluations they are expected to perform, and the results they need to generate. This report does not address these issues specifically, but does discuss steps toward the development and implementation of a risk-based program.

Implementing a Risk-Based Approach

The following tasks are necessary in order for FEMA to implement a modern risk-based approach successfully:

1. Develop, test, and deploy—with the same care and rigor with which standard procedures for hydrologic and hydraulic analysis and mapping have been developed, tested, and deployed—a procedure for the complete risk analysis. This procedure should be consistent in concept with that described in the *Summary Description of the Risk-Based Approach* section of this chapter, including all components shown. This includes algorithms and computational methods that consider the likely users of the risk analysis procedures.
2. Develop, test, and deploy a consistent procedure for assessing levee systems (rather than individual levee segments) and for describing levee performance with a fragility curve.
3. Design, develop, test, and deploy a software application that can be used for FEMA's risk analysis by agency staff, affected public agency staff, and contractors. This application may incorporate features of FEMA's Hazus application—particularly the consequence assessment features and databases.
4. Emphasize and expand FEMA's program of collecting and analyzing flood damage data, including insur-

TABLE 3-1 Products Useful for FEMA's Insurance, Risk Communication, and Mitigation Programs Yielded by a Modern Risk-Based Analysis .

Information Category	Risk Analysis Product	Use	Description
Flood hazard (floodplain)	Floodplain stage frequency distributions for locations in the floodplain; including at the individual property level. This result will also include an estimate of the uncertainty in the flood frequency estimate.	Risk communication	Report on likelihood of inundation to various depths for individual parcels, census tracts, or other subdivisions, better communicating risk to occupants, owners.
Flood hazard (floodplain)	Flood stage frequency distributions for locations in the floodplain; including at the individual property level. This result will also include an estimate of the uncertainty in the flood frequency estimate.	Mitigation	Provide information on depths, velocities, and other flood properties throughout floodplain, for specified events, to inform emergency evacuation and safe shelter planning by government agencies, industry, and public.
Flood hazard (floodplain)	Flood stage frequency distributions for locations in the floodplain; including at the individual property level. This result will also include an estimate of the uncertainty in the flood frequency estimate.	Risk communication, mitigation	Support wise land-use decision making by identifying annual exceedence probability of a parcel, tract, etc. in interior floodplain. This is consistent with earlier recommendations by NRC (2000) that "the Corps use annual exceedance probability as the performance measure of engineering risk . . . and the federal levee certification program focus <u>not</u> upon some level of assurance of passing the 100-year flood, but rather upon 'annual exceedence probability.'" For convenience, and in keeping with tradition, this information can be displayed on a map similar to existing FEMA maps (Figure 3-9).
Flood hazard (floodplain)	Flood stage frequency distributions for locations in the floodplain, including at the individual property level. This result will also include an estimate of the uncertainty in the flood frequency estimate.	Risk communication, mitigation	Similar to above, but conditional flooding can be displayed to answer question: Given channel flood flow or stage of specified annual exceedence probability, will a parcel, tract, etc. on the floodplain be inundated? For convenience, this answer too can be mapped. For example, properties with annual exceedence probability >0.01 can be identified and shaded, yielding a map similar to current FEMA maps, but taking into account performance of levees in other than a binary manner (Figure 3-9).
Deaggregation of the flood hazard (floodplain)	Conditional probability distribution on the relative contribution of different flood scenario types (levee overtopping, levee breach, breach locations, interior drainage and ponding, etc.) to the frequency and depth of flooding.	Risk mitigation, emergency preparedness	Report on the relative contribution of different events (flood events, levee failure events, etc.) that contribute to flooding in a community. This detailed information provides insight into the hazards to which a community is exposed and can guide the selection of mitigation measures and emergency preparedness and support risk communication.

TABLE 3-1 Continued

Information Category	Risk Analysis Product	Use	Description
Levee system performance	Levee system fragility estimates, including an estimate of the uncertainty in the system fragility.	Risk communication	A levee system fragility curve provides quantitative information on the performance of a levee system, which may comprise multiple levee reaches, gate structures, etc. It provides direct information on the reliability of infrastructure performance and the protection it is intended to provide for a community. This can be integrated with hazard information to provide information on combined probability that a certain flood level will occur and that the levee will protect from interior area flooding.
Levee system performance	Levee system fragility estimates, including an estimate of the uncertainty in the system fragility	Mitigation	Provides information useful for identification of deficiencies and for setting priorities on repairs or upgrades to existing infrastructure, especially when integrated with hazard information.
Frequency distribution on consequences	Frequency distribution on the consequences of flooding. This will also include an estimate of the uncertainty in the frequency of exceeding levels of consequences.	Insurance	Enables computation of expected annual damage for use with premium determination procedures described in Chapter 5. This information can be provided at the individual structure level and community level.
Frequency distribution on consequences	Frequency distribution on the consequences of flooding. This will also include an estimate of the uncertainty in the frequency of exceeding levels of consequences.	Mitigation	Enables computation of expected annual damage for comparison with cost estimated for flood management alternatives to assess feasibility of investment for federal, state, and local mitigation planning.
Frequency distribution on consequences	Frequency distribution on the consequences of flooding. This will also include an estimate of the uncertainty in the frequency of exceeding levels of consequences.	Risk communication	Provides property owner or floodplain occupant with information about flood losses that they may expect on average for any year or over a longer duration.

NOTE: Risk analysis products are results of or intermediate products of the risk analysis described in this chapter. Uses are limited to FEMA's main missions: risk communication, insurance, and mitigation.

ance claim data and corresponding flood properties, and adding to that damage and flood property data collected from other agencies, including USACE and state and local government agencies. Depth-damage models commonly used for assessing consequence as a component of risk analysis are best calibrated and verified with actual damage data. Completing this task will ensure that models for consequence estimates are accurate and reliable.

5. Develop, test, and deploy consistent procedures for describing the uncertainty in all inputs to the risk analysis, include uncertainty about flow frequencies, water surface elevations, levee performance models, and consequence estimates.

6. Refine and enhance the methods of communication of information on flood risk, including now information on consequence. With the revised risk analysis approach proposed here, the information to be communicated offers an estimate of direct, tangible, economic impact of flooding—the cost to a property owner and information on potential life-safety consequences. (The risk analysis conducted by IPET for New Orleans indicated the potential

deaths that could be encountered for various levels of flooding.) The revised analysis will also offer information about the probability that water will exceed specified depths in any year; this can be expanded easily to describe the probability that flood depths will exceed specified depths over the life of a mortgage or another informative duration.

7. Work closely with other federal agencies to ensure that techniques are consistent across agencies so that communities are not faced with using multiple approaches to evaluate a levee system. (See Chapter 8 for further discussion.)

New Technology and Cost

FEMA can leverage new technology to share the information now available. For example, in 2000 the state of North Carolina was designated a Cooperating Technical State by FEMA, a decision largely driven by the impact of Hurricane Floyd in 1999 highlighting the need for accurate flood maps for the state. North Carolina's Floodplain Mapping Program began updating and acquiring high-resolution flood hazard data, specifically light detection and ranging (Lidar) to generate digital elevation data, and creating new Digital Flood Insurance Rate Maps (DFIRMs) (North Carolina Floodplain Mapping Program, n.d.). This information is available on the Web through North Carolina's Floodplain Mapping Information System, which includes access to DFIRMs and other features such as data export capability and a property address search function to enable viewing of flood risk to a specific property (Figure 3-10).

The cost to FEMA of developing procedures for and implementing a program based on modern risk analysis cannot be predicted with certainty without information from pilot applications, which are a valuable pursuit. However, in the absence of pilot studies, a comparison between the costs for a modern risk analysis that would be above that incurred for a Flood Insurance Study (FIS) using current methods (i.e., the incremental cost) is drawn.[9]

Current FIS procedures require many of the same hydrologic and hydraulic inputs required for a modern risk-based analysis, which indicates that the incremental cost for pilot studies should not be significant. Similarly, information on exposure and vulnerability of floodplain property is available in FEMA's Hazus database, and so the incremental cost of gathering information should not be significant (unless a community prefers more recent information, in which case the cost of conducting a new structure inventory would be included). Topographic data are necessary for establishing the relationship between flood hazard and consequence; these are collected for current FIS studies, and so no additional cost will be incurred for this portion of the modern risk analyses.

The greatest cost, not currently incurred for a FIS, is the cost of developing system fragility curves—the functions that represent the likelihood that levees and appurtenant structures will provide the protection for which they were designed. Developing the curves requires information about the materials, methods of construction, operation and maintenance, performance history, and so on. Much of the information required to develop fragility curves is also required to meet the current requirements for certification and accreditation under 44 CFR §65.10. However, the manner of interpretation of the information is different when compared with the proposed modern risk analysis approach. No longer can an engineer compare the system or levee state to a standard, finding that the standard is or is not met. Instead, a qualified geotechnical engineer will evaluate the integrity of a levee system and incorporate this quantitatively in the conditional performance of the system. This can be done, as the State of California demonstrated with its Central Valley Flood Protection Plan. For risk analyses to support that study, engineers developed fragility curves at approximately 300 locations in approximately 1 year, following an extensive data collection effort (CA DWR, 2012). Similarly, USACE has developed levee fragility curves for use in ongoing planning studies, all of which now employ a risk analysis along with procedures and guidance for the development of appropriate levee fragility curves (USACE, 2006; Schultz et al., 2010).

---

[9] A FIS is an examination of flood hazards with a specific legislative definition (Appendix C). From the FIS, FEMA's FIRMs are produced. A flood hazard study, which is discussed in this chapter, is a consideration of the frequency of flooding included in the FIS.

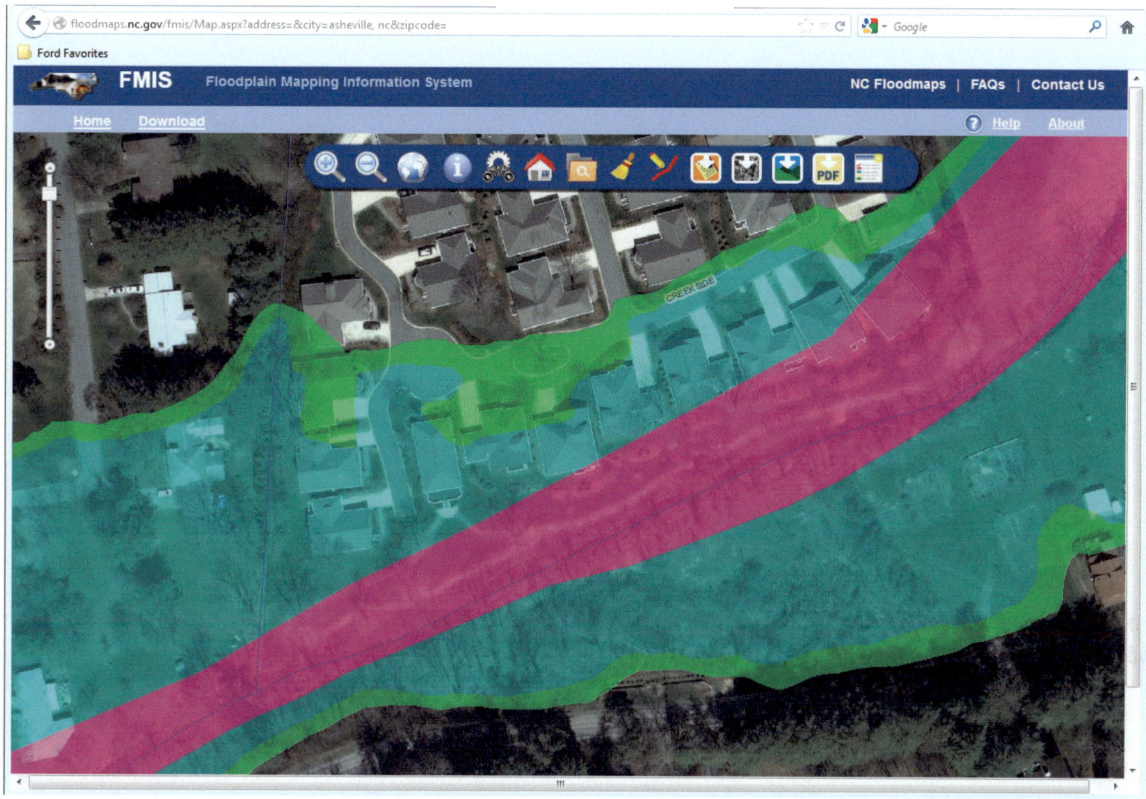

FIGURE 3-10 A screen shot of an area subject to flood risk in North Carolina according to high-resolution flood hazard data gathered. The menu bar at the top of the image allows the user to explore exposure of an individual structures to the flood hazard as well as other options such as a DFRIM export function. Pink depicts areas subject to one percent annual chance flooding (AE zone, floodway), blue areas are subject to 1 percent annual chance flooding (AE zone, has base flood elevation established), and green areas are subject to the 0.2 percent annual chance flooding (shaded X zone).
SOURCE: http://www.ncfloodmaps.com, accessed January 7, 2012.

A Living, Risk-Based Analysis: Maintenance and Agents of Change

A modern risk-based analysis will serve a very different role than FEMA's current analysis of the flood hazard, which yields a flood map that is not updated with regularity. The modern risk-based approach will be an integral part of all aspects of the NFIP: mapping, floodplain management, portfolio analysis, and insurance rate assessment. As such, the risk-based analysis will be a living, integral part of the NFIP; something that is used regularly by the program actuaries and analysts. By its very nature, the modern risk-based analysis will be a regularly updated, preferably online, resource for FEMA.

When a modern risk analysis is performed for a community, it would be based on information that is available at the time. Following its completion, there are factors that may change the risk profile of a community, and therefore require maintenance (updating of certain data) and possibly revision of hazard, performance, and consequence components and the resulting risk analysis products shown in Table 3-1. Given the role the risk analysis will play in the NFIP, it is important that FEMA develop and implement procedures to periodically review, maintain, and update the risk-based analysis, as necessary. There are a number of agents of change that may affect future flood risks: climate change, community development and changes in land use, changes in population, condition of flood mitigation systems (success of system maintenance, changes in system configuration, etc.), changes in the watershed, and so on. As such, FEMA's implementation plan needs to take these factors into account.

The processes for maintenance and review needs to include

- periodic review of the elements of the risk analysis (hydrologic analyses that describe the flood hazard, geotechnical engineering studies that yield description of levee system performance, inventory of properties, etc.);
- standards that define the steps and basis to revise, refine, or replace one or more elements of the risk analysis for a community, and that establish schedules for periodically reviewing and revising
    - individual property information,
    - levee and flood mitigation system properties and condition, and
    - hydrologic and hydraulic modeling.

Within its current program, FEMA has requirements for archiving hydrologic and hydraulic models used in the FIS (FEMA, 2003) and requiring that information on levees be maintained and certification be periodically updated. With the development and implementation of a risk-based analysis to evaluate community flood risks, FEMA needs to extend these practices to a community risk-based approach and faithfully maintain them.

FEMA policy with regard to levees is spelled out in 44 CFR §65.10. In addition to defining the base flood as the one percent annual chance exceedance flood, the regulation states that a community must provide evidence of a flood protection system's ability to provide protection from the base flood. As noted in Chapter 1, the requirements that determine whether FEMA recognizes a levee system on NFIP maps include (1) design criteria, (2) operation plan criteria, and (3) maintenance plan criteria. If the levee system satisfies the criteria and is certified, it is recognized on NFIP maps, with the protected floodplain "mapped out" of the hazard area. If the levee system does not meet the criteria, the levee system is assumed to provide no protection.

Within the context of a risk-based analysis to evaluate the floodplain and flood protection systems in particular, the analysis would quantify the residual risk in the area behind every flood protection system, and develop insurance pricing based on the level of protection that is provided. For instance, communities will be given credit for the level of protection that is provided, even if this level is below the base flood standard—the estimated risk analysis and corresponding rates will recognize this change.

REFERENCES

AIR (American Institutes for Research). 2012. AIR Models Overview. Available online at http://www.air-worldwide.com/Models/Overview/. Accessed November 28, 2012.

ASME/ANS (American Society of Mechanical Engineers and American Nuclear Society). 2009. Standard for Level 1/Large Early Release Frequency Probabilistic Risk Assessment for Nuclear Power Plant Applications, Addenda to ASME/ANS RA-S-2009. LaGrange Park, IL: ANS.

CA DWR (California Department of Water Resources). 2012. 2012 Central Valley Flood Protection Plan, Attachment 8E Levee Performance Curves. Sacramento, CA: CA DWR.

DOE (Department of Energy). 2002. Natural Phenomena Hazards Design and Evaluation Criteria for Department of Energy Facilities. DOE-STD-1020-2002. Washington, DC: DOE.

FEMA (Federal Emergency Management Agency). 2003. Guidelines and Specifications for Flood Hazard Mapping Partners. Available online at http://www.fema.gov/library/viewRecord.do?id=2206. Accessed December 12, 2012.

FLOODsite. 2004. Flood Risk Management. Available online at http://www.floodsite.net/html/flood_risk.htm. Accessed August 14, 2012).

ILPRC (Interagency Levee Policy Review Committee). 2006. The National Levee Challenge: Levees and the FEMA Flood Map Modernization Initiative. Available online at http://www.fema.gov/library/viewRecord.do?id=2677. Accessed August 8, 2012.

IPET (Interagency Performance Evaluation Task Force). 2009. Performance Evaluation of the New Orleans and Southeast Louisiana Hurricane Protection System: Final Report of the Interagency Performance Evaluation Task Force. Available online at https://ipet.wes.army.mil/. Accessed December 4, 2012.

Kaplan, S., and B. J. Garrick. 1981. On the quantitative definition of risk. Risk Analysis 1(1):11-27.

Morgan, G. M., and M. Henrion. 1990. Uncertainty: A Guide to Dealing with Uncertainty in Quantitative Risk and Policy Analysis. Cambridge, UK: Cambridge University Press.

Moser, D., T. Bridges, S. Cone, Y Haimes, B. K. Harper, L. Shabman, and C. Yoe. n.d. Transforming the Corps into a Risk Managing Organization. White Paper. Available online at http://corpsriskanalysisgateway.us/data/docs/ref/Explore%20 Resources/IWR%20Reports/White_Paper-Transforming_the_Corps_into_a_Risk_Managing_Org.pdf. Accessed February 5, 2013.

NOAA (National Oceanographic and Atmospheric Administration). 2012. Hydrologic Information Center—Flood Loss Data. Available online at http://www.nws.noaa.gov/hic/. Accessed February 27, 2013.

North Carolina Floodplain Mapping Program. n.d. LIDAR and Digital Elevation Data. Fact Sheet. Available online at http://www.ncfloodmaps.com/pubdocs/lidar_final_jan03.pdf. Accessed February 21, 2013.

NRC (National Research Council). 2000. Risk Analysis and Uncertainty in Flood Damage Reduction Studies. Washington, DC: National Academy Press.

NRC. 2010. Review of the Department of Homeland Security's Approach to Risk Analysis. Washington, DC: The National Academies Press.

NRC. 2012a. Dam and Levee Safety and Community Resilience: A Vision for Future Practice. Washington, DC: The National Academies Press.

NRC. 2012b. Disaster Resilience: A National Imperative. Washington, DC: The National Academies Press.

RMS (Risk Management Solutions). 2012. RMS Models. Available online at http://www.rms.com/models/. Accessed November 28, 2012.

Rogers, J. D. 2012. GE 301 Lectures—Evolution and Development of Flood Control Engineering. Available online at http://web.mst.edu/~rogersda/umrcourses/ge301/. Accessed November 15, 2012.

Sayers, P., G. Galloway, E. Penning-Rowsell, F. Shen, K. Wang, Y. Chen, and T. Le Quesne. 2012. Flood risk management: A Strategic Approach—Consultation Draft. UNESCO on behalf of World Wildlife Fund–UK/China and the General Institute of Water Design and Planning, China.

Schultz, M. T., B. P. Gouldby, J. D. Simm, and J. L. Wibowo. 2010. Beyond the Factor of Safety: Developing Fragility Curves to Characterize System Reliability. Washington, DC: U.S. Army Corps of Engineers.

SSHAC (Senior Seismic Hazard Analysis Committee). 1997. Recommendations for Probabilistic Seismic Hazard Analysis—Guidance on Uncertainty and Use of Experts. Rockville, MD: U.S. Nuclear Regulatory Commission.

USACE (U.S. Army Corps of Engineers). 1996. Risk-Based Analysis for Flood Damage Reduction Studies. EM 1110-2-1619. Washington, DC: USACE.

USACE. 2000. Design and Construction of Levees. EM 1110-2-1913. Washington, DC: USACE.

USACE. 2006. Risk Analysis for Flood Damage Reduction Studies. ER 1105-2-101. Washington, DC: USACE.

USACE. 2008. Improving Public Safety—From Federal Protection to Shared Risk Reduction. Letter to J. Nussle, Director, Office of Management and Budget, April 16, 2008. Washington, DC: USACE.

USACE. 2010. USACE Process for the National Flood Insurance Program (NFIP) Levee System Evaluation. EC-1110-2-6067. Available online at http://publications.usace.army.mil/publications/eng-circulars/EC_1110-2-6067.pdf. Accessed November 27, 2012.

USNRC (U.S. Nuclear Regulatory Commission). 2011. Practical Implementation Guidelines for SSHAC Level 3 and 4 Hazard Studies, NUREG-2117. Washington, DC: USNRC.

URS/JBA (URS Corporation/Jack R. Benjamin & Associates, Inc.). 2008. Delta Risk Management Strategy (DRMS), Phase 1, Risk Analysis Report (Draft). Sacramento, CA: California Department of Water Resources.

4

Levee Analysis and Mapping Procedure

As indicated in Chapter 2, the examination of levees under the Map Modernization Program identified many levees that no longer met the requirements of Title 44, Section 65.10 of the *Code of Federal Regulations* (44 CFR §65.10). As a result, the areas behind these levees in the one percent annual chance floodplain were to be remapped as being in the Special Flood Hazard Area (SFHA) and would be subject to the mandatory purchase requirements of the NFIP. In response to congressional requests, the administrator directed FEMA staff to develop an approach to mapping nonaccredited levees that would reflect any risk reduction and the commensurate rates that nonaccredited levees might afford the communities. In March 2011, FEMA issued a public notice stating that

> The new approach [being developed] will provide a more precise assessment of flood risk in areas impacted by levees. Because the new modeling methods may affect the treatment of a levee, FEMA will temporarily withhold issuing flood risk study final determinations for those communities whose levees do not meet accreditation requirements of 44 CFR §65.10. This temporary delay will allow FEMA to give proper consideration to levees under the new modeling procedures. (FEMA, 2011b)

Between, March 2011 and February 2012, FEMA developed and circulated for public comment the Levee Analysis and Mapping Procedure (LAMP) to identify where risk reductions might occur from evaluation of the performance of levees below the one percent annual chance standard. The LAMP was designed to be implementable without statutory or regulatory changes, or changes to the insurance program. The LAMP is currently being prepared for use in pilot analyses in communities around the country.

PROPOSED PROCEDURE

The proposed procedure is characterized by FEMA as being

> a cost-effective, repeatable, and flexible approach that (1) complies with all statutory and regulatory requirements governing the NFIP, most notably 44 CFR §65.10, (2) leverages local input, knowledge, and data through proactive stakeholder engagement; (3) aligns available resources for engineering analysis and mapping commensurate with the level of risk in the leveed area; and (4) considers the unique flooding and levee characteristic (solely from an engineering perspective) of each levee system. (FEMA, 2011a)

Under the existing approach, when a levee is no longer accredited, all land areas that would be subject to inundation by the one percent annual chance flood if the levee system was not present are designated as SFHAs. Under the proposed approach, based on analyses of levee reaches,[1] some lands areas may be determined to be still protected against inundation by the one percent annual chance flood. These land areas would be designated as Zone D, an existing designation that indicates undetermined, but possible flood hazard (Appendix E). Zone D areas are not subject to mandatory flood insurance purchase requirements, although structures in this zone may be subject to lenders' flood insurance requirements. Rates under the NFIP would be adjusted from the preferred rate to a rate that reflects the undetermined risk for structures in Zone D.

The proposed approach makes use of five "procedures." The Freeboard Deficient Procedure, Overtopping Procedure, and Structural-Based Inundation Procedure deal with common deficiencies in nonaccredited levee systems *as they affect specific reaches*. The Sound Reach Procedure deals with continuous reaches that were "designed, constructed, and maintained to withstand and reduce the flood hazards posed by a 1-percent-annual-chance flood" (FEMA, 2011a). The Natural Valley Procedure is applied to entire levee systems and assumes that "the levee system would not obstruct the river from flowing within the entire natural valley of the floodplain during the 1-percent-annual-chance flood event" (FEMA, 2011a).

The Freeboard Deficient Procedure is applied to levee reaches for which the only deficiency is inadequate freeboard. In this case, the Natural Valley Approach would be used to map the affected area. This area would be designated as Zone D, except for any portions that are affected by levee reaches with other deficiencies or are subject to flooding due to interior drainage.

The Overtopping Procedure is to be applied to levee reaches for which the one percent annual chance flood is higher than the crest, but overtopping will not result in structural failure. In this procedure the levee is modeled as a lateral weir, using a one percent annual chance flood plus a factor of safety. The inundated area is designated as an SFHA, and could be smaller than the inundated area that would result without consideration of the levee, as determined by the Natural Valley Procedure. If so, the remaining area is designated as Zone D, except for any portions subject to flooding due to internal drainage.

The Structural-Based Inundation Procedure is applied to levee systems that are not accredited because of structural integrity issues, such as insufficient foundation strength or vulnerability to seepage failure. In this procedure, possible failure locations are identified and evaluated to determine the inundated area. The modeling will be based on the one percent annual chance flood, without a factor of safety. Breaches will be assumed to occur at multiple locations. The inundated area will be designated as SFHA. This area could be smaller than the inundated area that would result without consideration of the levee, as determined by the Natural Valley Procedure. If so, the remaining area is designated as Zone D, except for any portions subject to flooding due to interior drainage.

The Sound Reach Procedure is applied to continuous sections of levee that have been designed, constructed, and maintained to withstand and reduce the flood hazards posed by a one percent annual chance flood. The areas landward of these levees are mapped using the Natural Valley Analysis and designated as Zone D, except where there is inundation due to internal drainage or to flows from the nonaccredited levee reaches.

LAMP also includes a stakeholder engagement process. Once FEMA has determined that a levee system does not meet the certification requirements, it will engage stakeholders to assist in the data collection phase. When appropriate, it will establish a Local Levee Working Group to provide additional data as well as feedback on the analysis and mapping. Depending on the nature of levee deficiencies, this data collection may involve considerable responsibility on the part of the local sponsors for technical data development, the costs of which may place a significant burden on their resources.

---

[1] A levee reach is a portion of a levee system (usually a length of a levee) that may be considered as a unit taken for analytical purposes to have approximately uniform representative properties. A levee reach will be the unique entity having properties different than other reaches of the levee system that may be used to evaluate the performance of a portion of the levee system. No maximum length is associated with a reach (USACE, 2010).

PUBLIC COMMENTS

As part of the development of the LAMP, FEMA solicited input on a preliminary version of the procedure and received over 1,400 comments. The main criticisms of the proposed method had to do with the identification of a Zone D and the probability that the insurance rates in Zone D would rise and thus discourage insurance purchase. Suggestions were made to use the Zone X designation which carries less of a negative connotation.[2]

Technical comments addressed such issues as the failure of the Natural Valley method to account for local backwater effects caused by flow impediments such as levee tiebacks, confusion over the definition of a reach, use of the Freeboard Deficient Procedure believing that it "would remove elevation and insurance requirements behind a levee that has only one inch of freeboard and no armoring," and failure in the Overtopping Procedure "to consider that the true one percent chance flood stage may be substantially higher than the median one percent stage," suggesting that the Overtopping Procedure be replaced by the Structural-Based Procedure which does not appear to use a factor of safety. Fears were expressed that the new approach might "mislead the public and mortgage lenders about their risk, result in more risky development and fewer insurance policies, and dis-incentivize levee investments." Other comments noted that Zone D lacks elevation data that could be useful to homeowners, builders, mortgage lenders, and communities.

EVALUATION OF LAMP

The committee examined LAMP, met with key FEMA staff to discuss its possible implementation, and reviewed its components. The LAMP is founded on sound algorithms with sound science and engineering behind them and follows established approaches to hydrology and hydraulics. It is a first cut at an approach to dealing with the hazards associated with levees that do not meet the standards for protection against the one percent annual chance flood.

LAMP represents a response to the issues raised by members of Congress that, given funding and personnel constraints, it will likely take several years to fully implement. It represents a diversion of NFIP effort away from the long-recognized need to move the NFIP to risk-based analysis. Because it would be highly unlikely for the NFIP to move forward concurrently with implementation of LAMP and development of a modern risk-based approach as described in Chapter 3, a move to LAMP would simply postpone the inevitable need to move the NFIP to the risk-based foundation that other agencies and nations have already endorsed (Kobayashi and Porter, 2012; Serra-Llobet et al., 2013).

The implementation of LAMP would be a time-consuming and expensive endeavor for FEMA and communities.[3] It will not yield the desired results of describing flood risks more accurately for the purposes of flood risk communication, mitigation, and insurance rate setting compared with a modern risk analysis. The LAMP approach does not identify or evaluate the properties at risk or the potential for losses and it does not consider levels of risk. In addition, it is not clear that the timescale for implementation of LAMP would contribute to resolving the flood mapping issues that currently exist.

Should the decision be made to implement LAMP, the implementation will not be simple or without controversy. It will require time for FEMA and communities to assess its viability and to establish the steps required for its systematic implementation. The steps in this process will necessarily involve

1. conducting a series of trial applications of LAMP at a number of locations to understand its impact, benefits, and the considerations in its broader implementation;
2. development of guidance for implementation of LAMP on a national scale, which might have a number of elements such as preparation of lessons from the trial applications, development of guidelines, preparation of training materials, Frequently Asked Question documents, holding of training seminars, etc.; and
3. implementation of a finalized LAMP across the nation, which may require
    a. new hydraulic and hydrologic evaluations for levee reaches under evaluation,

[2] Zone X designates an area of moderate to low risk (see Appendix F).

[3] Although no cost estimates were available for the implementation of LAMP, this statement is based on the judgment of the committee.

b. significant time and resource commitments by local communities and experts to develop local knowledge and data, and

c. resource-intensive FEMA review and evaluation of community submittals (acceptance or rejection).

It is not apparent to the committee how FEMA can reasonably and defensibly price flood insurance for the range of circumstances it will encounter with the LAMP approach. Criticisms that have been levied against the NFIP concerning high premiums may persist with such an implementation of LAMP. The implementation of LAMP and the concurrent actions that are required by the Biggert-Waters Flood Insurance Reform and Modernization Act of 2012 would likely lead to confusion among communities (e.g., the repackaging of Zone D) and will have a negative impact with respect to public engagement, risk communication, and NFIP viability and public acceptance of the program. **FEMA should move directly to a modern risk-based analysis for dealing with areas behind levees and not implement the Levee Analysis and Mapping Procedure.**

DEALING WITH THE CURRENT SITUATION

If LAMP is not implemented, what should FEMA do until a modern risk-based approach can be fully implemented? NFIP-related levees fall into four categories. The first group includes levees that are currently fully accredited. A second group encompasses deaccredited levees with deficiencies that the community intends to remediate in order to bring the levee back into compliance with the provisions of 44 CFR §65.10. Depending on the nature of the deficiency and the costs associated with remediation, such actions may take a few to several years. The third group of levees includes those with deficiencies that the community recognizes cannot reasonably be remediated and that the community does not intend to remediate. This situation develops when the levee deficiencies are severe or the potential costs for repair or upgrade are so substantial that it would be impossible for the community to bear such costs. The fourth group of levees includes those that are not now accredited or that are proposed to be built at less than 44 CFR §65.10 standards to provide some level of protection to those behind levees against more frequent than one percent annual chance floods (e.g., 2 percent annual chance or 50-year floods).

Should a decision be made not to implement LAMP and to develop and a modern risk-based analysis, dealing with the current situation is still necessary. **Pending development and completion of a modern risk-based analysis, FEMA should implement and promote these interim steps:**

1. Communities with currently certified levees not in the mapping process and not having been declared deaccredited would be expected to take necessary actions to remediate any deficiencies that have been brought to their attention as a result of USACE or FEMA programs so that when the community is remapped, the levee would meet standards.

2. If the community indicates that it will take steps to restore accreditation, the following action would be taken with respect to its levees:

a. Communities would commit to an aggressive but reasonable program to remediate all deficiencies in the levees using a program similar to the USACE System-Wide Improvement Framework (SWIF) policy (Appendix I) or an integration of the SWIF and an NFIP remediation program. The SWIF policy requires participating levee sponsors to

i. provide a "description of deficiencies or issues that will be included in the SWIF and discussion of how a system-wide approach will improve and optimize overall flood risk reduction";

ii. demonstrate "that significant nonfederal resources have been, or will be, committed for developing and/or implementing the SWIF (e.g., state legislative action, bond financing)";

iii. identify "risk reduction measures that will be implemented throughout the SWIF process, including [an] overall risk communication approach that addresses [how] the risk to life increased by system-wide deficiencies" is being mitigated during the remediation period"; and

iv. describe "existing or planned interagency collaborative efforts that will contribute positively to SWIF development, implementation and oversight" (USACE, 2011).

b. Issuance of preliminary maps would be suspended until remedial action was completed. These levees would be left in the current program and any actions to remove them from accreditation would be suspended temporarily. Areas behind levees would remain outside of the SFHA and would be considered in Zone X.

c. To ensure the protection of the health, welfare, and property of those in the leveed areas, communities would be required to carry out the following actions:

i. individually inform all residents living behind the subject levees of the failure of the levee to meet FEMA and/or USACE standards, the nature of the deficiencies, resultant risks that exist, and steps being taken by the community to remediate the deficiencies;

ii. review and exercise emergency plans to include plans for the evacuation of residents from behind the levees; and

iii. encourage those behind levees to purchase insurance policies.

3. If communities do not intend to seek reaccreditation of deaccredited levees:

a. Mapping efforts would continue through issuance of the effective maps indicating that areas behind the levees are in the SFHA and subject to mandatory purchase requirements and land development regulations. At the time the new maps go into effect, the rate charged would be that for the SFHA.

b. After FEMA completes the development of a new risk-based analysis approach, the rates for these areas would be reassessed to reflect the degree of risk reduction provided by the less-than-compliant levees.

c. To ensure the protection of the health, welfare, and property of those in the leveed areas, communities would be required to carry out the following actions:

i. individually inform all residents living behind the subject levees of the failure of the levee to meet FEMA and/or USACE standards, the nature of the deficiencies, resultant risks that exist and the intention of the community to not remediate the deficiencies;

ii. review and exercise emergency plans to include plans for the evacuation of residents from behind the levees;

iii. encourage the purchase by those behind levees of insurance policies at the preferred risk rate;

iv. provide FEMA with reports of locally funded periodic inspections of the assessed levees to ensure that they continue to maintain the degree of protection used in the risk analysis.

4. Communities wishing to build a new levee in an existing SFHA or in an area where there is no flood designation should be required to construct the new levees to NFIP standards. Where a participating community has within its geographic bounds and within the SFHA levees that do not currently meet NFIP standards and that have never been proposed for certification, it would be possible, on completion of the development of risk-based analysis, for these communities to request, at community expense, FEMA evaluation of the levees to determine risk-based SFHA insurance pricing for properties affected by these levees.

A CAUTION

Levees were introduced into the NFIP because, at the time, decision makers reasoned that because new construction within the SFHA would be required to be at or above one percent annual chance flood level in order to be removed from the mandatory purchase requirement, equity dictated that those protected by a one percent annual chance "flood control" structure should also be removed from the requirements of mandatory insurance and land management behind the structures. It was assumed that these structures would provide protection equal to or in excess of the protection provided by requiring the first flood of new development in the SFHA to be at or above the one percent annual chance flood level. As indicated in Chapter 2, because the decision that areas behind levees are not subject to the mandatory purchase requirement and land regulation, studies and internal deliberations between USACE and FEMA since the early 1970s have highlighted that the aggregate risk to those behind levees may be greater than the risk to those in front of the levees (FEMA, 2006; Galloway et al., 2006; NFIP Evaluation Final Report Working Group, 2006; ASFPM, 2007; CA DWR, 2007).

REFERENCES

ASFPM (Association of State Floodplain Managers). 2007. National Flood Policy Challenges Levees: The Double-Edged Sword. Madison, WI: ASFPM.

CA DWR (California Department of Water Resources). 2007. A California Challenge: Flooding in the Central Valley. Sacramento, CA: Department of Water Resources.

FEMA (Federal Emergency Management Agency). 2006. The National Levee Challenge: Levees and the FEMA Flood Map Modernization Initiative. Washington, DC: FEMA.

FEMA. 2011a. Analysis and Mapping Procedures for Non-Accredited Levees. Proposed Approach for Public Review. Available online at http://www.fema.gov/library/viewRecord.do?id=4954. Accessed November 28, 2012.

FEMA. 2011b. National Flood Insurance Program: Answers to Questions About the NFIP. FEMA F-084. Available online at http://www.fema.gov/library/viewRecord.do?fromSearch=fromsearch&id=1404. Accessed February 15, 2013.

Galloway, G. E., G. B. Baecher, D. Plasencia, K. G. Coulton, J. Louthain, M. Bagha, and A. R. Levy. 2006. Assessing the Adequacy of the National Flood Insurance Program's 1 Percent Flood Standard. Washington, DC: American Institutes for Research. Available online at http://www.fema.gov/library/viewRecord.do?id=2595.

Kobayashi, K., and J. W. Porter. 2012. Flood Risk Management in the Peoples Republic of China: Learning to Live with Flood Risk. Mandaluyon City, Philippines: Asian Development Bank.

NFIP Evaluation Final Report Working Group. 2006. The Evaluation of the National Flood Insurance Program: Final Report. Washington, DC: American Institutes for Research.

Serra-Llobet, A., J. D. Tàbara, and D. Sauri. 2013. The Tous Dam disaster of 1982 and the origins of integrated flood risk management in Spain. Natural Hazards 65: 1981-1998.

USACE (U.S. Army Corps of Engineers). 2010. USACE Process for the National Flood Insurance Program (NFIP) Levee System Evaluation. Available online at http://publications.usace.army.mil/publications/eng-circulars/EC_1110-2-6067.pdf. Accessed November 27, 2012.

USACE. 2011. Memorandum for Commanders, Major Subordinate Commands, and Districts—Policy for Development and Implementation of System-Wide Improvement Frameworks (SWIFs). Washington, DC: USACE.

5

Insuring Properties Behind Levees

Insurance is a method of prefunding (usually over time) the adverse consequences of a loss so as to have available funds to pay damages when they arise. It is a legal contract that transfers some of the financial consequences of an uncertain low-probability but high-consequence peril to another party (called an insurer) in return for periodic relatively low payments (premiums). If an event covered by the insurance contract occurs, and if coverage is triggered by conditions detailed in the contract, then the insurer will indemnify the insurance purchaser up to the limits of the insurance policy in accordance with the terms of the contract (e.g., the deductible, the policy limits, proof of loss, etc.). Thus, the value of insurance arises at the time of a covered loss, and can thus be viewed as a hedging instrument against the financial consequences of a loss.

In insurance nomenclature, the causes of loss is called a peril, and a condition or situation that increases either the likelihood or severity of the peril occurring is called a hazard. Risk, on the other hand, is the term used to designate the loss consequence of the realization of the uncertain peril, and it may be financial or nonfinancial in nature (e.g., reputational loss). Thus for example, fire is a peril, storing oily rags next to a gas water heater is a hazard for the peril of fire, and the loss of property and the financial losses due to fire-related property destruction is the risk. Concerning the peril of flooding, building unelevated property below the base flood elevation in a floodplain is a hazard, and the financial consequences of partial or total destruction of property due to flooding is the risk that the property owner faces. Risk is an adverse consequence of uncertainty concerning perils, and without uncertainty there can be no risk.

Not all risk can be insured in the private market (i.e., by private insurers) but when available, insurance can be part of an effective risk management strategy. Other parts of the risk management process include risk identification, risk assessment, risk mitigation or control, and risk communication. Insurance is not a risk mitigation technique, however, because the legal insurance contract does not change the physical probability or the severity of a realized peril, but rather it just allocates the cost to another party.[1]

From the perspective of the purchaser, insurance is a viable risk transfer technique when the probability of the event occurring, p, is relatively small and the consequence or loss, L, is relatively large (so as to be difficult to financially assume all at once by the purchaser without insurance), but such that the expected value in any given

[1] Insurance that is truly risk-based priced can motivate policyholders to mitigate risks to lower premium costs, but it is not that insurance itself that mitigates the risk, but rather the affirmative actions of the policyholder to lower frequency or severity.

year $p \times L$ is affordably small as a basic for an annual insurance premium calculation.[2] This expected value is called the "pure premium" in insurance and is subsequently "loaded" or modified to reflect insurance underwriting costs, claim adjusting and settlement costs, and returns to the providers of capital to the insurance organization. An additional loading is added to reflect the level of uncertainty in the estimation of the pure premium. Once loaded, the result is the actual premium charged to the purchaser of insurance.

As part of the National Flood Insurance Program (NFIP), the insurance component is one pillar to address catastrophic loss potential from floods (other pillars include motivating building and land-use restrictions in vulnerable areas and construction of dams, levees, and other structures). NFIP insurance is available to all property owners in communities that participate in the program. Coverage for flood damages extends up to $250,000 dollars for residential structures and $100,000 dollars for residential contents and $500,000 for business structures and $500,000 dollars for business contents.[3] Those that require greater amounts of coverage than available from the NFIP have access to additional flood insurance through private industry, e.g., their property insurance carrier.

BASICS OF INSURANCE PRICING

Pricing is done in essentially the same manner in all areas of insurance with some differences in models and assumptions corresponding to the characteristics of the different types of insurance. Property and casualty insurance (also known as hazard insurance or general insurance in some areas) generally has a shorter contract duration, usually exists in a more highly regulated environment than life insurance, and has more complicated claims structure (with more correlated losses, the potential for multiple claims within the policy period, etc.). These characteristics have led financial/actuarial pricing formulas specifically tailored to the property or liability arena, which are somewhat distinct from the models used for pricing life and health insurance. A brief introduction to various aspects of property insurance pricing including manual rate setting and individual risk rating is presented. (More detailed examination of pricing of general property insurance is given by Ai and Brockett [2008] and by the Casualty Actuarial Society [1990].)

An (oversimplified) essential representation of pricing of insurance policies is that prices (or charged rates) should equal the expected present value of the future losses on the contract augmented by a load for expenses, plus an insurer profit load, plus a load to compensate the insurer for risk bearing (see equation (5-1)). Hence, the price for insuring a risk exposure is given by

$$P = E[\sum_{i=1}^{N} L_i V(T_i)](1+k_1)((1+k_2)(1+k_3), \qquad (5\text{-}1)$$

where P is the price, E is the expectation operator representing the taking of the statistical expectation of the random variables involved in the brackets, and where

k_1 = the load for expenses (marketing underwriting, administration, taxes, claims adjusting, commissions, and general operating expenses);

[2] For insurance to be a viable alternative for an insurer to offer, there needs to be additional conditions that mitigate the financial hazard being assumed by the insurer so that the risk-assuming insurer itself does not go insolvent and can adequately manage the financial exposure they take on. These ideal conditions for a risk to be insurable include that (1) there be a large number of independent homogeneous exposure units to be insured (so the law of large numbers and the central limit theorem from statistics is available to the insurer for pooling and spreading risk among clients); (2) losses that occur are accidental; (3) a catastrophe cannot occur that affects a large number of exposure units simultaneously (this high correlation between losses thus defeating the independence of the exposure units and mitigating the insurers' use of the law of large numbers and the central limit theorem); (4) losses must be definite and measurable (so they can be priced); and (5) the probability distribution for losses should be determinable (so prices can be set). In addition to the above characteristics of a risk to be insurable, there are behavioral conditions that the customer can create which will make the insurer reluctant to offer insurance. One of these is adverse selection, which can occur when the individuals who have a higher likelihood of experiencing losses are disproportionately drawn to buying insurance defeating the risk pooling needed by insurers (e.g., if only high-flood-risk property owners seek flood insurance). Finally moral hazard also needs to be considered by insurers, where moral hazard is the term used to describe the behavioral property that can arise wherein the people who are insured will take fewer precautions than those who are not insured simply because they have insurance and are not fuly responsible for their losses. This escalates costs to insurers, and if severe moral hazard or adverse selection is anticipated, then insurance will not be offered (Baranoff et al., 2009).

[3] See floodsmart.gov.

k_2 = profit load (included so as to allow the capital suppliers to obtain a "fair" or competitive rate of return on capital);

k_3 = load for bearing uncertainty in future cash flows;

T_i = (possibly random) time until payment of the *i*th claim during the policy coverage period;

L_i = (possibly random) dollar severity realized in the *i*th claim, also called the severity (or size of loss) of the loss exposure;

N = (possibly random) number of claims incurred during the policy period, also called the frequency;

$V(T_i)$ = the discounted present value of $1 payable at time T_i (a priori this being a random stochastic process or a random variable).

Usually $V(T_i)$ is taken to have the form $V(T_i) = \text{v}^{T_i}$ where $\text{v} = 1/(1 + r)$ is the discount factor based on an interest rate r. Stochastic models taken for N often include Poisson (for rare events) and binomial distributions. Statistical distributions chosen for L_i generally depend on the particular hazard situation or peril involved. Indeed, the determination of appropriate statistical distributions and the dependence structure for N and L_i is a crucial part of property insurance pricing. (See Klugman et al. [2008] for explicit loss distributions models used in insurance pricing.)

Once calculated, these theoretical prices may be modified for political, social, or market competition reasons. For example, political or regulatory considerations may force the company to deviate from the theoretically derived prices and constrain or suppress rates (Witt and Hogan, 1993). Such regulatory intervention (and social influences) upon the mathematically derived pricing formulas occurs more in property insurance lines wherein purchase is mandated by contractual obligations (e.g., homeowners' insurance on mortgaged homes, comprehensive collision insurance on financed automobiles, etc.) and whose purchase is deemed .mandatory for social reasons (e.g., automobile insurance, workers' compensation, flood insurance[4]) (Ai and Brockett, 2008). Most state regulations impose the conditions that rates (or premiums) cannot be inadequate (to ensure solvency), excessive (to prevent excessive profits), or of an unfairly discriminatory nature (for social equity) (Ai and Brockett, 2008).

Pricing Property Insurance Using "Manual Rates"

The notion of "manual rating" is setting the rate (or price) based on a basic rating table (or tables) that produces a price per unit for different classes of risks. From such tables the user can obtain a numerical value (say, dollars to charge per $1,000 of coverage) that is then applied to obtain a price for any designated risk within the class of risks covered by the table. Such manual rates are used generally to price homogeneous groups of exposure risks and are not tailored to the idiosyncratic characteristics of any specific individual. For such manually rated policies, the issue is how to determine the table entries from which the appropriate rates can be read.

Manual rate encompasses two basic methods: the loss ratio method and the pure premium method (Casualty Actuarial Society, 1990). The pure premium method bases the rate on the expected fundamental loss exposure probability distribution itself. Mathematically,

$$R = \frac{E[L]+F}{1-V-Q}, \tag{5-2}$$

In equation (5-2), R denotes the rate per unit of exposure, $E[L]$ denotes the expected loss per unit of exposure (called the pure premium or actuarial value), F denotes the per-exposure unit fixed expense, V is a variable expense factor, and Q is a risk contingency factor that also incorporates anticipated profit. Essentially, this is analogous to equation (5-1) in that it says the rate should be sufficient to pay expected losses after allowing for required profits and fixed and variable expenses (Ai and Brockett, 2008).

The expected loss per unit of exposure is calculated as the discounted present value of the loss(es) per exposure

[4] According to the Federal Emergency Management Agency revised edition of the Mandatory Purchase of Flood Insurance Guidelines (FEMA, 2007), "The mandatory purchase law directed the Federal agency lender regulators and Government-Sponsored Enterprises (GSEs) to develop and adopt regulations requiring lenders subject to their jurisdiction not to make, increase, renew, or extend any loan on applicable property unless flood insurance is purchased and maintained to protect that property securing loans in high flood risk areas."

unit, $E[\sum_{i=1}^{N} L_i V(T_i)]$. After calculating this value for each possible rating class, a table (or set of tables) of values is created such that every possible exposure unit is uniquely found within a table. The coverage rate for a new exposure is ascertained by first determining the exposure's class, and then reading the multiplicative factor for this class from the corresponding table. Ultimately, by multiplying the number of exposure units encompassed by the risk exposure to be priced times this factor, a premium is determined. The difficulty in applying this method is appropriately stochastically modeling the frequency, N, and severity, L_i, variables and their dependencies. Many books on statistical analysis have been devoted entirely to this topic (cf. Klugman et al., 2008).

The second method, called the loss ratio method, is really a premium adjustment method as opposed to a premium-setting method. It assumes a current rate (R_0) and determines a rate change for the next year by multiplying by an adjustment factor (A) to obtain a new rate $R = AR_0$. The adjustment factor (A) is the ratio of the experienced loss ratio (Y) to a prespecified target loss ratio (T). Mathematically, $R = AR_0$, where R_0 is the current rate, $A = Y/T$, and $Y = L/[(R_0)(e)]$, where L denotes the current period experienced losses and e is earned exposure for the experience period. The target $T = (1 - V - Q)/(1 + M)$, where V denotes the premium-related expense factor, Q is the factor for contingencies and profit, and M is the ratio of nonpremium-related expenses to losses. The loss ratio method (by construction) is not applicable when a current rate is unavailable (such as for new lines of business). Still, it allows the insurer to obtain new rates for the next period by updating the previous period's rates. Brockett and Witt (1982) show that "when current prices are set by a regression methodology based on past losses, and the loss ratio method is used to adjust prices, then an autoregressive series is obtained, thus partially explaining insurance pricing cycles". Although the loss ratio method and the pure premium method are constructed from different perspectives, one can show that the methods yield the same prices given the same data (cf. Casualty Actuarial Society, 1990).

Individual Risk Rating

To obtain prices for individual exposure units, one can make modifications to the manual rates to accommodate for individual characteristics. A basic approach is to use a so-called credibility formula,[5] where individual loss experience is incorporated with an estimation of expected losses for the risk class derived from some other source to obtain a blended or averaged rate concatenating individual loss experience and exogenous derived expected losses (Ai and Brockett, 2008). Two such rating schemes are prospective rating and retrospective experience rating (Casualty Actuarial Society, 1990), where the rate for a future period is determined using a weighted combination of expected losses (from some other source such as industry data) an individual experienced losses.

In addition to the above methods, a somewhat different method for individual risk rating is called schedule rating (Ai and Brockett, 2008). It adjusts the manual rates to reflect individual risk characteristics that may affect future losses. Unlike the above individual risk rating methods, however, it does not necessarily use individual loss experience to make this adjustment, but instead creates a factor to apply to the manual rate to account for characteristics known to affect the likelihood of losses (e.g., a certain building standard, such as height of first floor, may result in a multiplicative factor less than one being applied to the manual table rates for flood insurance) (Ai and Brockett, 2008).

CURRENT RATE-SETTING PRACTICES WITHIN THE NFIP

FEMA's National Flood Insurance Program (NFIP) has approximately 5.6 million policies insuring over a trillion dollars in assets. Currently, 21,881 communities participate in the NFIP.[6] Broken down by structure type, this includes

[5] The concept of credibility theory, which dates back to Mowbray (1914), is said to be "the casualty actuaries' most important and enduring contribution to casualty actuarial science" (Rodermund, 1990).

[6] As of June 2012.

- 3,839,254 single-family policies,
- 281,508 nonresidential (i.e., commercial) policies,
- 341,607 multifamily (non-condominium) policies,
- 1,123,427 units (in 78,821 contracts or buildings) Residential Condominium Building Association policies.

The above numbers include all NFIP policies regardless of flood zone designation and 10,361 properties that have been designated as severe repetitive loss properties[7] (Andy Neal, FEMA, personal communication, July 19, 2012).

The purpose of this section is to explain the current process by which the NFIP sets rates and discuss the similarities and differences related to traditional private insurance processes for setting risk-based rates. Next, comments on the actuarial and fiscal soundness of the NFIP on the basis of the risk premium methodology used are presented. Finally, improvements and advancements that can be made to better serve the goals of the NFIP and the pricing of flood insurance are discussed.

Full-risk Class Rates

FEMA evaluates the flood risk of policy holders and determines rates for the "full-risk" class through a balance of elements including the extent and type of flood hazard, the base elevation of the insured structure and the structure type,[8] the contents location (first floor, second flood, etc.), and whether or not the community participates in the Community Rating System[9] that provides discounts for communities that actively manage their flood risk (Box 5-1).

At the core of FEMA's risk-based rate setting is the calculation of the expected loss for a property. FEMA calculates community losses using average annualized loss data (Box 5-2). In theory, this is calculated for each property by conditioning upon the floodwater level and the probability of that level of water inundation occurring during the year. Thus, they assess the frequency or probability that the floodwaters will reach an elevation of i (which they denote by probability of elevation or $PELV_i$). This number is then multiplied by the "loss severity" that would occur at the structure if the floodwaters reached level i (i.e., damage based on water depth in a given structure), which they denote by damage by elevation ($DELV_i$). This is then summed over all possible water elevation levels to arrive at an expected loss.[10]

$$\text{Expected Loss} = \left[\sum_{i=\text{Min}}^{\text{Max}} (\text{PELV}_i \times \text{DELV}_i) \right] \qquad (5\text{-}3)$$

This is consistent with a standard actuarial method for calculating the expected loss due to flooding similar to the pure premium method described in equation (5-2) earlier. In practice, properties having similar risk-related covariates (flood risk elevation, construction type, zone, etc.) are grouped together, and a single rate is given to all properties in this class nationwide. This is essentially similar to risk-based rate setting by private insurers in property insurance and automobile insurance where entities having the same set of underwriting classification characteristics are grouped together and given the same rate. Risk pooling in this regard is also used by private insurers to spread the risk.[11]

---

[7] Severe repetitive loss properties, as established in Section 1361A of the National Flood Insurance Act, as amended, 42 U.S.C. § 4102a) is defined as a residential property that is covered under an NFIP insurance policy and has at least four claim payments and for which two of these payments in total have exceeded the market value of the property. See http://www.fema.gov/severe-repetitive-loss-program.

[8] Condominium, single or multi-family home, residential or commercial use, number of floors, with or without a basement, ventilated crawlspace, etc.

[9] The discounts given under the Community Rating System are modeled back in to the expected aggregate premium calculation for the NFIP by adjusting all premiums upward so that the aggregate premium collected is sufficient to cover expected losses accounting for CSR discounts for the CSR discounts. Thus, CRS discounts constitute a relative discount which is adjusted for in the aggregation and is not an uncompensated for diminution of income as occurs with other types of non-full-risk class policies. (Andy Neal, FEMA, personal communication, August 22, 2012).

[10] See Hayes and Neal (2011, Appendix).

[11] Although risk pooling is used in private insurance, there have been questions about the appropriateness of the breadth (nationwide) of NFIP's grouping of all properties within the same zone classification for the purpose of obtaining a single rate applicable to all properties in the

BOX 5-1
FEMA's Community Rating System

FEMA's Community Rating System (CRS) is a voluntary incentive program that encourages floodplain management activities exceeding the minimum NFIP requirements. The CRS has 1,211 participating communities among the approximately 21,000 communities participating in the NFIP. However, policies held within CRS communities represent 68 percent of the policies in force (Bill Blanton, FEMA, personal communication, July 10, 2012).

Typically, communities receive between a 5 and 15 percent discount for participating in the program with a maximum discount of approximately 45 percent (FEMA, n.d.a; Bill Blanton, personal communication, July 10, 2012). Discounts are calculated based on a credit point system for actions that help save lives and property in the event of a flood—and that exceed the minimum NFIP requirements. There are four categories of activities that are credited: (1) public information programs that advise the public about the hazard, flood insurance, and strategies for reducing the hazard; (2) mapping and regulation programs that increase protection to development that postdates existing maps; (3) flood damage reduction programs, such as relocating or retrofitting vulnerable structures); and (4) flood preparedness programs, such as flood warning and dam and levee safety programs (FEMA, n.d.a). For additional insight regarding the CRS, see Chapter 7.

BOX 5-2
Average Annualized Flood Losses

Average annualized losses (AALs) due to flooding represent the estimated value of flood losses in any given year in a selected region. Analysis of such data provides information on changes in losses by geographic area that are useful in identifying loss trends and geographic anomalies in loss patterns. The data are also useful to leaders in establishing policies and priorities and developing flood risk management strategies.

FEMA's Hazus-Multi Hazard (Hazus-MH) offers analysts the opportunity to quantify annualized flood losses in a given community over multiple flood recurrence intervals. The Hazus-MH program, which can calculate damages, casualties, and economic losses, is based on national data from U.S. government agencies. It may also use, when available, higher resolution data available within the communities. For flood AALs, it calculates losses under each of five flood recurrence intervals (10, 25, 1.0, 0.5, and 0.2 annual chance floods or 10, 50, 100, 200, and 500 years, respectively) and then multiplies the loss values by their respective annual frequencies of occurrence, then summing the values for an annual average. This approach is focused on the recurrence intervals within the bounds of FEMA FIRMs' upper delineation, the 0.2 percent flood. Additional calculations using recurrence intervals for extreme events (>0.2 percent) would provide a more robust portrayal of the potential AALs but would likely increase the costs of the AAL process.

FEMA began the calculation AALs in 2009. Pilot data on flood AALs have been released for initial review within the floodplain management community but release of the entire report is awaiting final FEMA review. The use of Hazus-MH provides a relatively straightforward and simple but time-consuming method of making such calculations. Because the quality and resolution of the data used are subject to considerable variation across the country, the resultant AAL figures should be taken as general estimates and not high-resolution information. Such an approach is appropriate for the management uses envisioned by FEMA.

The determination of the components in the above expected-value calculation is as follows: The event frequency (probability) or PELV is obtained by reference to a collection of PELV probability curves created to describe the probability of various water depths relative to the 1 percent flood depth, that is, at this property the probability of a flood during the year reaching the height +2 (2 feet above the base flood level of the one percent annual chance flood) is a certain point on the PELV curve.[12] The PELV curves provide probabilities up to the 0.2 percent annual chance event (flood recurrence interval of 1 in 500 years). For all flood events greater than the 0.2 percent level, the program assumes a "catastrophic" water depth level by doubling the 0.2 percent depth. Because of the low likelihood of this "catastrophic" water depth level and because of the generally small marginal increase in the damage assumption between the 0.2 percent water depth and the damage at the assumed catastrophic depth, this approximation does not affect the rate significantly. Currently, "residual risk" resulting from the likelihood of overtopping a levee, levee failure, human development resulting in other topographic changes affecting inundation probability, and so on are not mapped or considered in the PELV determinations (FEMA, 2006).

The second component in the above expected-value calculation, the damage estimate as a function of the water depth elevation, or DELV, is based on historical damage data obtained by FEMA associated with different water depths in the given zone, and varies with structure content and location. When the NFIP's historical damage data are sufficiently credible[13] based on the number of claims and the variability of those claims, the NFIP data are used to develop the statistical damage estimate. When historical NFIP data are absent, U.S. Army Corps of Engineering (USACE) data are used. When some historical NFIP data are available but are is not fully credible (in an actuarial sense), the NFIP program uses a blend of NFIP data and USACE data to determine the DELV variable. This is consistent with standard property and casualty actuarial practice in private insurance.[14]

Once the expected loss is determined, the actual premiums or rates for the NFIP are determined by "loading" the expected losses in a manner similar to equation (5-3). Specifically, the charged rate for the group charged at the full-risk costs is determined by the formula

$$\text{Rate} = \left[\sum_{i=\text{Min}}^{\text{Max}} (\text{PELV}_i \times \text{DELV}_i)\right] \times \frac{\text{LADJ} \times \text{DED} \times \text{UINS}}{\text{EXLOSS}} \tag{5-4}$$

where LADJ is the loss adjustment expense loading, similar to the factor k_1 in equation (5-1), and DED is a factor that adjusts the DELV variable to compensate for the amount of the deductible chosen (so that the loss actually paid reflects the deductible amount) (Hayes and Neal, 2011). The variable UINS represents an adjustment to account for the underinsurance amount because not all properties can (or do) insure to their full value. The UINS factor is determined by the NFIP using a review of past insurance claims data. Incurred losses are a nonlinear function, with most losses being smaller and relatively fewer losses being much larger (i.e., the loss distribution is skewed); hence the variable UINS adjusts the loss for the percentage coverage to account for this nonlinearity (Hayes and Neal, 2011). This is similar to traditional homeowners insurance. Currently the value used for LADJ is 1.05 and

---

zone nationwide, without regard to state or topography. For example, the comparison of losses to premiums over time differs dramatically across states even within similar zones. This suggests that some properties in one state are cross-subsidizing other properties in other states because they are charged the same rate but have much different loss probabilities or risks (GAO, 2008). In a competitive insurance market with many participants, such cross-subsidization would be arbitraged away because an insurer who can identify those policyholders being overcharged relative to their risk (the subsidizers) would recruit such policyholders by offering lower premiums until perceived discrepancies disappear. Because the NFIP is essentially a monopolistic insurance provider, this cross-subsidization may continue to exist, and innovation to identify better pricing methods and cross-subsidization opportunities may not be pursued with the same vigor as in the private competitive market.

[12] The curve for a given flood zone (A01-A30) is based on the difference between the 1 percent flood elevation and the 10 percent flood elevation with each policyholder being assigned a PELV curve. The original PELV curves are based on studies conducted at the time the program was established. FEMA is currently collecting data to reevaluate the original PELV curves (Andy Neal, FEMA, personal communication, November 7, 2011).

[13] In actuarial practice "creditability theory" delineates when there is sufficient historical data to give a confident assessment of losses for estimation purposes, and how to augment the estimate of loss using auxiliary datasets when there are not sufficient historical data. See Casualty Actuarial Society (1990).

[14] The accuracy of the loss data used for determining the values of DELV has been questioned since many entries in the NFIP data set obtained from claims history have loss values but are missing data on elevation at the site. In these cases NFIP uses a zero elevation in their calculation, which will bias the losses experienced at zero elevation, and distort the DELV curve values. See GAO (2008) for details and references to further studies validating this criticism.

the value used for DED is 0.98 (Andy Neal, FEMA, personal communication August 22, 2012). Finally, the factor EXLOSS represents the expected loss ratio and contingency loading[15] and adjusts the rate to accommodate commissions, acquisition costs, and other costs such that the rate times the expected loss ratio is sufficient to cover the expected loss accounting for the loss adjustment expenses as well as the idiosyncratic choices by the purchaser of the deductible and underinsurance amount. Again, this formula is actuarially appropriate given the constraints and attributes of the program (absence of bankruptcy risk, ability to borrow from the government, etc.).

Discounted Rates

By statute, Congress mandated that the NFIP make flood insurance available to certain properties at less than their full-risk rates, that is, at discounted rates below those specified by equation (5-2). This occurs. for example. if (1) a structure was built before the flood insurance maps were available, that is, it was a "pre-FIRM" (Flood Insurance Rate Map) structure or built before December 31, 1974; (2) a structure was built in a V zone before 1981 and before maps that consider wave height were adopted in setting flood insurance rates; (3) a structure is in an AR or A99 zone for levees in the course of reconstruction or construction (so their current actuarial risk does not correspond to their current risk charge); or (4) the policyholder participates in a group flood insurance policy.

Discounts for structures that were built pre-FIRM or where the NFIP does not have any elevation information are by far the largest group—with 1,082,201 policies or a little less than 20 percent of policyholders. Only 7,508 policies are discounted because of entering the program prior to including wave height in rate setting, that is, pre-1981 V zones. Discounted rates are applied to approximately 24,907 policies for property owners behind levees in construction or deaccredited levees but where the community is showing a good-faith effort to construct levees or repair the levees (AR/A99).

Pre-FIRM Policies

The largest group of discounted policies consists of those that predate the existence of the flood insurance maps. By legislative directive, pre-FIRM policies were given a discount in order to encourage community participation in the NFIP, to help maintain property values for homeowners who might not have known of their flood risk at the time of purchase, and to encourage NFIP participation among those with a higher risk of floods (Michel-Kerjan, 2010). Currently, of the 5.6 million NFIP policies, there are 1,082,201 or approximately 19.3 percent of the policies that are pre-FIRM discounted policies (Andy Neal, FEMA, personal communication, July 19, 2012). These can be further decomposed as

- 736,066 single-family policies (150,674 of those are nonprincipal residences),
- 82,387 nonresidential policies (commercial) policies,
- 93,669 multifamily (non-condominium) policies, and
- 170,079 Residential Condominium Building Association policies (corresponding to 11,516 contracts or condominium buildings).

By statute, highly discounted premiums have been made available for pre-FIRM buildings in the Special Flood Hazard Area (SFHA). Not all structures that were built in SFHAs prior to the FIRM are discounted, however, because some structures meet more stringent building codes and qualify for better rates on a full-risk basis because of elevation above the base flood level than they would receive taking the discount off the higher SFHA rate. For the discounted older buildings, the average full-risk premium is about five times greater than the average full-risk premium for compliant buildings. With the discounted premium level, the noncompliant NFIP discounted buildings only pay between 40 percent and 45 percent of what they should pay would they be charged using the actuarially determined full-risk premium for being in the SFHA. Even so, their discounted premiums are still much

[15] Currently, the contingency loading incorporated into the EXLOSS factor includes a 10 percent additional load in nonvelocity zones and 20 percent in velocity zones (cf. Hayes and Neal, 2011).

higher than those that would be paid by actuarially rated policyholders for buildings constructed in compliance with the building codes.

The status of the pricing subsidies given to certain policyholders will change in the future. The Biggert-Waters Flood Insurance Reform Act of 2012 included provisions that were intended to help the fiscal soundness of the NFIP by eliminating certain pre-FIRM discounts and increasing the ability of the NFIP to raise rates to achieve fiscal soundness (Nechamen and Inderfurth, 2012). Effective July 1, 2012, pre-FIRM discounts are being removed from certain types of properties, with permissible increases in their flood insurance rates of up to 25 percent per year until the actuarial rate level is achieved for the property. Discounted policies losing their discount include nonprimary residences (e.g., second homes), severe repetitive loss properties,[16] all business properties, homes that have had substantial damage or improvements (of over 30 percent of the market value of the property) after the implementation of the Act, and several other classes of properties (Nechamen and Inderfurth, 2012).

Group Flood Insurance Policies

A group flood insurance policy is a discounted policy offered after a presidentially declared flood event for property owners who qualify for federal individual assistance. Qualifications include low-income property owners who do not qualify for a Small Business Administration disaster loan. On a one-time basis, owners are provided by FEMA with a 3-year policy with a coverage amount equal to FEMA's Individual and Households Assistance Program maximum (currently $31,400). The policy premium is set according to 44 CFR §61.17 at $600 per year, which is typically subtracted from the insurance adjustment payment received in the event of a loss (Andy Neal, FEMA, personal communication, January 15, 2013). Approximately 42,000 group flood policies were issued in the wake of Hurricane Katrina. These policies will eventually become nondiscounted and will be charged actuarial rates, and so they constitute a short-term imbalance, but new group flood insurance policies will be created as new floods occur (continuing to affect the fiscal soundness of the NFIP).

Grandfathered Rates

Grandfathered policy rates are applied to certain policies when revisions to FEMA's FIRMs result in increased premiums. Unlike the discounts described above, grandfathering is achieved through cross-subsidies with other policyholders. Grandfathered rates are available to structures that were built in compliance with a particular map or to structures that purchased a policy prior to the revised map's effective date, and whos purchasers have maintained continuous coverage since the map change. Grandfathering can apply to both the flood zone changes, resulting in zone grandfathering, and the one percent flood elevation changes, resulting in elevation grandfathering.

The most common form of zone grandfathering occurs when a policyholder was once outside the SFHA and paying a lower rate, but is now included in the SFHA because of to remapping. In this case, non-SFHA to SFHA zone grandfathered policies do not pay the lower preferred risk policy (PRP) non-SFHA rate. Instead they pay an average rate, called the X zone standard rate that is sufficient for the spread of risk for all non-SFHA to SFHA zone grandfathered structures. Effective October 1, 2010, FEMA introduced a 2-year PRP extension, allowing structures that were recently mapped into the SFHA to use PRP rates for a limited, 2-year period before moving to the higher X zone standard rate. This was achieved through a cross-subsidy by increasing the premium for all PRP policyholders. Approximately 90,000 PRP 2-year extension policies have been issued. There are approximately 1.6 million non-SFHA PRP policyholders (Andy Neal, FEMA, personal communication, August 9, 2012).

Zone grandfathering can also occur when a policy moves from a lower risk zone to a higher risk zone, for example, from an AE zone to a VE zone. In this case, the grandfathered structure pays the lower AE zone rate. Elevation grandfathering applies when new maps increase the elevation of mapped 1 percent flood without chang-

[16] The Biggert-Waters Act defines severe repetitive loss properties as those that have "incurred flood-related damage (i) for which 4 or more separate claims payments have been made under flood insurance coverage under this title, with the amount of each claim exceeding $5,000, and with the cumulative amount of such claims payments exceeding $20,000; or (ii) for which at least 2 separate claims payments have been made under such coverage, with the cumulative amount of such claims exceeding the value of the insured structure."

ing the zone. For example, a property that was 3 feet above the 1 percent flood elevation according to the previous FIRM and is now only 1 foot above the 1 percent flood elevation according to the revised FIRM would be eligible to use the +3-feet rate. FEMA is currently engaged in a study to quantify the number of zone and elevation grandfathered policies (Andy Neal, FEMA, personal communication, August 9, 2012).

The use of grandfathered rates does allow the grandfathered properties pay a lower rate than they would pay if the property were truly risk rated, and hence grandfathering implies that not all properties having the same NFIP premium are actually of the same risk level. This is a move away from risk-based pricing. Because properties receiving a higher risk classification under the Map Modernization program continue to receive insurance at the lower, pre-Map Modernization rate in spite of their now-recognized higher risk, the effect is similar to subsidization in that some properties are not charged their risk-based price, but rather the price that would have occurred if they had remained in their previous risk classification. Moreover, the grandfathered status continues indefinitely, even upon sale of the property (GAO, 2008).

Expected Annual Income from Premiums from the NFIP

To ensure an adequate aggregate expected annual income from premiums for the NFIP as a whole, the premiums are adjusted for the discounted risk classes. First, the aggregate expected full-risk premium anticipated to be received using the rating formula for the nondiscounted classes is calculated, the calculated base rates for the subsidized classes is added in, and then this value is subtracted from a calculated target "average historical loss year" value. The resulting difference is then used to determine an upward adjustment of rates among the discounted classes so as to accomplish a balance between the total anticipated income received and the targeted average historical loss year cost set by the NFIP. Note that by this process, some properties will be paying more than their actuarial rate.

There is a problem with this method, however. As noted by the Government Accountability Office (GAO) the"average historical loss year value is calculated using loss data over a limited loss experience time frame coupled with extreme loss values, such as the 2005 loss year (Box 5-3), which are heavily discounted when input into the historical loss year calculation (GAO, 2008).[17] Thus, this historical loss year average value may balance the books in an historically typical year, but this value is neither a reasonable estimate of either actual historical average yearly losses (which now do include catastrophic loss years such as 2005), nor a reasonable long-term estimate of the actual expected future loss because it diminishes the impact of real or anticipated catastrophic loss events that have happened in the past and can be expected to happen again in the future.[18] **The method used to calculate the average historical loss year ensures that in the long run there will be inadequate premiums collected to cover costs in significant flood years.**[19] Even with the mandate in the Biggert-Waters Act that catastrophic loss years be fully incorporated into the NFIP calculation of the "historical loss year average," there is still a potential long-run shortage becausee larger but less frequent catastrophic floods (500-year floods, 1,000-year floods, etc.) may not have been recorded in the flood record at a particular location (although their likelihood might be modeled), yet these events might occur in the future. Hence, the historical average does not reflect the real expected long-term average loss.

NFIP's Approach to Insurance Compared with Private Insurer Approaches

Insurance companies started to appear in the United States in the mid-1700s but did not expand rapidly, and encountered many bankruptcies at first. Early insurers were small, localized in geographic coverage, and usually

---

[17] For example, since 2007 the catastrophic losses from the 2005 year were given only a 1 percent weight in the historical loss averaging process in "an attempt to reflect the events of 2005 without allowing them to overwhelm the pre-Katrina experience of the Program" (Hayes and Neal, 2011).

[18] The full-risk class and the risk classes outside the 100-year flood level have premiums calculated using formula (5-2) and already incorporate the possibility of catastrophic risk.

[19] FEMA itself recognized the inadequacy of using the mandated "historical average loss year" goal in premium setting even prior to the dramatic 2005 losses. From the NFIP's 2004 Annual Rate Review, "[t]he underwriting experience period has, to date, included 7 heavy-loss years. Despite these heavy-loss years, the absence of extremely rare but very catastrophic loss years leads to the conclusion that the historical average is less than what can be expected over the long term" Hayes and Sabade (2004).

BOX 5-3
The 2005 Loss Year

The record-breaking hurricane season of 2005 included five Category 4 and 5 hurricanes—Dennis, Emily, Katrina, Rita, and Wilma—that battered the Gulf Coast of the United States along with neighboring countries. In addition, the Los Angeles River experienced flooding in January 2005, and there was flooding in New England and the Mid-Atlantic during October 2005 caused by excessive rainfall due to a subtropical depression, Tropical Storm Tammy, and another tropical disturbance. The NFIP experienced an unprecedented amount of paid losses, 212,235 claims, totaling over $17 billion (Figure 5-1).

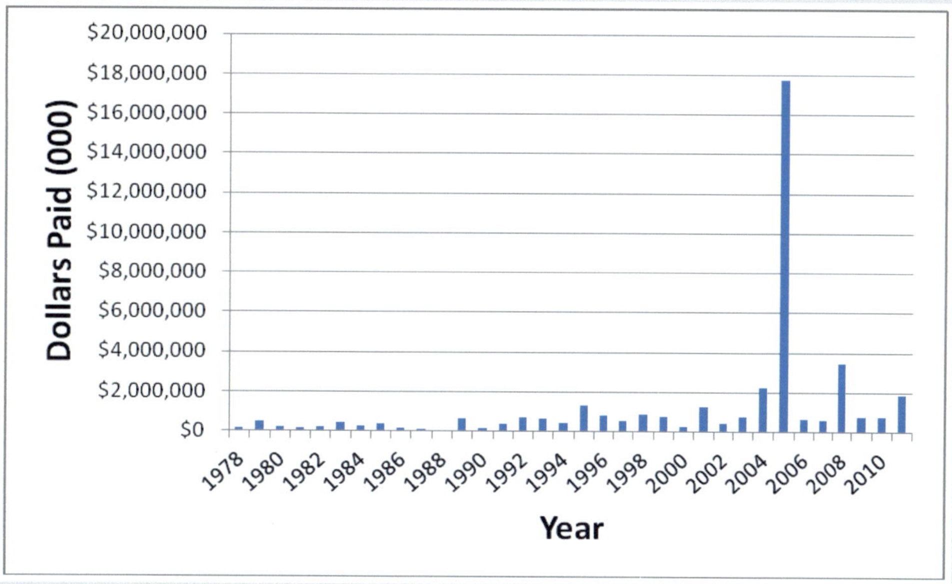

FIGURE 5-1 Loss dollars paid by the NFIP per calendar year.
SOURCE: http://www.fema.gov/policy-claim-statistics-flood-insurance/policy-claim-statistics-flood-insurance/policy-claim-13/loss.

specialized in a single type of insurance (e.g., fire insurance). Thus, initially, private localized insurers were reluctant to include flood coverage because of potential highly correlated losses and catastrophic total loss payment possibility (floods would simultaneously affect many of their customers, undermining the law of large numbers and central limit theorem for advantageous risk pooling) and because of the lack of ability to calculate appropriate actuarial rates that were protective of the insurer and at the same time affordable to the policyholder. Additionally, moral hazard and adverse selection considerations would imply that in a voluntary market the higher risk properties would disproportionately seek insurance and those that had insurance would be less motivated to take precautionary measures if they had insurance coverage. Nevertheless, in the late 1920s, there were still several dozen insurers willing to sell flood insurance, perhaps in part because USACE had declared in 1926 that improvements had been made to the levee system sufficient that they could now "prevent the destructive effects of floods" (Daniel, 1977; see also Moss, 1999; King, 2005).

The problems with private companies writing flood insurance policies were brought home dramatically by the great Mississippi River flood of 1927 and other riverine flooding in 1928, so that all of these several dozen insurers stopped writing flood policies (King, 2005). Private flood insurance has been essentially unavailable in the United States for residential property since then (Browne and Halek 2010). The previously described ideal characteristics needed for a risk to be privately insurable are simply, to a large extent, missing in flood risk (Baranoff et al., 2009). According to William Hoyt and Walter Langbein in 1955, floods "are almost the only natural hazard not now insurable by the home — or factory owner, for the simple reason that the experience of private capital with flood insurance has been decidedly unhappy" (Moss, 1999).

The lack of availability of flood insurance did not stop floods from occurring, nor did it stop development in flood-prone areas.[20] Such flooding motivated the federal government to provide aid to the victims as a humanitarian action. Over time the post-disaster aid moved increasingly away from being primarily provided by private charities (e.g., the Red Cross provided most of the relief in the 1927 Mississippi River flood) to a more central role for federal assistance (Moss, 1999). Indeed, the federal government has had an ever-increasing role in providing disaster relief after floods, and this increased involvement in after-the-fact emergency aid to disaster victims prompted Congress to investigate creation of a federal flood insurance program that could be considered as a form of preloss financing (Moss, 1999). Whereas private insurers feared the catastrophic and highly correlated nature of flood losses to local insurers, a federal program could spread risk nationwide and could make premiums more affordable by not charging (and reserving) for periodic very large disasters. Moreover, the federal government has the ability to mandate purchase, mitigating the adverse selection problem, and enforce building codes, mitigating some of the moral hazard problems. In 1968, Congress passed the National Flood Insurance Act[21] with the goal of creating a preloss (insurance) financing arrangement that would lower post-loss disaster relief, encourage responsible development in floodplain areas, and encourage maintenance of flood control operations (e.g., levees).

Browne and Halek (2010) give a comprehensive review of the developments in the federal flood insurance marketplace since the passage of the 1968 law creating the NFIP. Moreover, the challenges reaching the original legislative goals of the NFIP and the various legislative flood insurance law changes enacted to address the shortcomings are detailed in other sections of this report. It remains true, however, that for the reasons previously stated, the private market for flood insurance is still problematic. Private flood insurance is primarily available to commercial businesses that can demand flood coverage as part of a larger insurance program that includes "all-risk" coverage.[22] The NFIP remains the primary source of flood insurance for almost all homeowners and small businesses. However, there are a limited number of private insurance companies that offer flood insurance coverage for homeowners and small businesses as excess protection above the NFIP coverage limitations.

There have been many criticisms of the NFIP based on the perception that it does not operate like a private insurer, and impressions of how it should operate are based on notions of how private insurance operates. For this reason, and to set expectations of what is and is not possible for the NFIP in perspective, it is useful to compare the NFIP operation with those that occur in a private insurance setting. Toward this end, it is important to note that there are several ways in which the NFIP operates and prices risks that differentiate it from how a commercial insurer would price similar flood risks. These are in part due to the governmental nature of the NFIP which has different goals and constraints than private insurers, and these differences are briefly detailed below.

First, and importantly, commercial insurers must manage their capital reserves to ensure that they are able to honor their commitment to claims payable for policies in force at any one moment and to remain solvent for paying future claims. This is especially a challenge for any peril that has a potential for a major loss aggregation, of which flood is one of the greatest challenges. The private insurance industry is heavily regulated to ensure that policyholders can be confident that insurers will be able to honor claims made for the coverage they have purchased.

Commercial insurers also consider the time value of money and discount losses from the time of occurrence (T_i) to the time of premium payment using a discount function, $V(T_i)$ that reflects their cost of capital, per equation (5-1). In the NFIP, the time value of money is not considered, and no adequately large contingency reserve has been set up to pay future claims because FEMA is backed by the U.S. Treasury Department should the premiums

[20] Along this line, Moss (1999) quotes a congressional task force (U.S. Congress, 1966) as recognizing this problem when they observed "Floods are an act of God; flood damages result from acts of men."

[21] Public Law 90-448.

[22] "All-risk" or "all-peril" policies cover any fiscal loss unless it is specifically excluded in the policy.

collected prove insufficient to pay claims. Indeed, if the NFIP had been able to set up a contingency reserve fund that grew large enough during noncatastrophic loss years to be able to fund the losses in catastrophic loss years, the "perceived excess" might be reclaimed and spent elsewhere in the federal budget. By contrast, private insurers *must* be self-sufficient and hence must consider the time value of money and the cost of capital and set up a reserve to forestall insolvency.[23]

Next, while the NFIP sets premium rates (for the full-risk class) sufficient to pay its own expected claims (including catastrophic claims), the classes of subsidized policyholders paying discounted prices makes no allowance for extreme nonhistorical-average-loss years in the premium calculation for the discounted classes. As noted above, the method used to calculate the average historical loss year for the NFIP ensures that in the long run there will be inadequate premiums collected to cover costs in significant flood years. A private insurer would not use this financially unsound, downwardly biased risk premium method without facing the extreme uncertainty of returns and danger of future insolvency.

Another difference between NFIP and a private insurer is that there is no profit factor loading in the formula for premium setting by FEMA. It can be argued that elimination of the profit loading factor might make sense for governmentally run programs because the objective function of a governmental program is different from that of a for-profit insurance firm. Additionally, NFIP does not need to compete in the capital markets to attract investor funds,[24] and the issue of NFIP loss-induced insolvency and continued survival are not issues for the government-run program because of the financial backing of the government and its taxing authority. Moreover, neither is the profit or return on capital investment an objective of the government-run program, but rather the government has a societal welfare objective function (rather than a "profitability" objective function).

Since its formation the NFIP has had an objective of encouraging participation in the program, motivating community building standards, and making premiums affordable. Affordability of premiums has been and continues to be a prime concern of FEMA and is one important motivation for allowing discounts and subsidies in the program and for not adjusting premiums upward to reflect periodic catastrophic losses. To "load" the pure premium sufficiently to allow the NFIP to withstand reasonably anticipated, low-probability catastrophic losses (such as those in 2005 and worse) would require substantially elevated premiums. Setting aside such catastrophic contingency reserve capital, as would be necessary to endogenously withstand the financial consequences of catastrophic flooding, could undermine the societal welfare objective of encouraging community participation because premiums would necessarily rise, sometimes significantly, making some policyholders drop out of the market.

Moreover, given the parallel involvement of the government in flood disaster relief (separate from insurance), from a governmental perspective there is a natural trade-off between insurance pricing and disaster relief costs. Either way, the government pays (Czajkowski et al., 2012). Suppressing current insurance premiums and not allowing the buildup of large contingency reserves for future claims payments will lower current premium costs, encourage participation in the NFIP, and reduce the amount of disaster relief needed from the government after a flood. On the other hand, such a strategy of suppressed rate setting runs the inevitable consequence of periodic subsidization of the NFIP in catastrophic loss years. This is another issue that makes the NFIP rate-setting methodology differ from private insurers' rate-setting methodology, in part due to the governmental backing and differing objective function.

Further compounding the issue, the dual role of the government in providing both flood insurance and flood disaster relief (even to those who did not purchase insurance) presents what Browne and Hoyt (2000) refer to as "charity hazard" wherein the individual does not have the motivation to take proactive steps (such as risk financing or insurance purchase) when they believe that after an event they can depend on the charity of others (the government, charitable organizations such as the Red Cross, etc.). Although in truth the amount of money received by

---

[23] Under provisions of the Biggert-Waters Act, FEMA is to establish a National Flood Insurance Reserve Fund that contains at least 1 percent of the total potential loss exposure funded by putting in 7.5 percent of the required amount each year until fully funded (Nechamen and Inderfurth, 2012). A reserve level of 1 percent of expected annual losses would still not be considered prudent in private insurance.

[24] It could be cogently argued that the requirement that the NFIP pay back with interest the U.S. Treasury for the funds borrowed after the catastrophic 2005 loss year implies that they do indeed have a cost of capital. Interest rates on this debt are now relatively low (0.25 percent) but have been as high as 4.875 percent (Hayes and Neal, 2011).

the flood victim is much larger and more rapidly delivered with flood insurance, the perception of many is that if a flood occurs, the government will fully compensate victims.

Arguing in favor of building a catastrophic contingency reserve and/or putting a profit loading factor into the rating formula is the reality that, although FEMA was able to borrow money from the U.S. Treasury to finance the losses experienced by the NFIP in the catastrophic loss year 2005, the program is now being required to repay this "loan" back to the Treasury with interest. This directive could mean raising premium levels on new and current policyholders to pay back the loan and interest, which is again contrary to the objective function of the NFIP to encourage participation and make premiums affordable. Moreover, should interest rates rise from their current very low levels, it may be unlikely for NFIP (i.e., policyholders) to pay the interest on the loans from current revenue.

Actuarial Soundness Versus Fiscal Soundness of the NFIP

There has been much discussion concerning the actuarial soundness of the NFIP since the NFIP is required by law to promulgate rates that have actuarial soundness (Mathewson 2011). Some, such as the GAO (2008) , specifically state that "[t]he program is not actuarially sound." Even actuaries at the NFIP (Hayes and Neal, 2011) state: "Because the NFIP, as explained above, charges highly discounted premiums for many older buildings, it is currently impractical for the NFIP to be actuarially sound in the aggregate." The notion of actuarial soundness in the context of the NFIP, however, is somewhat subtle and elusive because of the differences described previously between a government-run program with mandated discounts, no profit requirement, no significant interest income, and reliable, significant borrowing power and the private sector from which the notions of actuarial soundness arose and are routinely applied.[25]

Indeed, even within the private insurance market, a precise definition of "actuarial soundness" in the Property and Casualty Insurance industry could not be found by the Actuarial Soundness Task Force of the American Academy of Actuaries,[26] although hints at what might be meant can be gleaned from the use of the term in the context of actuarial statements of principles. These principles include the following:

1. an actuarial rate should reflect the expected value of the insured risk;
2. a rate should be an estimate of the expected value of future costs, accounting for all costs associated with the transfer of risk;
3. a rate should provide for the costs associated with an individual risk transfer; and
4. the rate should be reasonable and not be excessive, inadequate, nor unfairly discriminatory (Mathewson, 2011).

These principles imply, in particular, that revenues collected should be sufficient to pay future losses for the class from whom premiums were collected. Moreover, the rates charged should be determined in such a manner as to be reasonable, and be calculated in a statistically valid manner reflecting the expectation of future loss costs for the classes used as groups for pooled risk underwritten. In the context of private insurers, these statements make good sense, especially since insurance is a future performance-contingent claims contract, with governmental regulations directed especially toward ensuring the ability of the private insurer to remain solvent and pay future claims. This is much less of (or perhaps not even) a consideration in the governmentally backed NFIP where the ultimate payment of the claims is not in question, profit is not a motive, and other objectives dominate.

To achieve actuarial soundness, predictability, and affordable rates, private insurers use risk pooling (over either policyholders such as occurs in automobile insurance or commercial general liability insurance, over space as occurs in property insurance, or over time such as occurs in life insurance, or some combination of these techniques) to hedge, spread, diversify, or pool risk so that in the aggregate the premiums collected would be sufficient to pay

---

[25] The NFIP can collect interest when there is a surplus in the fund, but its borrowing authority is legislatively set and was significantly raised in the wake of hurricanes Katrina and Sandy.

[26] "Unlike the Health actuarial practice . . . , the Actuarial Standards of Practice applicable to property/casualty practice do not directly define the term actuarially sound or actuarial soundness" (American Academy of Actuaries, 2012).

expected losses. The NFIP on the other hand operates as a governmental entity in a more political environment. One result of this more political influence is that the NFIP does not adopt any of these approaches. The program does not adopt risk pooling across properties (as some are offered discounted rates and the lost discount dollars are not recovered from other policyholders in the risk pool), and risk is not pooled over time because no sufficient reserve fund is set up using surplus in 1 year to compensate for losses in a subsequent year.[27]

Dictated by legislation, the NFIP created a dual system of rate promulgation that is not adequately pooled over time, space, or policyholder. The NFIP has essentially a rate classification methodology created by statute that creates a "full-risk" rate class (consisting of about 80 percent of the policies) that are charged risk-based (or actuarial) premiums and a "discounted risk" class (about 20 percent of the NFIP policies). Because the discounted policies are charged premiums that are lower than their expected losses and expenses, and because the difference between the full-risk premium for a discounted policy is simply revenue lost to the NFIP (i.e., this lost revenue is not compensated for in any other part of the premium structure), the NFIP is, by legislated mandate, fiscally unsound in the long run. Congress could have appropriated funds to compensate FEMA for the lost revenue dictated by the statutory discounts, and had they done so, the NFIP would be in a stronger fiscal condition.

The statements of actuarial principles do allow actuaries to use differing models with different assumptions, provided they can be reasonably justified and estimate future loss costs in a defensible manner, so the notion of actuarial soundness in the context of legislatively mandated discounts can be defended in a governmental context apart from strict fiscal soundness that is appropriate for private insurers. For catastrophic governmental insurance programs, such as the NFIP, the California Earthquake Authority (CEA), the Texas Windstorm Insurance Association, and others, there are additional mitigating considerations concerning the adequacy of rates since, in the event of insufficient funding via the premium collection mechanism, the program may have recourse to alternative funding mechanisms (e.g., borrowing from the U.S. Treasury or taxpayers in the case of the NFIP). As stated in the *Special Report on Actuarial Soundness* (American Academy of Actuaries, 2012),

> While not all publicly based catastrophe programs rely on outside sources of funding (e.g., taxpayer dollars or assessing a broader policy base), when they do, additional examination is needed to evaluate actuarial soundness.

For example, the setting of discounted rates so as to make the entire NFIP system able to obtain sufficient income that the premiums collected are sufficient to cover the "average historical loss year" value described previously can be construed as a form of directed actuarial soundness appropriate for this government program with significant borrowing power and mandated discounted rate classes. It will not, however, be financially sound, nor was it ever intended to be.

An additional consideration that must be added to general actuarially sound rate-setting principles is that the rates must be set consistent with the law. Even if one class of insureds has a documented higher expected loss cost than another, if the law specifies that the insurance company handle their premium rate setting in a specific manner, then the insurer's rates may not be *able* to legally reflect these expected loss cost differences. The rates obtained following legal dictates and not using this class variable, and the resulting cross-subsidy (mandated by law) would not generally be construed as making the new rates actuarially unsound even as they could result in cross-subsidies and not strictly risk-based premiums. For example, 60-70 years ago, African American males were charged higher rates for life insurance because of having (actuarially or statistically verified) lower life expectancy than white males. This process was legally outlawed in the 1960s, but this does not mean that life insurance premium rates that now follow the dictates of the law and ignore race are actuarially unsound. Overall prices are adjusted so that the entire book of business is balanced, and this is acceptable actuarially.

Thus, the concept of actuarial soundness must be contextualized in the framework of the laws restricting how rates may be set. A system may be fiscally unsound because of the inability of the actuaries to charge rates that reflect risk in an appropriate manner, but this is different from being *actuarially* unsound. This is relevant when

[27] Private insurers that write flood risk coverage for commercial clients often use reinsurance contracts to spread the risk, or use catastrophe derivatives available in the capital markets to hedge the financial impact of flood risk. However, the NFIP does not use reinsurance or capital markets to hedge flood risk.

discussing the actuarial soundness of the NFIP because, as the *Special Report on Actuarial Soundness* (American Academy of Actuaries, 2012) states:

> State and national catastrophe insurance programs present additional points of discussion regarding what constitutes actuarially sound cost estimates or rates. Rates may be, by design, subsidized. The programs are generally not designed to generate a profit, and large losses may be funded by alternative mechanisms.

Indeed, in the case of the NFIP, legal constraints on the premiums that can be charged to certain groups of policyholders constitute an economic externality on the cash flow such that the program's financial viability would be threatened were it to rely solely on its own generated premium income. Financial soundness and actuarially soundness are coincident in private insurers, but they need not be in governmental insurance programs. In private insurance when regulations on pricing suppress prices so as to constrain the insurer to charge insufficient rates, they have the option to drop out of the market and stop writing business.[28] The NFIP and other governmental programs cannot make this decision on their own. Thus, the NFIP uses a general rate-setting formula given by equation (5-3) that is consistent with standard actuarial models such as equation (5-1), calculates this rate for all properties, but can legally only apply it to the full-risk class. A form of overall balance is targeted via an adjustment of the discounted rates to obtain funding for the average historical loss year discussed previously.

For estimation and actual rate-setting purposes, insured properties are cross-classified according to a smaller but fixed number of characteristics, grouped into a smaller number of hazard zones and property attributes (with basement, multiple story, etc.), and a common nationwide rate determined for each class. According to Hayes and Neal (2011), "[t]here are 30 numbered A Zones for which different sets of *PELV* values may be assigned." However, the risk zones are combined for rating purposes. Combining exposure units according to common underwriting characteristics for rating purposes is an actuarially sound manner for setting premiums, consistent with methodology used in other insurance settings (although there is always the question concerning oversimplification involved in having too few risk classes and consequently too heterogeneous exposure units within a given class, creating induced inequities in premiums). In the case of the NFIP today it is argued that improved and more accurate graduated risk-based premiums could be achieved by further improving estimates of inundation level probabilities (using improved maps or hydrologic studies, incorporating residual risk into the probability of inundation of a property behind a levee, etc.), and that improved estimates of the severity of the loss accuracy may be obtained using Lidar data (or something similar) to obtain better estimates of structure elevation and damage characteristics (Chapter 3). However, these are improvements to an already actuarially sound rating method.

The major structural drawback to the fiscal soundness of the NFIP is *not* actuarial unsoundness due to inappropriate or inadequate rate-setting methodology or inadequacies in using a formula such as equation (5-4), but rather the fiscal unsoundness caused by not being able to apply equation (5-4) uniformly across all classes. The NFIP-mandated "discounted risk" class was anticipated to eventually decrease substantially over time, and this has indeed been the case. Still, there are a substantial number of discounted policies in the NFIP, approximately 20 percent of the 5.6 million policies. The difference between the full-risk premium for a discounted policy and the discounted premium for the policy is revenue lost to the NFIP (this lost revenue is not compensated for in any other part of the premium structure). FEMA is legally unable to raise premiums in a manner sufficient to allow the NFIP to be financially sound or to build a contingency reserve fund sufficient to pay for a catastrophic future loss. Without this ability, the NFIP is destined by law (not by actuarial methodological inadequacy) to run short of money in the long run, with occasional dramatic shortfalls (such as occurred in 2005). Furthermore, repetitive loss properties continue to drain the fiscal soundness of the NFIP.[29]

---

[28] Examples include restrictions on rates in high-risk coastal areas in Texas and Florida that drove insurers from the market, or in automobile insurance in Massachusetts when it was very heavily regulated.

[29] A 1995 Senate Task Force Report, *Federal Disaster Assistance: Report of the Senate Task Force on Funding Disaster Relief,* noted that in 1993 repetitive loss properties accounted for 2 percent of the properties covered by flood insurance but accounted for 53 percent of the claims filed and 47 percent of the dollars paid from the Flood Insurance Fund (Moss, 1999).

The NFIP is constructed using an actuarially sound formulaic approach for the full-risk classes of policies, but is financially unsound in the aggregate because of constraints (i.e. legislative mandates) that go beyond actuarial considerations. Parts of the Biggert-Waters Act that relate to adjustment of fiscal practices, discussed above, move the program toward a financially sounder approach.

Improving the NFIP Rate-Setting Process: Incorporating Technological Improvements to Better Obtain Risk-Based Premium Structures

While the mathematical construction used by the NFIP in determining their risk rates is based on actuarially sound principles as applicable to a government program, these processes and procedures could still be enhanced and aligned with a more rigorous and risk-based methodology by using finer graduation of risk propensity at the individual policyholder's structure level. Although the NFIP currently uses a small number of zones for nationwide pricing purposes, the suggested finer graduation would not be foreign to the NFIP. According to GAO (2008):

> NFIP implemented the nationwide class-rating system because of the nature of the program and the desire to make it less complex and easier for agents and customers to understand. In the early years of the program, rates were set on a community-by-community basis. But as the number of communities participating grew, this system became unwieldy and costly to maintain.

The complexity, costliness, and unwieldiness referred to above have been alleviated by technological and computer advancees that could allow agents and customers to interactively obtain structure-specific premium quotes online in real time. Additionally, existing advances in meteorologic and geotechnical modeling can allow the underwriter to model the likelihood and severity of flood damage individually so the complexity of the calculation or the number of zones is no longer a compelling factor at the agent or customer level. These advancements were not available in the 1960s when the core structure of the risk-based pricing framework was originally developed but can be utilized now to improve the system.

Although the general form of the rates given in equation (5-3) is actuarially based, in practice the implementation using a small number of zone categories and homogeneity of premiums within any given zone across the entire nation ignores differences within contiguous zone properties and makes the premium determination cruder and less directly risk-based than they need to be given today's knowledge and technology. Currently, the following zone categories are used:[30]

1. coastal, V Zone, within the 100-year flood frequency zone and subject to wave action or velocity-induced damage;
2. inland, A Zone, within the 100-year flood frequency zone but not subject to wave action or velocity-induced damage; and
3. all other, X Zone, which includes other flood zones and those outside identified 100-year flood zones (considered to be of low or moderate flood risk).

Properties that are located behind levees are placed in one of the last two categories. If the levee is accredited by FEMA to the one percent annual chance flood level, the "all other" or X zone category is applied. If not, the property is considered in the category that would apply without the levee being there. It can be argued that this approach to properties behind levees is appropriate based on the size of flood needed to create damage (over the one percent annual chance flood level). However, when viewed from a risk assessment perspective, it is apparent that the range of potential scenarios behind a levee and the loss corresponding to properties behind 100-year levees is very different from the loss that occurs from floods on the adjacent natural floodplain above the 100-year flood level (Figure 5-2).

[30] There is also a preferred risk policy available to properties in lower-risk areas with a minimal loss history.

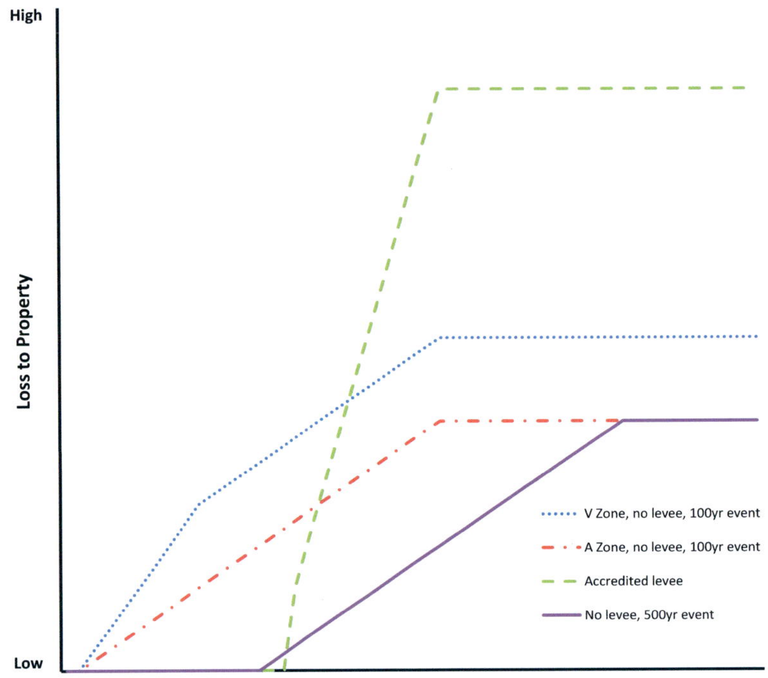

FIGURE 5-2 Notional schematic illustrating the damageability (or vulnerability) curves associated with example properties. The properties in the 100-year zone are prone to initial damage more frequently (the V Zone has increased damage due to wave action which has the potential to cause structural damage). The property in the 500-year zone does not experience loss as frequently, but the impact of flooding results in a similar profile of loss as the water depth increases. The property behind the levee has the advantage of no damage until the levee is overtopped or fails (similar to a property located outside the 100-year flood zone). However, once the water depth increases above the 100-year level, the potential loss to properties could be far more severe than if the levee was not there, and certainly more severe than that to properties located at elevations above the 100-year flood level. This is due to the fact that when the levee is overtaxed, water flow is likely to be rapid (similar to the coastal V zone) and the water level may be significantly deeper if the area behind the levee is at a low elevation compared with the levee.

The current process can sometimes result in dramatic premium changes occurring for rather small perturbations in actual risk (e.g., changing the protection level of a levee from the 100-year level to the 99-year level, or the determined elevation level of a structure from −1 foot to +1 foot above the base flood level). Two houses next door to each other can have one just above the one percent annual chance flood level (and have no mandatory flood insurance mandate, and a low-priced policy if the property owners do decide to buy) while the house next door is just inside the 100-year floodplain and is charged much higher rates (and is required to buy insurance). In reality, both houses are exposed to essentially a similar hazard, but are treated differently. This abrupt change in rates and mandates that corresponds to being inside or outside the one percent annual chance floodplain as specified by the FEMA flood map has created problems as new maps are produced that change a community or set of properties from a moderate-risk zone to a high-risk zone. Even within the full-risk class the actuarial accuracy of the pricing within the program could be improved by incorporating more exact and detailed probability-of-inundation calculations instead of using a small number of rating zones and assuming the flood hazard to be uniform across the entire zone across the entire country. Better estimates of loss at stated water inundation levels is also possible utilizing more detailed economic damage models that are now available (and used in the private risk market). The use of more detailed and better probability models and better damage estimates will result in a more risk-based premium at the structure level.

One problem resulting from the very small number of zones and the special attention given to the 100-year floodplain is that in many respects the pricing mechanism in the NFIP today is based largely upon whether the structure's location is inside or outside the SFHA on the NFIP's flood maps. Premiums outside the 100-year floodplain on the map are much less expensive than those inside the 100-year floodplain, even if outside by only a foot. Requirements are different too. In general, if a structure is located inside a 100-year floodplain, there is a mandatory flood insurance purchase requirement associated with being inside the SFHA. On the other hand, if a structure is located outside the 100-year floodplain, mandatory purchase is waived.

For NFIP insurance pricing purposes, an area within a 100-year floodplain that is located behind a 100-year levee is treated riskwise (and premium-wise) as equivalent to being outside the 100-year floodplain area. Thus, the flood insurances premiums are lower (and purchase is voluntary) when behind a certified 100-year levee. If the exact same property were protected by a 99-year levee, it would be treated for NFIP purposes as if the levee did not exist (i.e., that the levee is providing zero protection), and purchase of flood insurance at the higher "without protection and within the 100-year floodplain" rate is mandated. Rates and mandates change dramatically depending upon whether a structure is behind a 99-year levee (the X Zone for pricing purposes) or a 100-year levee (the A Zone for pricing purposes).

An additional problem with the current NFIP structure is that, because of the small number of zones, there can be substantial internal variation in risk and losses within the zone that is not reflected in rates, since all similar properties are charged the same rate within the same zone classification. Structures having similar characteristics located at different locations in a particular designated flood zone (say, the X500 zone of moderate flood risk between 100 years and 500 years) are all given the same NFIP insurance rate nationwide, even though it has been shown that there can be substantial risk differences across the zone within a given geographical area (such as a county) (Czajkowski et al., 2012). Even more expected loss variation exists across similar structures in the same zone classification between counties within the same state (Czajkowski et al., 2012) although the NFIP would charge all of the structures the same rate. Differences in loss ratios for policies within the same NFIP rating zones across states can also vary greatly (GAO, 2008). Thus, the current system oversimplifies the rating structure (by forming a nationwide zone classification) resulting in rates that are not truly individual risk-based, and instead have premiums set that have been aggregated over properties having dissimilar loss structures. This results in imbedded cross-subsidies between properties within the same risk classification. Additionally, the dramatic changes in rates that arise from moving between the few rating zones does not recognize the continuous nature of the flood risk. **Rate setting for properties behind levees, accredited or not accredited, should be improved by using the modern risk analysis method employing advances in hydrology, meteorology, geotechnology, and engineering to more precisely calculate the probability of water inundation levels and the associated damage estimates throughout the area behind the levee in a graduated fashion.** This will replace the current approach and more appropriately reflect the risk. One then gets a gradation of probabilities that can differ from property to property even within the same geographic area or flood zone (the PELV variable in equation (5-2).

Likewise, modern Lidar imaging technology can be used to obtain a property-specific assessment of the elevation of each property rather than using a similar value for all structures in the zone being priced. This Lidar information, together with other currently used underwriting information, such as existence of a basement, material structure of the building, and contents located above the first floor, can be used to obtain a better estimate of the expected loss or damage to the specific property at each location (the DELV variable in equation (5-2)). In this case, the resulting premiums determined for a property behind a 99-year levee could be very similar to the premiums determined for a property behind a 100- or a 101-year levee, and premiums could be allowed to vary continuously across properties within the same zone. In essence, this process moves the premium setting from a binary model (inside or outside the 100-year flood zone) to a graduated model with premiums varying more continuously as risk profiles change continuously. Many of the challenges that occur when new FIRMs are created that are due to the without-levee approach of the rating process could be alleviated by implementing a continuous premium structure of the type described above using modern assessments of probability of water inundation levels at the specific property and modern assessment of the elevation of properties as described in Chapter 3. A graduated and risk-based probability assessment and severity assessment would move the process from a without-levee approach (in the X zone, out of the X zone) to a dial approach with a large array of probabilities and severities that are determined for each property, allowing a property-specific calculation of the expected loss per property by flooding. In this context, not only can floodwater inundation levels be incorporated at the individual property level, but also residual risk such as levee failure due to overtopping, seepage, and so forth can be incorporated. This detailed level of probability assessment and damage assessment at the insured-structure level is where NFIP needs to head in the long term.[31]

Probabilistic risk assessment (PRA) involving Monte Carlo–simulated flood result scenarios is being done by private firms such as Swiss Reinsurance and CoreLogic (Czajkowski et al., 2012), and can be adapted to work nationwide to price flood insurance in the NFIP. The state of North Carolina has already demonstrated its ability to estimate inundation-level probabilities at the structure level, using Lidar or similar technology, and estimate structure damage consequences at each water inundation level for any given structure, eliminating the need to treat all properties across the nation in the same zone as having equivalent risk and being charged the same premium. These are the critical inputs into the actuarial rating formula (5-3), which can now be performed even more accurately at the individual-structure level than before, and can provide a more accurate alternative to simply aggregating (or averaging) expected loss costs across disparate entities within a grouped NFIP zone classification. Indeed, Czajkowski et al. (2012) have shown that substantial expected loss differences are discernible even within a single-priced NFIP zone inside a concentrated geographic location (e.g., Travis County, Texas), and this variability is exacerbated when comparing across disparate geographic areas having the same zone classification (e.g., Galveston, Texas). Because the same rate is charged by the NFIP to all entities nationally having the same characteristics and which are classified as being in the same zone, this can lead to substantial over- and undercharging of premiums on a risk basis (GAO, 2008). This can be improved (corrected) using PRA methods, resulting in more accurate risk-based (actuarial) pricing within NFIP. Moreover, not all properties in a similar zone classification pose the same risk, and charging the same premium across these properties encourages adverse selection and does not enhance actuarial soundness of the program.

To the extent possible to better achieve fiscal soundness, the properties that have been grandfathered into the NFIP or that receive discounts due to statutory considerations should have their pricing moved to the actuarially based prices using the detailed probability-of-inundation estimates and detailed economic damage models. This would strengthen the program and further the goal of risk-based pricing.

ROLE AND UTILITY OF MANDATORY INSURANCE BEHIND LEVEES

When the NFIP was created, property owners could purchase flood insurance voluntarily. In 1973, Congress passed the Flood Disaster Protection Act (Public Law 93-234) finding that,

> (1) annual losses throughout the Nation from floods and mudslides are increasing at an alarming rate, largely as a result of the accelerating development of, and concentration of population in, areas of flood and mudslide hazards; (2) the availability of Federal loans, grants, guaranties, insurance, and other forms of financial assistance are often determining factors in the utilization of land and the location and construction of public and of private industrial, commercial, and residential facilities; (3) property acquired or constructed with grants or other Federal assistance may be exposed to risk of loss through floods, thus frustrating the purpose for which such assistance was extended; (4) Federal instrumentalities insure or otherwise provide financial protection to banking and credit institutions whose assets include a substantial number of mortgage loans and other indebtedness secured by property exposed to loss and damage from floods and mudslides . . . .

The purpose of this Act was to

> require State or local communities, as a condition of future Federal financial assistance, to participate in the flood insurance program and to adopt adequate flood plan ordinances with effective enforcement provisions consistent with Federal standards to reduce or avoid future flood losses; and require the purchase of flood insurance by property owners who are being assisted by Federal programs or by federally supervised, regulated, or insured agencies or

[31] In the short or intermediate term the individually assessed expected losses can be used for aggregation to form a smaller set of risk classes, each of which is charged an average price. This is different from what is currently being done.

institutions in the acquisition or improvement of land or facilities located or to be located in identified areas having special flood hazards.[32]

This requirement is called the Mandatory Purchase Requirement (MPR). Because of the low levels of participation among eligible property owners and the low NFIP financial reserves, enforcement of the MPR in the SFHA was strengthened in the National Flood Insurance Reform Act of 1994 (Public Law 103-325). The National Flood Insurance Reform Act increased the responsibilities of lenders by requiring them to escrow flood insurance payments (if other mortgage-related expenses were required to be escrowed) and notify property owners when their policies lapsed. It also instituted penalties for lenders who failed to carry out these requirements.[33] In 2012, the Biggert-Waters Reform Act strengthened monetary penalties on lenders, including removing the cap on the total amount of penalties imposed on a lender in any calendar year. See Table 5-1 for a summary of legislative changes related to the purchase of flood insurance from 1968 to 1994.

As indicated in Chapter 2, following congressional hearings in 1973 that included discussion of the role of flood control structures in protecting those behind or below them from flooding, the Federal Insurance Administration determined that areas behind levees providing protection from the one percent annual chance flood would be considered outside of the SFHA, and the MPR and associated land-use restrictions would not apply to properties in these areas. This remains true today; property owners living behind an accredited levee (i.e., providing protection from the one percent annual chance flood) are not subject to the MPR. Areas behind a levee are shown on the FIRMs as outside the one percent annual chance floodplain or SFHA, although exceptions do exist (Box 5-4).

Figure 5-3 illustrates the growth in NFIP policies in force over time. Of note, after the 1994 legislation, approximately 1 million new policies were purchased between 1994 and 1997. Also, the great Mississippi and Missouri rivers flood of 1993 heightened awareness of flood risk across the nation and most likely had an impact on the number of policies. However, after the immediate impact of these two events, the growth in policies leveled off from 1999 through 2004, instead of achieving the desired, long-term continuous growth (Figure 5-3).

In 2004, in the aftermath of Hurricanes Charley, Frances, Ivan, and Jeanne, the NFIP experienced the largest claim year in the history of the program at that time, soon to be eclipsed when Hurricanes Katrina, Rita, and Wilma in 2005 resulted in the most significant loss year in the history of the program to that date (Box 5-2). The total policies in force grew through 2008, likely as a result of awareness of flood risk after the 2004 and 2005 events (Figure 5-3).

The number of NFIP policies in force also varies from year to year as a result of the mix of dropped and new policies. A study by Michel-Kerjan et al. (2012) indicates that, based on analysis of policy identification numbers, the NFIP adds new policies each year but also loses policies. The analysis indicates that an immediate drop in policy retention occurs 1 year after purchase, ranging from 20 to 30 percent between 2001 and 2009 (Table 5-2). Cancelations of policies after the first year continue, but are relatively fewer. The average tenure of policies issued between 2001 and 2006 was 2 to 4 years (Table 5-2). Policy retention in the SFHA, where the MRP is in place, and outside the SFHA, is reasonably similar (Table 5-2). Because of shifts in mortgage ownership and reissuance of policies in states such as Florida and Louisiana where there have been significant post-disaster rebuilding and policy migrations, policy identification numbers as opposed to policy addresses may overstate the number of structures for which policies were dropped (e.g., when policies are dropped from one company and shifted to another.) Of interest, between 2001 and 2009, of 8.36 million new policies purchased,[34] 41 percent were outside of the SFHA, with the highest number of purchases taking place after Hurricane Katrina. Since the properties outside the SFHA are not subject to the MPR, it is likely that these purchases represent an awareness of flood risk and the lower cost of policies not in the SFHA.

---

[32] Two other purposes of the 1973 Act are articled, dealing with increasing the amount of coverage authorized under the NFIP and dissemination of information concerning flood-prone areas.

[33] Section 524.

[34] This number is the summation of all the new flood policies in force according to Figure 5-4.

TABLE 5-1 Summary of Legislative Changes Related to the Purchase of Flood Insurance, 1968-1994

	Legislation		
Topic	National Flood Insurance Act of 1968	Flood Disaster Protection Act of 1973	National Flood Insurance Reform Act of 1994
Availability of flood insurance	Available in participating communities	Available in participating communities	Available in participating communities
Purchase of Flood Insurance	Voluntary	Mandatory for certain property owners: (a) building or manufactured home in or to be located in an SFHA in a participating community and receiving a loan from a regulated lender; (b) building for which any federal financial assistance is used for acquisition or construction of a building or a manufactured home in an SFHA of a community participating in the NFIP	Mandatory for certain property owners: (a) building or manufactured home in or to be located in an SFHA in a participating community and receiving a loan from a regulated lender; (b) building or manufactured home in or to be located in an SFHA in a participating community and whose loan is purchased by Fannie Mae or Freddie Mac; (c) building for which any federal financial assistance is used for acquisition or construction of a building or a manufactured home in an SFHA of a community participating in the NFIP
Responsibilities of regulated lenders	Not applicable	Require flood insurance for buildings or manufactured homes located in the SFHA of a participating community Ensure that flood insurance is maintained on such structures for the life of the loan	Determine whether the building or manufactured home is, or will be, located in an SFHA Complete a hazard determination form Provide notice of availability of flood insurance to borrower, lending institution, and loan servicer Require flood insurance for structures located in SFHAs of a participating community Ensure that flood insurance is maintained for the life of the loan Escrow for flood insurance if other mortgage-related expenses are required to be escrowed Force place flood insurance in the event of nonrenewal or insufficient coverage

SOURCE: Tobin and Calfee (2005).

BOX 5-4
SFHA May Exist in Areas Behind Accredited Levees

Flooding can occur from sources other than rivers and storm surge, for example, from heavy rainfall or low elevation. A community's interior drainage system, which can consist of storm sewers, pumping stations, and canals, is designed to prevent this type of flooding but can become overwhelmed. Generally, this type of flooding is less of a catastrophic problem than, for example, levee failure.

When FEMA's FIRMs are developed, an assessment of susceptibility to this type of flooding is performed. Areas behind the levee where susceptibility exceeds the one percent annual chance flood level are designated as inside the SFHA even though the area might have an accredited levee. For example, much of New Orleans, Louisiana, is susceptible to flooding from interior drainage despite having accredited levees. The pumping system in New Orleans is one of the largest in the world, boasting nearly 100 pumping stations in the greater New Orleans area designed to handle approximately 9 inches of rainfall in a 24-hour period (ASCE, 2007).

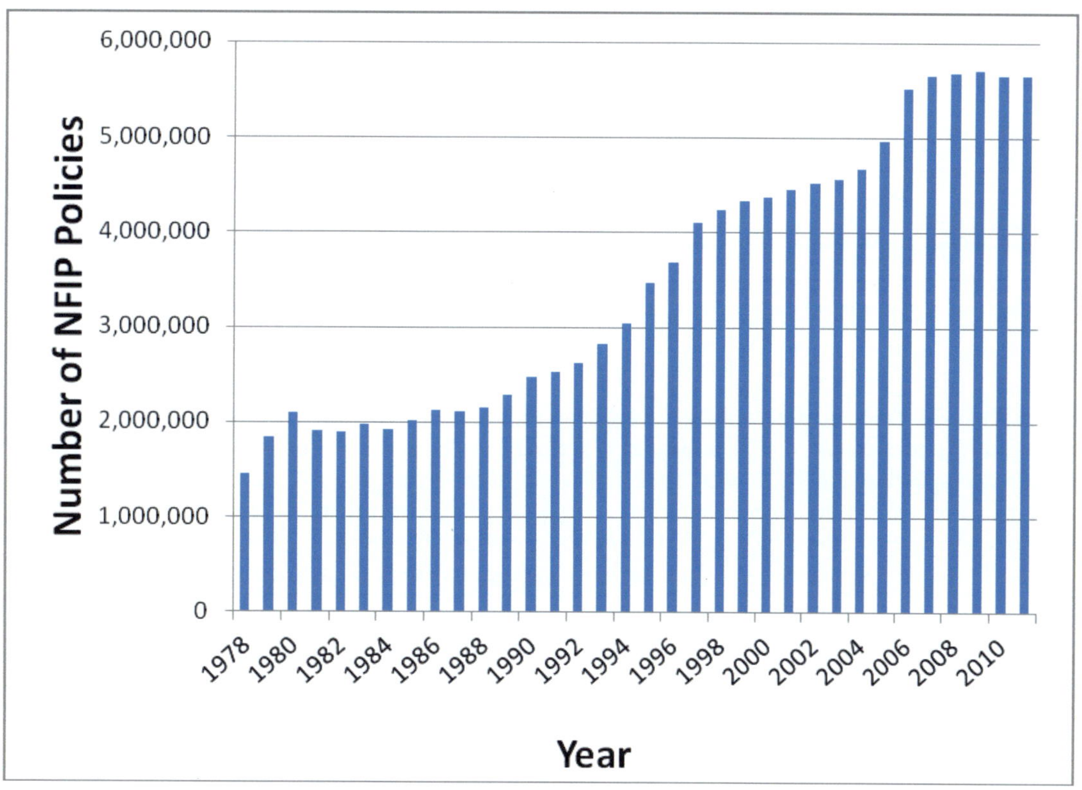

FIGURE 5-3 The total number of NFIP policies in force by calendar year.
SOURCE: Data from http://www.fema.gov/policy-claim-statistics-flood-insurance/policy-claim-statistics-flood-insurance/policy-claim-13-14.

TABLE 5-2 Number of new policies and duration from 2001 through 2009.

	2001	2002	2003	2004	2005	2006	2007	2008	2009
New flood policies in force (000s)									
All	**841**	**876**	**1,186**	**986**	**849**	**1,299**	**974**	**894**	**1,051**
SFHA/non-SFHA	*542 299*	*613 264*	*880 306*	*696 291*	*529 320*	*635 664*	*542 432*	*487 407*	*595 456*
Tenure longer than:									
1 year	73%	67%	77%	78%	76%	73%	74%	73%	
SFHA/non-SFHA	*74% 71%*	*67% 67%*	*78% 76%*	*77% 80%*	*75% 78%*	*74% 72%*	*74% 74%*	*75% 70%*	
2 years	49%	52%	65%	65%	63%	59%	58%		
SFHA/non-SFHA	*48% 52%*	*52% 50%*	*66% 64%*	*64% 67%*	*62% 64%*	*59% 60%*	*58% 59%*		
3 years	39%	44%	57%	55%	53%	48%			
SFHA/non-SFHA	*37% 41%*	*44% 43%*	*57% 56%*	*54% 57%*	*53% 54%*	*47% 49%*			
4 years	33%	38%	50%	48%	44%				
SFHA/non-SFHA	*32% 36%*	*39% 38%*	*50% 48%*	*47% 49%*	*43% 44%*				
5 years	29%	33%	44%	38%					
SFHA/non-SFHA	*28% 31%*	*34% 33%*	*44% 42%*	*38% 38%*					
6 years	25%	30%	33%						
SFHA/non-SFHA	*24% 28%*	*30% 29%*	*34% 32%*						
7 years	22%	26%							
SFHA/non-SFHA	*21% 25%*	*26% 25%*							
8 years	20%								
SFHA/non-SFHA	*18% 22%*								

Market Penetration and Compliance

Under the NFIP, flood insurance is available to both those subject to the MPR and to any property owner living within the NFIP community, whether in or out of the SFHA or in front of or behind a levee. From a flood risk management perspective, the NFIP encourages all who live in areas subject to any level of flooding to purchase insurance so as to provide loss coverage to them if their covered property is damaged from a flood event and to reduce the cost to the federal government for their recovery. Therefore, FEMA is interested in knowing what percentage of those living in flood-prone areas have insurance, that is the market penetration:[35]

$$\text{Market Penetration} = \frac{\text{Number of buildings in SFHA eligible for insurance purchase that have policies}}{\text{Number of buildings in SFHA eligible for insurance purchase}}$$

Compliance with the NFIP MPR is a subset of the market penetration:

[35] "Insurance may be written on any building eligible for coverage with two or more outside rigid walls and a fully secured roof that is affixed to a permanent site. Buildings must resist flotation, collapse, and lateral movement. The structure must be located in a community that participates in the NFIP . . . . Buildings entirely over water or principally below ground, gas and liquid storage tanks, animals, birds, fish, aircraft, wharves, piers, bulkheads, growing crops, shrubbery, land, livestock, roads, machinery or equipment in the open, and most motor vehicles are not insurable through the NFIP" (FEMA, 2011).

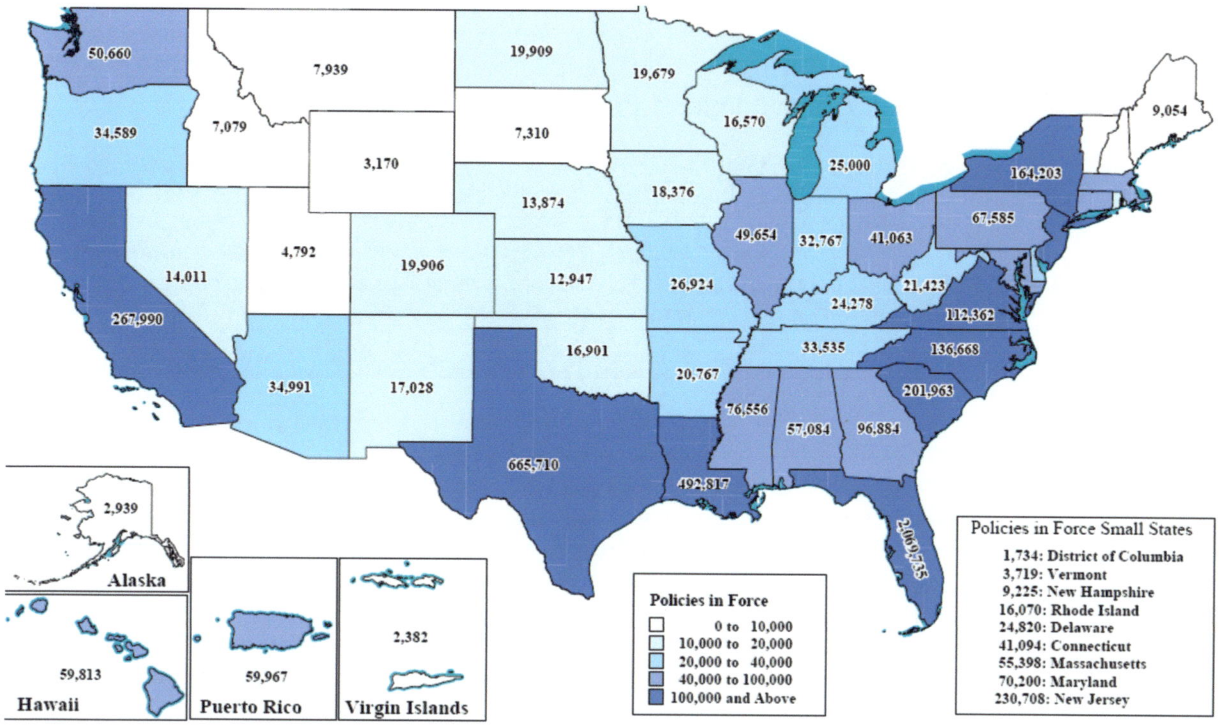

FIGURE 5-4 National Flood Insurance Program total number of policies in force, by state, as of September 30, 2012. SOURCE: http://www.fema.gov/policy-claim-statistics-flood-insurance/policy-claim-statistics-flood-insurance/policy-claim-13-2, accessed January 11, 2012.

$$\text{Compliance} = \frac{\text{Number of buildings in SFHA eligible for insurance with federal mortgages that have flood insurance}}{\text{Number of buildings in SFHA eligible for insurance with federal mortgages}}$$

In a scenario with full lender application of the regulation, compliance with the MPR would be approximately 100 percent.

There have been several analyses over the last two decades of the market penetration of the NFIP. According to Dixon et al. (2006), a 1999 report by PricewaterhouseCoopers estimated that 28 percent of *residential structures* in the SFHA carried flood insurance. In 2006, RAND estimated that 49 percent of *single-family homes* carried flood insurance (Dixon et al., 2006). Preliminary results from a J. Walter Thompson (JWT) analysis currently under way estimate that nationwide only approximately 30.6 percent of *identified structures* in the SFHA are insured.[36] Because each of the three analyses used a different set of the total structures in the SFHA, it is difficult to compare the results, although the ongoing JWT analysis appears to provide the highest degree of resolution.

Market penetration of flood insurance policies across the country varies with, for example, geography (coastal versus inland communities) and frequency of major flooding events in an area (Figure 5-4). The JWT analysis indicates that South Carolina had the highest penetration rate in the SFHA at 46.69 percent. Other coastal and flood-prone states have high penetration in the SFHA rates, such as Louisiana (43.69 percent), which is consistent with its history of high penetration particularly in the New Orleans area (Box 5-4, Louisiana), Florida (37.75 percent), New Jersey (37.94 percent), and California (33.49 percent).

The development of statistics for compliance with the MPR is confounded by several factors. As indicated, some penetration studies focused on single-family homes whereas others included all residential structures or

[36] A market penetration and mortgage compliance analysis, by state, in preparation for FEMA by J. Walter Thompson (JWT) advertising company. The committee used the version from January 2013, which is heretofore referred to as the "JWT analysis" in the report.

BOX 5-5
New York's Market Penetration Experience

Although detailed analyses of market penetration have not been conducted in areas impacted by Hurricane Sandy, preliminary figures indicate that approximately 62,000 of the owners of the 116,000 residential structures damaged by Sandy in New York City held NFIP policies. This equates to approximately a 54 percent market penetration rate. In the devastated Breezy Point area of Queens, New York, owners of 1,394 of the 2,150 structures damaged submitted claims, approximately a 64 percent market penetration rate (Jeffrey Woodward, FEMA, personal communication, February 27, 2013).

addresses, parcels, etc.[37] Many structure owners do not carry mortgages and others do not insure with federally regulated lenders. Both of these groups are exempt from the MPR and, combining the results of several studies, represent somewhere between 15 and 40 percent of the residential structure ownership in the SFHA. Tobin and Calfee (2005) reported that 77 percent or more of the debt in the SFHA was associated with lenders subject to the MPR, whereas Dixon et al. (2006) assumed that 85 percent of the single-family homes in the SFHA were subject to the MPR.

Dixon et al. (2006) also assumed that policies sold by private insurers, in addition to those sold under the NFIP, represented an additional 7 percent in policies, and that 38 percent of those not required to buy polices did purchase them. By applying these assumptions to the probable number of single-family homes with mortgages, by region, Dixon et al. identified compliance rates ranging from 43 percent in the Midwest to 88 percent in the West with a national average of 78 percent. The ongoing JWT analysis indicates, based on a sample of structures in the SFHA, that national compliance may be in the 50-60 percent range, with several states having compliance rates lower than 20 percent and other, high-risk states with compliance higher than 80 percent (Louisiana) (Box 5-5). According to Dixon et al. (2006), a 1999 PricewaterhouseCoopers study of purchase of flood insurance, assuming that 10 percent of homes compliant in 1 year are not compliant in the following year, resulted in MPR compliance rates between 55 to 60 percent. However, more recent retention data by Michel-Kerjan et al. (2012), noted above, found that the drop rate was considerably higher and would thus drive down the compliance rates.

Review of all these studies indicates that estimates of both market penetration in the SFHA and compliance with the MPR have varied considerably and are very sensitive to the assumptions made in the study process. National market penetration most likely lies somewhere between 30 and 50 percent, with the most recent JWT analysis, which is seemingly the most definitive study, indicating that national market penetration is the lower value. National MPR compliance has been estimated to be between 50 and 78 percent. The JWT study, again which appears to be the more definitive study because of a higher degree of resolution, presents the values at the lower end of this range.

For those who live in areas subject to flood risk both in front of (SFHA) and behind levees, insurance provides an effective and efficient means of transferring financial risk and thus reducing the exposure of individuals and businesses to catastrophic consequences and the need to call on the government for assistance. In 1973, the Congress found that for the NFIP to maintain fiscal integrity and protect federal interests, it needed to mandate insurance for those in the SFHA. However, in spite of four decades of effort, the program has yet to achieve high levels of compliance. Given the emphasis that has been placed on market penetration and compliance with the MPR in FEMA's risk communication activities, the efficacy of the mandatory purchase requirement is called into question. **The current policy of mandatory flood insurance purchase appears to be ineffective in achieving widespread purchase of NFIP flood insurance policies.**

[37] Dixon et al. (2006) found that residential structures make up 87percent of parcels in the SFHAs examined, and 57 percent of the total were single-family homes.

Multiyear policies tied to the property rather than the individual, thus avoiding policy cancellation, have been proposed as one solution (NRC, 2012). FEMA views the fault to rest with the enforcement by those agencies charged with lender oversight as problematic (GAO, 2002). Discussions with personnel in the trade indicate that budget and manpower realities limit the number of oversight inspections. In addition, in the committee's judgment, enforcement does not appear to have high priority within the lender community because it is seen as a tangential aspect of mortgage oversight. However, lenders assert that they are increasing their success (GAO, 2002). Retention statistics would also indicate lack of progress, in either oversight or enforcement. **The rate of compliance with the mandatory purchase requirement indicates challenges with lender enforcement and federal oversight of this lending.** On the other hand, statistics indicate that where residents are able to identify their own risk, some voluntarily acquire insurance. The fact that 41 percent of new policies between 2001 and 2009 were outside the SFHA, and were not mandated, gives support to this view (Table 5-2).

Mandatory Purchase Behind Levees: The Current Approach Verses a Modern Risk Analysis

Beginning in the early 2000s, FEMA took steps to modernize FIRMs via "Map Mod." During Map Mod, FEMA discovered that because FIRM maps carried forward certification from previous years, many levees previously thought to be able to withstand the one percent annual chance flood were not able to withstand this event. Thus, increased attention was placed on ensuring that the accreditation status for all levee systems in the NFIP was up to date and accurate through Procedure Memorandum 34 and the Provisionally Accredited Levee (PAL)[38] designation (Chapter 2). The process of ensuring that all one percent annual chance levees are properly constructed, certified, and accredited continues today, as affected communities are faced with the possibility of movement into the SFHA and the MPR. Often this is the catalyst for controversy as communities and levee owner/operators repair levees to the one percent annual chance standard to avoid MPR (Box 5-6). Thus, the one percent annual chance flood has become a de facto design standard that promotes the idea that if you live behind a levee you are "safe"—an unintended consequence of the MPR. During Hurricane Sandy, coastal V zone flows along open shorelines resulted in significant losses. Also, many residents and insurers learned that many areas outside the SFHA were subject to catastrophic loss. It is important to realize that overtopping of levees can produce flood flows behind levees whose velocity can match the coastal V zone flows on open shorelines, and results in damages similar to those associated with V zone flows. **Mandatory purchase requirements have led many property owners to perceive that if they are not mandated to have insurance they are not susceptible to damage from floods.**

The committee examined the extension of mandatory purchase to include areas behind levees and took into consideration several factors. In the current NFIP approach to analyzing risk behind levees, areas behind accredited levees are considered to be at a low to moderate risk and insurance rates are priced accordingly. However, this evaluation is an incomplete reflection of flood risk (Chapter 3). Thus, extending the mandatory purchase requirement behind accredited levees would be extending the federal government's exposure without a complete understanding of the risk that exists and the possible consequences. Furthermore, without the benefit of information provided in the modern risk-based analysis, those living behind levees will, like those in the SFHA, find it difficult to understand the reasoning behind the mandatory purchase requirement and the premiums associated with this requirement; a difficulty that is promoted by the idea that when you live behind an accredited levee, you are "safe."

The post-Katrina rise in insurance purchases outside of the SFHA may, if it continues, indicate that public awareness of flood threats is increasing and that given credible risk information, owners may voluntarily move to such purchases, thereby avoiding the necessity to extend the administrative challenge of the mandatory purchase requirement. Experience since Hurricane Katrina indicates that where residents are able to identify their own risk, some voluntarily acquire insurance (Figure 5-3). Furthermore, 41 percent of new policies between 2001 and 2009 were outside the SFHA and thus were not mandated. This gives support to the view that credible risk information

[38] PAL is applied to a previously accredited levee on an effective FIRM for which FEMA is awaiting data or information that will show compliance with NFIP regulations.

BOX 5-6
Challenges in East St. Louis

The American Bottom floodplain is home to approximately 156,000 people and businesses that employ 56,000 workers in the St. Louis area. The five levee systems (Figure 5-5) protecting the region were built in the 1940s and 1950s and, at the time of construction, were designed to provide protection against a 500-year flood. The levees were last upgraded in the 1960s. Since the flood of 1993, underseepage and sand boils have been recognized as a problem for these levees (FEMA, n.d.b).

In 2004 and 2005, FEMA began revising and updating FIRMs that include levees in and around the St. Louis area. In 2007 and during the revision process, FEMA notified the impacted communities that five of the levee systems in the eastern suburbs (Metro East St. Louis) could no longer be depicted as accredited on revised flood maps without demonstrated compliance with the design requirements set forth in 44 CFR §65.10 (Chain of Rocks Levee, Fish Lake Drainage and Levee District, Metro East Sanitary District [MESD], Prairie du Pont Levee and Sanitary District, and Wood River Drainage and Levee District). If no longer accredited, the area was to be included by FEMA as in the SFHA (i.e., mapped as without a levee) and the mandatory purchase requirement would apply to the eastern suburbs.

Inclusion in the SFHA and the MPR can have a significant impact on housing cost, economic development, and insurance rates. The East-West Gateway Council of Governments estimated mandatory insurance costs paid by residents and businesses in the American Bottom floodplain as a result of implementation of the MPR to be approximately $50 million dollars per year (East-West Gateway Council of Governments, 2010). Thus, the community faced the question of how best to restore the levee systems to a level that provided adequate protection and FEMA accreditation as well as how to finance this restoration. This was even more challenging given the high poverty rate in the East St. Louis area—an estimated 41 percent of persons in East St. Louis were below the poverty level from 2006 through 2010 (U.S. Census Bureau, n.d.).

In 2009 a local sales tax was passed, to collect the estimated $160 million dollars to repair the levees. As a result of legal action taken by local sponsors, FEMA withdrew issuance of a preliminary FIRM and is awaiting information from the sponsors as to what actions they propose to take to remediate identified deficiencies. The sponsors, using the $160 million dollars raised for the project, have proposed remediating the deficiencies to the point where the levee system can successfully pass the one percent chance flood and meet FEMA's requirements for accreditation. However, because the levee was originally built by USACE and then transferred to the local sponsors as a 500-year system, the sponsors must obtain approval from USACE to carry out any local work on the levee.[1] USACE has indicated that the levee deficiencies should be remediated in a federal–local cost-shared project to meet the standards for a 500-year levee and is taking steps to obtain the federal funding. If the project were to proceed as a local project, the sponsors must assure USACE that any work done to meet the one percent annual chance flood accreditation requirement will not interfere with the ability to eventually repair the levee system to the 500-year level. If federal funding can be obtained, USACE would move ahead in a process that would carry out the required work for a 500-year levee and, as part of this work, obtain accreditation from FEMA for one percent annual chance flood protection (Colonel Christopher Hall, USACE, personal communication, February 7, 2013)

---

[1] 33 U.S.C. §408 (Section 408) provides that any proposed modification to an existing USACE project must obtain permission from the Secretary of the Army by demonstrating that such proposed alteration or permanent use and occupation of the federal flood control project is "not injurious to the public interest and will not impair the usefulness of such work."

BOX 5-6 continued

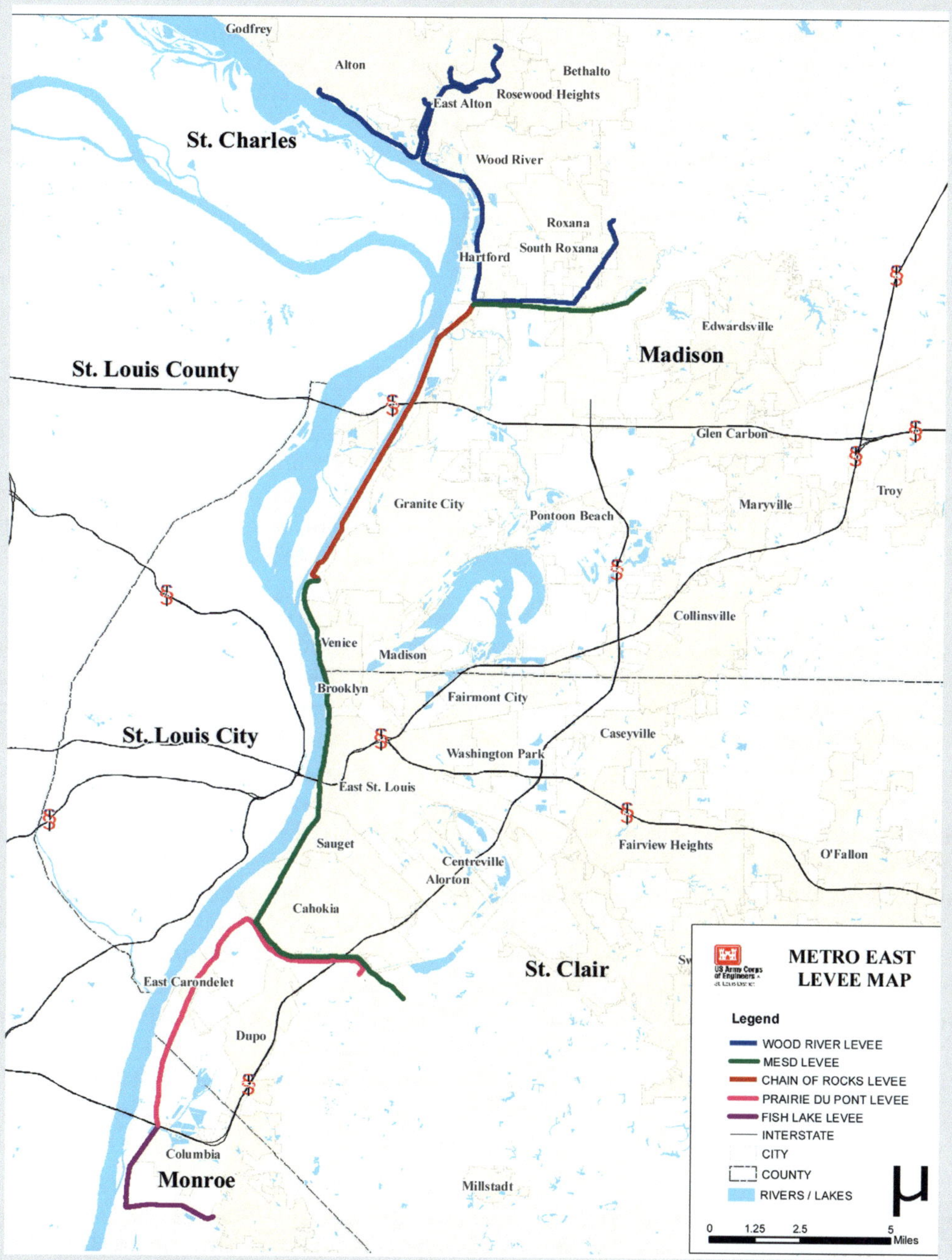

FIGURE 5-5 Map of the Metro East Levee System.
SOURCE: USACE (2010).

will prompt voluntary policy purchase, although retention rates also raise questions about property owner long-term commitment (Table 5-2).

To date, relying on federal supervisory agencies to oversee and lenders to require purchase has only marginally worked in the SFHA where the flood hazard is relatively obvious. There is little to suggest that most property owners in areas behind levees, if mandated to purchase insurance, would move to obtain it. Given the number of people behind levees, this would create an even greater challenge for lenders and oversight agencies.

In summation, instituting this requirement would be imprudent where there is (1) limited and conflicting evidence to support the effectiveness of the enforcement mechanism, (2) a communication impediment to overcome that is associated with the without-levee approach (when you live behind a levee you are "safe"), and (3) an incomplete understanding of risk behind a levee.. **At this time, there is no sound reason to institute mandatory purchase of flood insurance in areas behind accredited levees.** This conclusion is not to be interpreted as the committee not supporting the purchase of flood insurance behind levees and throughout the SFHA—quite the opposite. Other paths to achieve increased insurance purchase are critical: for example, effective communication of the risk behind levees; strong community support for voluntary acquisition of insurance behind levees; meaningful engagement with insurance companies and agents, lenders, and real estate professionals; and a modern risk-based flood risk analysis generating more accurate pricing and flood risk information.

Regardless of location (behind a levee or not behind a levee), because of the catastrophic nature of flooding, it is in a property owner's and, when applicable, lender's best interest to have the financial security provided by flood insurance coverage in the SFHA, assuming the cost of coverage is not an issue. As noted in Chapter 3 and above, a modern risk-based analysis will lead to a better informed, more differentiated assessment of properties' susceptibility to flood damage. If insurance rates are priced accordingly, this would provide for more granular policy pricing that better reflects the actual risk to the property. This includes areas behind nonaccredited levees. For example, an area with a 50-year levee will be given some credit for risk reduction and the insurance rates will be priced accordingly.

A perception exists that if the NFIP simply sold more policies the program's fiscal issues (e.g., debt to the U.S. Treasury resulting from the 2005 loss year) would be solved.[39] However, simply adding new policies doesn't automatically improve the program's fiscal soundness. Adding new polices could lower the program's overall risk *if* the policies increase the program's diversification of risk acquired through those policies. However, adding new policies very well may increase the NFIP's overall risk through concentration of policies in, for example, high-risk areas such as the SFHA. Furthermore, adding new policies would also increase the program's exposure to risk accumulation, or the additive risk of multiple events occurring at the same time. If NFIP policies continue to be concentrated in the same high-risk areas around the country, the diversification of risk and the fiscal soundness of the program will remain relatively unchanged.

Upon implementation of a modern risk analysis, insurance rates will more accurately reflect flood risk behind a levee. If the insurance rate goes up, it indicates that the property is at more peril than previously understood to be; if the insurance rate goes down, it indicates that the property is in less peril than previously understood to be. Thus, FEMA's moving to a modern risk-based analysis and pricing premiums accordingly will have an impact on NFIP policy purchase. More accurate pricing and more information created by a modern risk analysis has the potential to prompt additional policy purchases. Property owners would be more favorably inclined to buy flood insurance if individual risk is well known and insurance rates are priced to match the probability of flooding and the financial impact of flooding events. This includes areas behind levees, both accredited and not accredited, in the SFHA at large, and also non-SFHA areas. It is important to note that the Biggert-Waters Act directs full, risk-

[39] As noted by Baranoff (2009), the assumption that a large number of independent exposure units in a risk pool leads (via the law of large numbers and the central limit theorem) to an increased ability of the insurer to handle more risk at greater profitability. This is because the contingency loading for risk bearing added to the expected loss can decrease as the number of exposure units increase by the central limit theorem. This common perception, which is the mathematical underpinning of private insurance, can lead one to believe that a larger number of policies in the NFIP would lead to a similar result. Major differences, however, are that the NFIP premiums are not set in a fiscally sound manner (because of discounts, subsidies, no contingency reserve, etc.), and so, increasing the scale of an already cash-losing operation will not improve its position. Moreover, if the increase in policies is due to higher risk properties entering the NFIP (termed adverse selection, see Baranoff [2009]), then more policies can actually worsen the fiscal soundness of the NFIP.

based premiums and the elimination of certain discounts, also promoting a change to the NFIP that will impact policy purchase.

Lenders may choose to protect their portfolio regardless of the role of the NFIP (Buckley, 2010). It is very likely that if a lender has more information through a risk-based analysis, the lender is more aware that requiring flood insurance purchase is in line with their fiduciary responsibility. Regardless of the MPR, lenders currently can require insurance regardless of any regulatory responsibility as a function of their normal underwriting process. Furthermore, the recent NFIP reauthorization includes a provision requiring that lenders provide a disclosure of the availability of flood insurance to all purchasers at loan closing or settlement[40]—real estate agents might realize that disclosure of flood risk and promoting the purchase of a flood insurance policy is in their best interest.

The reasons why residents in flood-prone areas do not buy flood insurance from the NFIP have been studied both theoretically and empirically. Both behavioral and economic reasons for nonpurchase include the availability of aid from FEMA and other sources (i.e., charity hazard), underestimation of the risk probability, lack of knowledge of availability of the product and its price, the feeling that the coverage is not extensive enough, and several other rationales including experience with previous floods that did little damage, leading them to feel safe for the future (Kunreuther 1984, 1996; King, 2009). Effective communication of the risk behind levees is critical, for example, to overcome the impression that when you live behind an accredited levee you are "safe," an impression that is likely linked to policy purchase. Nuances in risk communication confound the premise that greater risk information will prompt additional policy purchases. For example, despite the positive association between risk perception and protective actions, awareness does not necessarily lead to actions to mitigate risk because of reasons judged to be more important, such as personal experience and evaluation of protective actions themselves. (See Chapter 7 for an expanded discussion.)

An increase in the purchase of policies, particularly in non-high-risk areas, will diversify the program's risk and ultimately generate a more fiscally sound NFIP. Thus, the necessity of the MPR in the SFHA may change, along with the decision to waive the MPR behind an accredited levee. **A modern risk-based analysis has the potential to impact the purchase of flood insurance, diversify the NFIP's exposure to flood risk, and generate a fiscally sounder program. Once the risk-based approach has been put in place and matures, FEMA should review and study the necessity of the mandatory purchase requirement, behind levees and throughout the SFHA.**

REFERENCES

Ai, J., and P. L. Brockett. 2008. Insurance pricing/nonlife. Pp. 922-933 in Encyclopedia of Quantitative Risk Assessment and Analysis, E. Melnick and B. Everitt, eds. Chichester, UK: John Wiley and Sons, Ltd.

American Academy of Actuaries, Actuarial Soundness Task Force. 2012. A Public Policy Special Report: Actuarial Soundness. Available online at http://www.actuary.org/files/publications/Actuarial%20Soundness%20Special%20Report%20FINAL%205%2010%2012.pdf. Accessed September 21, 2012.

ASCE (American Society of Civil Engineers). 2007. The New Orleans Hurricane Protection System: What Went Wrong and Why. Washington, DC: ASCE.

Baranoff, E. G, P. Brockett, and Y. Kahane. 2009. Risk Management for the Enterprise and Individuals. Available online at http://www.flatworldknowledge.com/. Accessed February 27, 2013.

Brockett, P. L., and R. C. Witt. 1982. The underwriting risk and return paradox revisited. Journal of Risk and Insurance 49: 621-627.

Browne, M. J., and M. Halek. 2010. Managing flood risk: The National Flood Insurance Program and alternatives. Chapter 5 in Public Insurance and Private Markets, Jeffrey R. Brown, ed. Washington, DC: American Enterprise Institute for Public Policy Research.

Browne, M. J., and R. Hoyt. 2000. The demand for flood insurance: Empirical evidence. Journal of Risk and Uncertainty 20: 291-306.

[40] Notice of flood insurance availability is required under the Real Estate Settlement Procedures Act (RESPA); see Section 100222 of the Biggert-Waters Flood Insurance Reform and Modernization Act.

Buckley, M. 2010. The NFIP Outside the Box. Background Reading for the Gilbert F. White National Flood Policy Forum, 2010 Assembly, Managing Risk to Humans and to Floodplain Resources. Available online at http://www.asfpmfoundation.org/pdf_ppt/2010_GFW_Forum_Background_Reading.pdf. Accessed February 21, 2013.

Casualty Actuarial Society. 1990. Foundations of Casualty Actuarial Science. New York: Casualty Actuarial Society.

Czajkowski, J., H. Kunreuther, and E. Michel-Kerjan. 2012. A Methodological Approach for Pricing Flood Insurance and Evaluating Loss Reduction Measures: Application to Texas. Available online at http://opim.wharton.upenn.edu/risk/library/WhartonRiskCenter_TexasFloodInsurancePricingStudy.pdf. Accessed April 4, 2012.

Daniel, P. 1977. Deep'n as It Come: The 1927 Mississippi River Flood. New York: Oxford University Press.

Dixon, L., N. Clancy, S. A. Seabury, and A. Overton. 2006. The National Flood Insurance Program's Market Penetration Rate: Estimates and Policy Implications. Washington, DC: American Institutes for Research.

East-West Gateway Council of Governments. 2010. Additional Flood Insurance Costs Resulting from the Federal Emergency Management Agency Flood Insurance Rate Map Modernization Program in the American Bottom. Available online at http://www.floodpreventiondistrict.org/wpcontent/uploads/2011/03/InsuranceCostsReport.pdf. Accessed February 21, 2013.

FEMA (Federal Emergency Management Agency). 2006. Designing for Flood Levels Above the BFE. Available online at http://www.fema.gov/library/file?type=publishedFile&file=fema549_apndx_e_ra8.pdf&fileid=143da3a0-0316-11dc-a1f1-000bdba87d5b. Accessed September 5, 2012.

FEMA. 2007. Mandatory Purchase of Flood Insurance Guidelines. Washington, DC: FEMA.

FEMA. 2011. National Flood Insurance Program: Answers to Questions About the NFIP. Available online at http://www.fema.gov/library/viewRecord.do?fromSearch=fromsearch&id=1404. Accessed February 15, 2013.

FEMA. n.d.a. National Flood Insurance Program Community Rating System. A Local Official's Guide to Saving Lives, Preventing Property Damage, and Reducing the Cost of Flood Insurance. FEMA 573. Available online at: http://www.fema.gov/library/viewRecord.do?id=3655. Accessed December 7, 2012.

FEMA. n.d.b. Mapping the Risk, Flood Map Modernization, Metro East—Frequently Asked Questions. Available online at https://www.fema.gov/pdf/about/regions/regionv/faq_east_stlouis.pdf. Accessed February 21, 2013.

GAO (Government Accountability Office). 2002. Flood Insurance: Extent of Noncompliance with Purchase Requirements Is Unknown. Washington, DC: GAO.

GAO. 2008. Flood Insurance: FEMA's Rate-Setting Process Warrants Attention. Available online at http://www.gao.gov/products/GAO-09-12. Accessed August 27, 2012.

Hayes, T. L., and D. A. Neal. 2011. National Flood Insurance Program Actuarial Rate Review in Support of Recommended October 1, 2011 Rate and Rule Changes. Available online at http://www.fema.gov/library/viewRecord.do?id=4853. Accessed June 12, 2012.

Hayes, T. L., and S. S. Sabade. 2004. National Flood Insurance Program Actuarial Rate Review. Available online at http://www.fema.gov/library/viewRecord.do?id=2377. Accessed May 28, 2012.

King, R. 2005. Federal Flood Insurance: The Repetitive Loss Problem. Congressional Research Service Report RL32972. Available online at http://www.fas.org/sgp/crs/misc/RL32972.pdf. Accessed February 26, 2013.

King, R. O. 2009. National Flood Insurance Program: Background, Challenges, and Financial Status. Congressional Research Service 7-5700, R40650. Available online at https://opencrs.com/document/R40650/.

Klugman, S. A., H. H. Panjer, and G. E. Willmot. 2008. Loss Models: From Data to Decisions, 3rd Rev. Ed. Chichester, UK: John Wiley and Sons, Ltd.

Kunreuther, H. 1984. Causes of Underinsurance Against Natural Disasters. Geneva Papers on Risk and Insurance 31: 206-220.

Kunreuther, H. 1996. Mitigating Disaster Losses Through Insurance. Journal of Risk and Uncertainty 12: 171-187.

Mathewson, S. B. 2011. Actuarial Rates in the Context of National Flood Insurance Program (NFIP) Hazard Mapping. Available online at http://www.actuary.org/files/publications/NCOIL_Simons_NFIP_Mapping_20110717_final.pdf. Accessed September 5, 2012.

Michel-Kerjan, E. O. 2010. Catastrophe economics: The National Flood Insurance Program. Journal of Economic Perspectives 24: 165-186.

Michel-Kerjan, E., S. L. de Forges, and H. Kunreuther. 2012. Policy tenure under the U.S. National Flood Insurance Program (NFIP). Risk Analysis 32: 644-658.

Moss, D. A. 1999. Courting Disaster? The transformation of federal disaster policy since 1803, Chapter 8 in The Financing of Catastrophe Risk, Kenneth Froot, ed. Chicago: University of Chicago Press.

Mowbray, A. H. 1914. How extensive a payroll exposure is necessary to give a dependable pure premium. Proceedings of the Casualty Actuarial Society 1: 24-30.

Nechamen, B., and M. Inderfurth. 2012. Summary of Contents: Biggert-Waters Flood Insurance Reform Act of 2012, H.R. Rep. No. 112-4348, Title III (Pages 521-576). Available online at http://www.floods.org/ace-files/documentlibrary/2012_NFIP_Reform/2012_NFIP_Reform_Act_ASFPM_Summary_of_Contents.pdf. Accessed February 7, 2013.

NRC (National Research Council). 2012. Dam and Levee Safety and Community Resilience: A Vision for Future Practice. Washington, DC: The National Academies Press.

Rodermund, M. 1990. Introduction in Foundations of Casualty Actuarial Science. New York: Casualty Actuarial Society.

Tobin, R. J., and C. Calfee. 2005. The National Flood Insurance Program's Mandatory Purchase Requirement: Policies, Processes, and Stakeholders. Washington, DC: American Institutes for Research.

U.S. Congress, House. 1966. A Unified National Program for Managing Flood Losses. 89th Cong., 2d Sess. H. Doc. 465.

USACE (U.S. Army Corps of Engineers). 2010. Metro East Levee System Maps. Available online at http://www.mvs.usace.army.mil/me_levees/maps.html. Accessed November 14, 2012.

U.S. Census Bureau. n.d. State and County Quick Facts, East St. Louis, Illinois. Available online at http://quickfacts.census.gov/qfd/states/17/1722255.html. Accessed February 27, 2013.

Witt, R. C., and A. M. B. Hogan. 1993. Economic, legal and social factors influencing insurance costs and prices. Journal of Insurance Regulation 11(3):314.

6

Implementing Flood Risk Management Strategies

Flood risk management seeks to reduce the risk from flood events to the people who are located in flood-prone areas. As indicated in earlier chapters, there is some level of risk to all locations within the floodplain. The magnitude of that risk is a function of the flood hazard, the characteristics of a particular location (its elevation, proximity to the river or coast, and susceptibility to fast-moving flows and surges, etc.), measures that have been taken to mitigate the potential impact of flooding, the vulnerability of people and property, and the consequences that result from a particular flood event. The initial risk is represented by the area's characteristics without consideration of mitigation and risk transfer measures and the vulnerability of the population. Each mitigation and risk transfer measure reduces the overall risk to some degree, but it is impossible to completely eliminate risk. A flood risk management strategy identifies and implements measures that reduce the overall risk and what remains is the residual risk (Figure 6-1). In developing the strategy, those responsible judge the costs and benefits of each measure taken and their overall impact in reducing the risk. This chapter describes measures that can be used to reduce the risk behind levees.

Levees represent one method of reducing the impacts of flooding on a community or a region. Levees keep the floodwaters away from the area behind the levee until the point at which the levee is overtopped or fails and the area behind the levees is inundated and the people and property are affected. The risk to those behind levees is a function of the characteristics of the levee (height, strength), their location, and the mitigation and risk transfer measures and vulnerability reduction actions that they have taken or have been taken on their behalf. As has been previously discussed, every location within a floodplain, regardless of the presence or absence of a levee and whether or not the levee is accredited, is subject to some level of risk.

It is important for those located in the floodplain and those responsible for activity in the floodplain (public officials, investors, and those relying on activities in the floodplain, etc.) to ensure that those in the floodplain understand the nature of the risks they face and the steps that may be taken to reduce this risk. In communities that are part of the National Flood Insurance Program (NFIP), those portions of the community located in the Special Flood Hazard Area (SFHA) are subject to mandatory insurance purchase and special land-use requirements including minimum first-floor elevations for new construction. FEMA's Community Rating System (CRS) identifies actions that can be taken by the community to reduce their risk and gives insurance premium reductions for communities that take appropriate mitigation actions.

Structure owners and occupants in NFIP communities who are not location in the SFHA, either outside of the one percent annual chance floodplain or behind an accredited levee, have no such federal restrictions even though

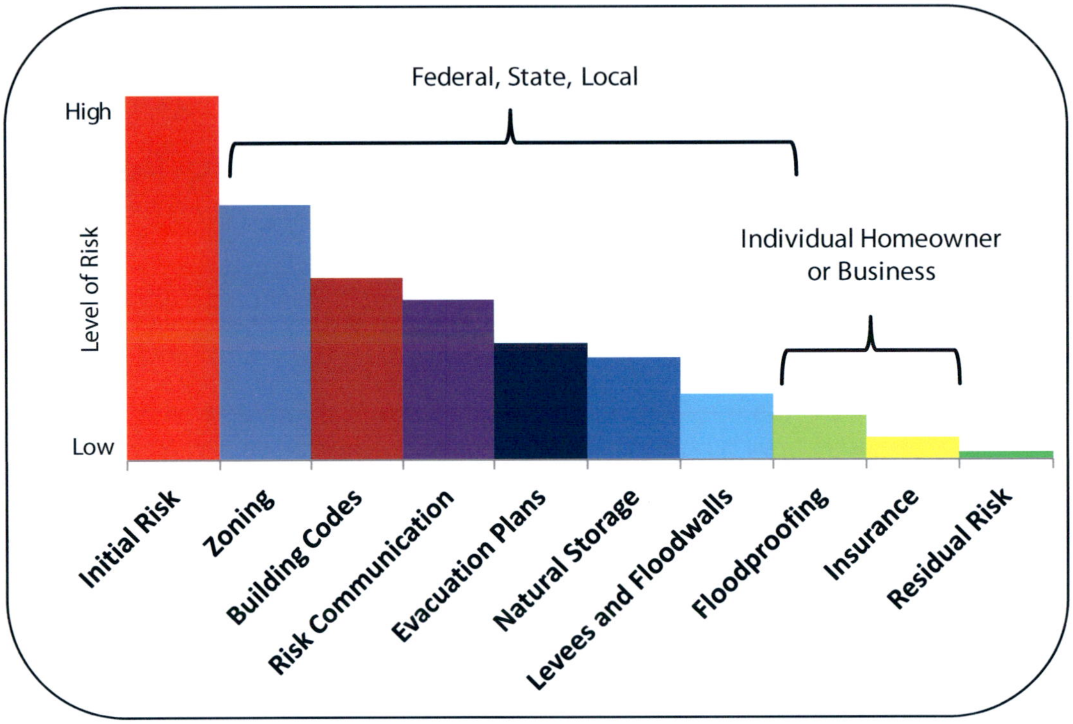

FIGURE 6-1 Examples of tools and measures to reduce and transfer flood risk in flood-prone areas. The remaining risk after these techniques are employed is the residual risk. The bar on the far left indicates the initial, unmitigated risk that is faced by a community. Actions taken through the methods indicated in the subsequent bars, which are illustrative, reduce the unmitigated risk. Some of these actions are taken at the federal, state, and local levels, whereas others are taken by the homeowners and businesses at risk. The risk that remains after these actions are taken (bar on the far right) is the residual risk.
SOURCE: Modified from USACE (2006).

in many cases they face significant risks. As Figure 6-2 illustrates, a structure whose first-floor elevation is at the one percent annual chance level, might suffer only minor consequences should a greater than one percent annual chance flood occur, whereas those behind a one percent annual chance levee might suffer significant consequences. Because there are no federal requirements for land-use restrictions or mitigation in the non-SFHA area, many owners and public officials erroneously assume that the absence of requirements for action can be equated with absence of risk, when in reality, the risk may actually be greater in some areas of the SFHA.

All NFIP communities, because of the land-use regulation provisions for the SFHA, have given some attention to the development of risk management strategies. However, once a levee is accredited, the area behind that levee is considered by most communities to be outside the floodplain and not subject to land-use regulation or requirement for communities to consider the consequences of failure or overtopping. Depending on the topography of the area behind the levee, these strategies consider interior drainage systems to ensure that water "trapped" behind the levee during a storm will not create internal flooding and a SFHA.[1]

One measure of the potential risk behind levees is measured by the number of structures protected by levees. FEMA currently estimates that 8.6 million housing units (6.5 percent of units in the United States) are located in SFHAs (FEMA, 2012a). The U.S. Army Corps of Engineers (USACE) estimates that there are 14 million structures behind the 14,000 miles of levees for which it has oversight; many of these levees are also accredited under the NFIP (Chapter 8) (Bryan Baker, USACE, personal communication, January 10, 2013). USACE is currently

[1] See 44 CFR §65.10 (b)(6).

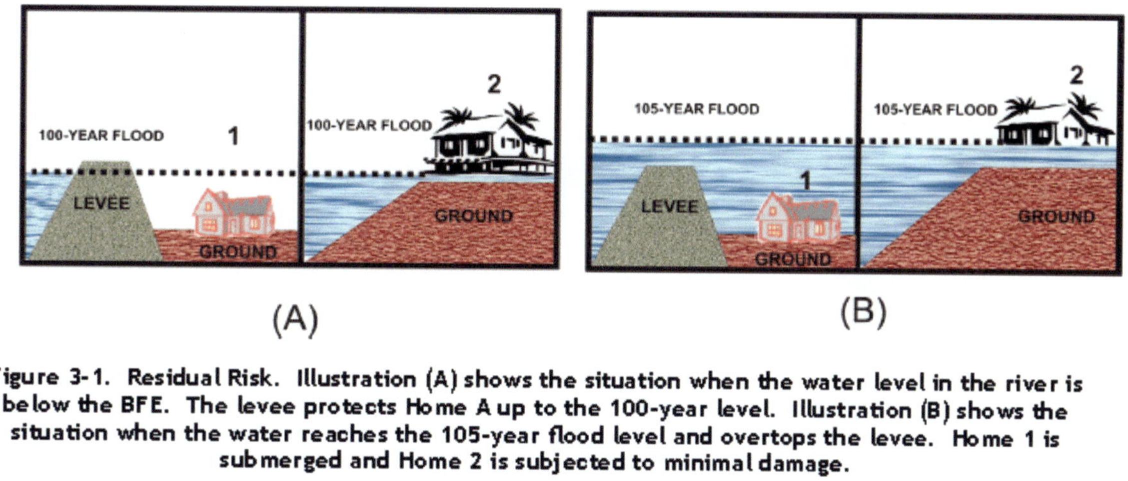

FIGURE 6-2 Scenarios of residual risk. In scenario (*A*), the levee protects the home up to the one percent annual chance flood. In scenario (*B*) the water is higher and the one percent annual chance flood overtops the levee. The consequences to Home 1 are more significant than to Home 2: Home 1 is submerged; Home 2 is flooded.
SOURCE: FEMA (2006).

developing a Levee Safety Action Classification (LSAC) for use in screening the results of its inspection of levees under its oversight. Information from the LSAC will be provided to levee owners (USACE, 2012a).

MITIGATION AND RISK TRANSFER MEASURES

Mitigation is defined by the Federal Emergency Management Administration (FEMA) as "sustained action taken to reduce or eliminate long-term risks to people and property from hazards" (FEMA, 2010b). Traditionally, mitigation is divided into structural and nonstructural options. USACE, n.d.a defines structural and nonstructural mitigation measures as follows:

> Structural measures such as dams, levees, and floodwalls alter the characteristics of the flood and reduce the probability of flooding in the location of interest. Nonstructural measures alter the impact or consequences of flooding and have little to no impact on the characteristics of the flood.

A portfolio of structural and nonstructural mitigation measures can reduce the likelihood and/or impact of flooding (Table 6-1). For nearly two centuries, the nation relied principally on structural measures to control floods.

TABLE 6-1 Examples of Structural and Nonstructural Flood Mitigation and Risk Transfer Measures

Structural	Nonstructural
Levees	Structure elevation
Floodwalls	Natural systems
Seawalls	Risk mapping
Dams	Hazard forecasting, early warning systems, and emergency plans
Floodways and spillways	Dry and wet floodproofing
Channels	Land-use planning and zoning
Controlled overtopping	Construction standards and building codes
Levee armoring	Acquisition and relocation
Seepage control	Insurance

In the mid-20th century, in recognition that even with significant use of structural approaches, flood damages were still on the rise, governments initiated multifaceted floodplain management and expanded use of nonstructural means and risk transfer (flood insurance) mechanisms (Sayers et al., in press).

Structural Mitigation Measures

A brief description of the structural measures most often used in flood control follows, along with discussion about how each of these measures relates to levees. This discussion was adapted, in part, from NRC (2012b).

Levees, Floodwalls, Seawalls, and Other Appurtenant Structures

These structures are designed to prevent floodwaters and storm surges from reaching areas that are at risk. Consequences of failure can be catastrophic because those behind the structure can be subject to rapid inundation and flooding conditions more severe than if the floodwaters had risen gradually.

Dams

Barriers that impound hydrologic flows, dams retain floodwaters before they reach areas at risk. For example, during high-precipitation periods, dams hold upstream floodwaters that are released gradually to minimize the likelihood of damage to downstream communities. However, during exceptionally large events, the storage capacity of a dam can be exceeded and uncontrolled flood flows are passed downstream. Under these circumstances, downstream levees may not be able to contain floodwaters and will fail. This condition occurred in 2011 during spring floods on the Missouri River (USACE, 2012b). Under exceptional circumstances, dams can fail and send significant quantities of water downstream, resulting in damage or destruction of levees and communities below the dams.

Floodways, Spillways, and Channels

Floodways, spillways, and channels are constructed to carry floodwaters around a community or region where the capacity of a river to pass a large volume of floodwaters past a critical location is limited. Under some circumstances, river channels can be modified to increase their flood carrying capacity. During the 2011 flooding of the Mississippi River, USACE opened floodways near New Madrid Missouri to take the pressure off upstream and downstream levees in Illinois, Kentucky, and other locations in Missouri, and three floodways in Louisiana to relieve pressure on structures in the New Orleans area. A similar floodway provides relief when needed to relieve pressure on levees surrounding Sacramento, California.

Structural Mitigation Through Improved Levee Design or Modification

During a flood, levees are under continuous stress that threatens their integrity. The most serious challenges result from wave action against a levee face, the erosion of the land side of a levee as the levee is overtopped or subject to waves breaking over its top, and seepage under the levee that destroys the levee from within. Forms of these occurred during Hurricane Katrina and caused the failure of levees in the New Orleans region in 2005. Use of controlled overtopping, armoring, and underseepage control can greatly reduce the potential for catastrophic failure.

Controlled Overtopping and Breaching of Levees

During a flood event, the risk of a levee overtopping can be significant and the consequences can be catastrophic. Controlled overtopping of levees or engineered overtopping involves designing a levee to force overtopping in the least hazardous location (USACE, 1986). This can be done by using different levee heights, known as superiority, or notches or openings in a desired location (Figure 6-3).

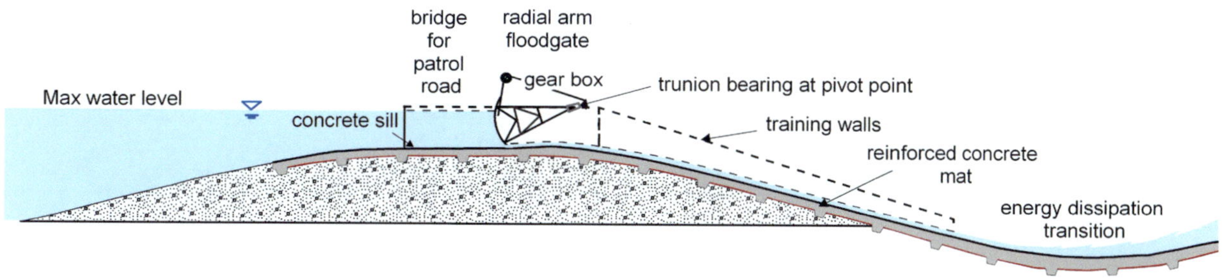

FIGURE 6-3 A gated overflow spillway that would control flows into a desired area.
SOURCE: Courtesy of Jae-won Chung.

The advantages of controlled overtopping in a designated area are (1) reducing the impact of overtopping failure in the selected area and in other parts of the levee system, (2) reducing the likelihood of overtopping in less desirable areas (i.e., areas with more development), and (3) reducing levee maintenance and repair costs after the flood event.

Controlled breaching of levees is occasionally carried out during sustained high flow events (floods), when the benefit of the intentional diversion is deemed to be justified, economically, to mitigate more extensive damage that would likely otherwise occur. Controlled breaches are relatively rare, but have been used in most major flood events since 1927 as a last ditch means to lower flood stages threatening high-value areas, such as New Orleans (in 1927), Cairo, Illinois, and the St. Francis Basin in Missouri and Arkansas 1937 and 2011 along the Mississippi River, or historic towns such as Prairie du Rocher, Illinois, during the 1993 Missouri–Middle Mississippi River flood. These are not to be confused with so-called "forced breaches," which are routinely employed during floods, usually to help drain flooded tracts of urban or agricultural land. Forced breaches are necessary if there is no other ready means of draining flooded lands that are ringed by intact levees about their lower, or "downstream" perimeters. Forced breaches are usually made using tracked backhoe excavators and are commonly employed whenever a diked tract of land is compromised by flooding. The natural breaches and forced breaches must then be repaired before the succeeding flood season. Since records began being tabulated within the NFIP in 1969, forced breaches have historically accounted for 40 to 50 percent of the post-flooding repair costs sustained by local agencies (Storesund et al., 2009).

Levee Armoring

Armoring a levee involves making a levee less susceptible to erosion induced by floodwaters and overtopping. It involves the use of a variety of materials, from concrete to vegetation. Three key factors in determining levee survival in a significant flood event and overtopping are depth and duration of flow, flow velocity (a function of slope inclination, height of the drop, and flow friction), and the erosive resistance of the vegetation mat and underlying soils (Briaud et al., 2008; Storesund et al., 2009).

The earliest methods of armoring employed in the 1800s, involved timbers reinforcing the submerged wall of the levee. Today, particularly after levees failed in Hurricanes Katrina and Rita in 2005, considerable attention is being given to new methods and technologies that might make earthen levees more survivable during short-term overtopping, as occurs during hurricane-whipped storm surges. These concepts include the use of embedded soil reinforcement (high-density polyethylene mesh and fiber reinforcement), turf reinforcement, and paving land-side slopes with soil admixtures, such as clay, soil cement, and other forms of tensile reinforcement (Rogers, 2009; Xu et al., 2012). An example of one such armored levee is shown in Figure 6-4.

Seepage Berms and Cutoff Walls

Depending on the nature of the material used in the construction of the levee and the foundation of the levee, water may flow through or under a levee creating the potential for collapse of the levee or its foundation (Figure

FIGURE 6-4 Example of an armored levee in Japan.
SOURCE: Courtesy of Gerald E. Galloway.

6-5, *Top*). Cutoff walls and trenches can be used to stop the flow through and below a levee (Figure 6-5, *Lower Left*). Construction of land-side seepage berms can add sufficient weight to counteract the upward seepage forces (Figure 6-5, *Lower Right*). Pressure relief wells can be placed on the land side of the levee to deal with seepage and are part of the levee repair effort for the East St. Louis, Illinois levees mentioned in Chapter 5.

Nonstructural Mitigation Measures

Nonstructural measures vary in cost and effectiveness and the physical and political effort required to implement them. The selection of which measure to use is a function of the location in which it will be employed (topography, expected flood levels, etc.), the availability of funds, and public acceptance of use.

Structure Elevation

By raising a structure above the expected flood level, flood damages can be prevented (Figure 6-6). Behind an accredited levee, if a flood greater than the one percent annual chance flood occurs, there could be some damage to such elevated properties but considerably less than if the structures had been at the base flood elevation (BFE). In the SFHA, buildings are elevated to the BFE in the case of new construction, substantial improvements are made to existing buildings, and repairs are made to substantially damaged buildings to comply with the NFIP (FEMA, 2000). NFIP insurance rates for structures in the SFHA are discounted for elevations above the BFE, providing that supporting documentation (i.e. an elevation certificate) is provided.

Elevation includes moving key or essential equipment from low-lying elevations within a structure to areas that would not be subject to flooding. Having building support systems such as computers, heating and air-conditioning units, and electrical stations located in the basement areas puts the functionality of the entire building

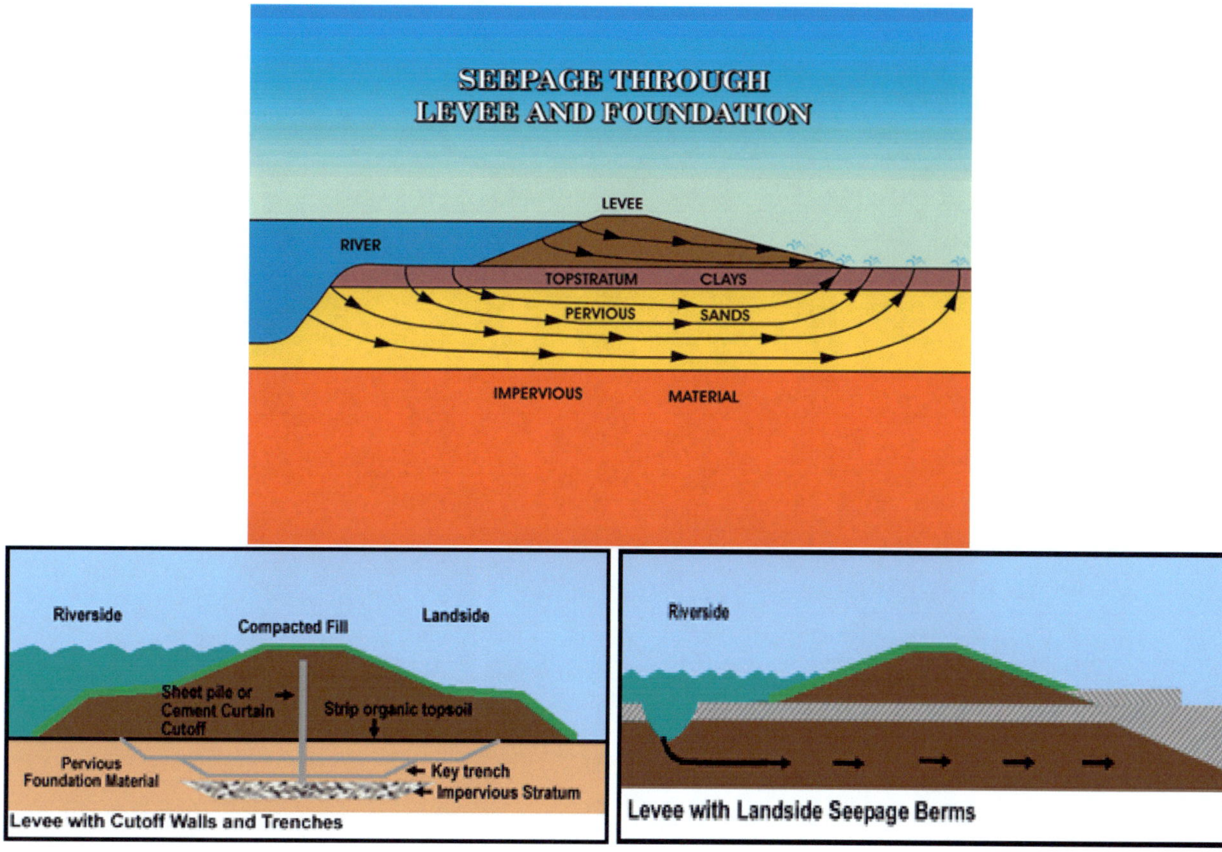

FIGURE 6-5 Levee seepage and control methods. (*Top*) Seepage paths under and through a levee. (*Lower Left*) Use of cutoff walls and trenches to block seepage. (*Lower Right*) construction of a land-side seepage berm to provide additional weight and length to counteract upward seepage forces.
SOURCE: USACE Management Measures Digital Library. Available online at http://www.iwr.usace.army.mil/docs/MMDL/FLD/Feature.cfm?ID=5, accessed January 25, 2013.

at risk during a flood event. In 2006, a heavy storm in downtown Washington, D.C. flooded much of the federal triangle including the National Archives, the Federal Bureau of Investigation Headquarters, and the basement of the Internal Revenue Service (IRS), damaging essential mechanical and electrical equipment. The IRS was closed for 6 months to allow for repairs (GAO, 2007; National Capital Planning Commission, 2008).

When building a structure on the floodplain, the cost of incorporating freeboard in a pile or a masonry pier foundation averages approximately 1 to 2 percent of the at-BFE building cost for 4 feet of added freeboard. For a masonry wall with interior pier (crawlspace) foundation, the cost averages 3 to 6 percent of the at-BFE building cost (Jones et al., 2006). Jones and colleagues (2006) concluded that it is financially reasonable to spend between 103 to 106 percent of the at-BFE building cost to elevate a structure, depending on local circumstances.

Challenges in elevating structures do exist. For example, properties that service people with disabilities may require either an elevator or a long ramp, the cost of which could make elevation economically infeasible. In addition, there is growing concern that having an entire community elevated creates the potential for small islands (houses) in a flood sea that cannot be accessed during the flood event. In the case of fire or a medical emergency, first responders would find it very difficult to reach many of these homes. If flooding were to last for an extended period, the elevated homes would rapidly become uninhabitable.

FIGURE 6-6 Elevated homes behind levees in New Orleans, Louisiana.
SOURCE: Courtesy of Gerald E. Galloway.

Natural Systems

Naturally existing, restored, or developed wetlands, as well as land in periodic cultivation, can store overflow waters from riverine flooding and help reduce downstream impacts. Wetlands also provide a natural barrier to storm surge inundation. Coastal sand dunes protect structures built behind them and help slow down coastal erosion and also the immediate impact of rising storm surge. Floodways built to relieve flood pressure can also provide ecosystem benefits. These natural systems can be effective in reducing flooding behind levees with interior drainage problems (Galloway et al., 2009; Opperman et al., 2009). Use of natural systems typically requires some form of real estate acquisition (fee simple, easement, payment for use, etc.) because the benefits from such measures normally do not accrue to the individual who may own the property required. In some cases, there will be a need for construction of appropriate inlet and outlet works to permit the entry and exit of stored waters.

Risk Mapping

Accurate mapping of risks provides those living or working in flood-prone areas, in front of or behind levees, with the information necessary to make rational decisions in developing their personal or corporate flood risk management strategies. Risk mapping is discussed further in Chapter 7.

> Combining the natural hazard risk assessment with quantitative consideration of mitigation measures yields expected outcomes that can be graphically portrayed in a manner that facilitates public understanding of the risk and its implications for them.
>
> NRC (2012b)

Hazard Forecasting, Early Warning Systems, and Emergency Plans

Detailed weather forecasts of the path and severity of a tropical storm, and accurate predictions of stages (heights) of flooding rivers enable government officials and the public to make decisions to evacuate or move valuable property from high-hazard areas. Advance information about potential failures of levees or dams can significantly reduce the consequences should failures occur. Preparation of emergency action and evacuation plans can similarly reduce or eliminate casualties and property losses.

Dry and Wet Floodproofing

Damage to structures in the SFHA and behind levees can be greatly reduced through effective dry and wet floodproofing. Dry floodproofing seals structures to prevent floodwaters from entering; wet floodproofing makes uninhabited portions of a structure resistant to floods by allowing water to enter and flow through the structure during a flood. FEMA defines floodproofing as

> a combination of adjustments and/or additions of features to buildings that eliminate or reduce the potential for flood damage. Examples of such adjustments and additions include anchoring of the building to resist flotation, collapse, and lateral movement; installation of watertight closures for doors and windows; reinforcement of walls to withstand floodwater pressures and impact forces generated by floating debris; use of membranes and other sealants to reduce seepage of floodwater through walls and wall penetrations; installation of pumps to control interior water levels; installation of check valves to prevent the entrance of floodwater or sewage flows through utilities; and the location of electrical, mechanical, utility, and other valuable damageable equipment and contents above the expected flood level (FEMA, n.d.a).

The Federal Alliance for Safe Homes (FLASH) indicates that dry floodproofing can be accomplished by measures such as

- Applying a waterproof coating or membrane to the exterior walls of the building;
- Installing watertight shields over doors, windows and other openings;
- Anchoring the building as necessary so that it can resist floatation;
- Installing backflow valves in sanitary and storm sewer lines;
- Raising utility system components, machinery and other pieces of equipment above the flood level;
- Anchoring fuel tanks and other storage tanks to prevent flotation;
- Installing a sump pump and foundation drain system;
- Strengthening walls so that they can withstand the pressures of floodwaters and the impacts of flood borne debris;
- Construct nonsupporting, break-a-way walls designed to collapse under the force of water without causing damage to the house or its foundation.[2]

Wet floodproofing includes construction of veneers to seal potential water entry into or under a structure, installation of vents to allow water to move through crawl spaces under homes (avoiding different water levels outside and under a home).[3] Floodproofing can also include permanent or temporary installation of barriers such as modular dams and small levees designed to keep floodwaters away from one or more structures for limited periods of time (Figure 6-7).

There is an increasing demand for products and techniques that can be used in existing properties to reduce the potential damage when flooding occurs. To ensure that floodproofing products are well designed and manufac-

---

[2] See http://www.flash.org/peril_inside.php?id=59.
[3] See http://www.flash.org/peril_inside.php?id=60.

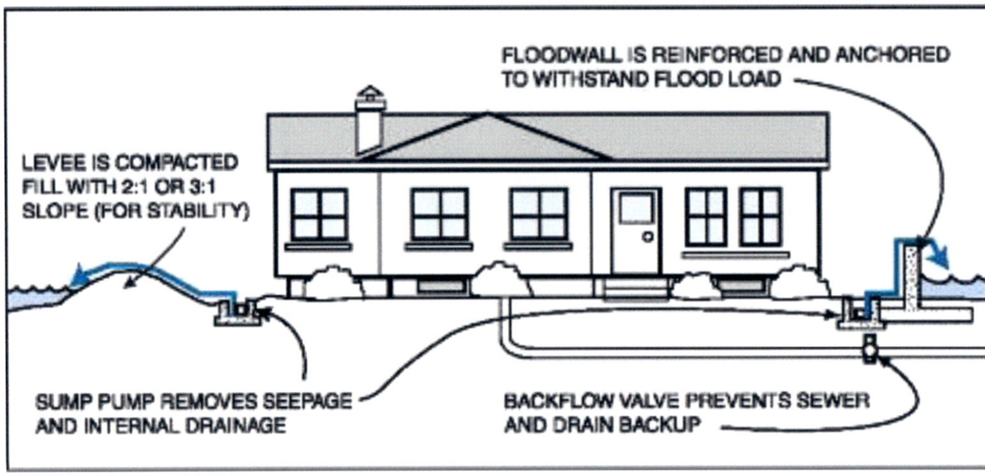

FIGURE 6-7 Barriers and small levees floodproof a home from low-level flooding.
SOURCE: FEMA (2007).

tured to perform the task for which they are being developed, both the Association of State Floodplain Managers (ASFPM) and FM Global have developed certification standards to support this need (FM Global, 2006, n.d.).[4]

Land-Use Planning and Zoning

Wise land use is at the center of nonstructural flood mitigation activity and is an effective tool for reducing risk at the community level. Land-use planning implements public policy to direct how land in a given area is used. It is executed through zoning ordinances and takes place on multiple levels of government, from national policy to local policy where there may be designation of parcels for a specific use at the local level. When appropriate, wise land use may mean a lack of investment in a particular area (Box 6-1).

Few issues have gained attention and controversy like land-use planning. The tension between promoting development in order to foster growth in the community and the potential long-term liabilities (i.e., flood risk) is significant. Unfortunately, when individuals or businesses develop an area that may be subject to flooding and are unaware of the potential risk, these individuals suffer when a flood event occurs. Although the NFIP requires regulation of the land within the SFHA, it does not require flood-prone communities to regulate areas beyond the one percent annual chance flood level or areas behind levees even though both areas face flood risks. When a community's land-use plans treat all areas behind a levee the same way even though some areas are substantially below the BFE and would be fully inundated in the event of a major levee failure, these land-use plans ignore the risk in placing occupants of the lower areas at an economic and safety disadvantage.

Construction Standards and Building Codes

Construction standards and building codes can be developed at any level of government but they are enforced at the local level. While codes provide for public safety, they also prescribe practices and measures that directly address known causes of disaster damages. Damages can be significantly reduced by attention to modern construction standards and building codes. Kunreuther (1996) found that one-third of the damages associated with the 1992 Hurricane Andrew could have been avoided had Florida enforced its building codes.

[4] See http://www.floods.org/index.asp?menuid=421&firstlevelmenuid=183&siteid=1.

BOX 6-1
The Coastal Barriers Resources Act

Coastal barriers possess many characteristics that make them attractive building sites. These include their rich biological diversity, their status as popular vacation destinations, and their role as large drivers to local economies. However, these areas pose substantial risks to both developers and homeowners. They are often the location of first landfall by tropical storms, bearing the full force of storm surges and hurricane winds, and are the victim of a constantly fluctuating landscape due to chronic coastal erosion (FWS, 2002). The federal government historically subsidized and encouraged development in these coastal areas until the late 1970s and early 1980s, when it was realized that this subsidization and encouragement had been resulting in the loss of natural resources; presented a threat to human life, health, and property; and cost American taxpayers millions of dollars each year (FWS, 2012).

In an effort to remove any federal incentive to develop on coastal barriers, Congress passed the Coastal Barrier Resources Act (CBRA) in 1982. Reauthorized in 1990, the CBRA designated relatively undeveloped coastal barriers along the Atlantic and Gulf coasts, the Florida Keys, the Great Lakes, Puerto Rico, and the U.S. Virgin Islands as part of the newly created John H. Chafee Coastal Barrier Resources System. This designation renders these coastal barriers ineligible for most new federal expenditures and financial assistance, including, most significantly, access to federal flood insurance through the NFIP (FWS, 2012). Although no longer encouraged by the federal government, development in these hazard-prone areas is not prohibited, provided any work and investment is made by private developers and other nonfederal parties. In return for a lack of restrictions on development, any individual or developer choosing to live and invest in these areas agrees that they will bear the full cost of development and rebuilding (in the case of a flooding event or other natural disaster) instead of relying upon federal funding for roads, wastewater systems, potable water supply, and disaster relief (FWS, 2012). This approach has saved over $1 billion in federal dollars between 1982 and 2010 and is expected to continue saving federal dollars in the future (FWS, 2012).

Acquisition and Relocation

Acquisition or relocation of properties that are repetitively flooded, substantially damaged, or need flood-related project construction have been supported by FEMA and USACE, allowing many communities to deal with the challenges of these frequent flood losses. When land is not suitable or only marginally suitable for development or agricultural activity, that land can be acquired and placed into public use for either environmental or recreational purposes to enhance the overall quality of life in the community. Buyouts of properties that are repetitively flooded have been supported by FEMA and have allowed many communities to deal with frequent flood losses.

Although a few sections of communities have been relocated, more relocations or removals usually take place where specific parcels are identified as being at risk. When property is acquired for environmental purposes. this land can frequently be used during flood periods for off-river storage of floodwaters. This flood storage reduces the downstream impact of the flood and the area is restored after the flood to its previous condition. During the 2011 Mississippi River floods, over 500,000 acres of land in the lower Mississippi Valley was used for flood storage, which dramatically reduced river stages that otherwise would have affected large communities such as New Orleans and smaller ones such as Cairo, Illinois.

Insurance

Through the purchase of insurance, some of the financial risk of living on a floodplain in or outside the SFHA or behind a levee is transferred, for a premium, from the individual to the entities that provide insurance. Floodplain managers tend to consider this a mitigation technique, when in fact unlike physical measures that reduce damages, insurance does not reduce damages but simply compensates the affected parties for these losses. In most flood-prone areas in the United States, this insurance is provided to residential properties and small businesses through the NFIP. Commercial flood insurance is available and is issued by numerous major insurers and is issued by a worldwide network of reinsurance organizations, providing coverage for property value in excess of the NFIP maximum of $250,000 for residential property and $500,000 for businesses. However, unlike the NFIP, commercial insurance rates and rates for homeowner coverage in excess of the NFIP cap are based on more detailed analysis of properties as they apply for coverage and may reflect more sophisticated risk analysis and requirements for mitigation measures (Chapter 5).

When property owners take steps to mitigate potential losses through use of mitigation measures, commercial insurers typically will reduce the premiums to reflect the risk mitigation. The NFIP provides premium reductions in the SFHA in two cases, for elevation of property higher than the BFE and some types of commercial property floodproofing. However, the NFIP does not provide premium reductions in areas behind levees (preferred risk policies). When risk-based pricing is implemented, discounts for mitigation measures behind levees need to be considered because it is likely that, in many cases over the life of a mortgage, the premium discounts could exceed the costs of the mitigation and would be beneficial to the mortgage holder and the mortgagee.

The perception exists that flood insurance is unnecessary for individual homeowners in the floodplain because, after a flood event, government and private agencies assist the property owners in getting back on their feet. Media statements and political rhetoric regarding disaster aid imply generous aid to repair and replace property. Analysis has indicated that this is not necessarily the case (Kousky and Shabman, 2012). Possession of an insurance policy provides rapid payment to the owner of claims for property damage and gives those with insurance the ability to recover much faster than those seeking to piece together other forms of assistance, which, in the long run, will not make up the entirety of losses.

DEVELOPING COMMUNITY STRATEGIES FOR AREAS BEHIND LEVEES

Circumstances produce a variety of flood risk scenarios behind levees. These circumstances can include anything from proximity to the coast or river, local weather patterns, existing infrastructure, and community preparedness. Some risk scenarios are associated with long time horizons, such as flooding associated with climate change, and might be less perceptible to the at-risk community compared with, for example, a design deficiency in a levee discovered during the Map Modernization process. Complicated scenarios arise as communities grapple with unique situations (Box 6-2). A mix of flood risk management measures tailored to the risk of a particular

BOX 6-2
Challenges to Mitigation and Flood Management: Norfolk, Virginia

Living in an area that is vulnerable to flooding, the community of Norfolk, Virginia, has spent years wrestling with flood risk. Home to the world's largest naval base and 1.7 million people, even high tides or heavy rains can cause problems for this low-lying port at the mouth of the Chesapeake Bay. Challenging issues such as the combination of sea level rise and subsidence dramatically changing the level of the ocean in Norfolk face this community (Figure 6-8; Boon et al., 2010).

The current flood management system of Norfolk consists of pumping stations and floodwalls. Each year, the city invests in improving drainage and elevating roads. Norfolk has taken additional steps to prepare for flooding, including hiring a Dutch coastal engineering firm to conduct a vulnerability analysis to enhance mitigation and inform development. This new strategy includes, for example, new floodgates, pumping stations, and floodwalls at an estimated cost of $1 billion dollars (NPR, 2012). This strategy also includes something called "managed retreat" by Mayor Paul Fraim, saying that the cost of flood mitigation is sometimes too steep and giving up homes makes more sense (NPR, 2012).

community paired with an appropriate communication strategy is the most effective flood risk reduction strategy (Burby and Dalton, 1994; Brody et al., 2007; Sayers et al., in press; see also Figure 3-3 in Chapter 3).

Flood Risk Management Strategy: Goals at the Local Level

To successfully develop a flood risk management strategy, a community, in collaboration with those entities that may be called upon to assist in the implementation of the strategy, needs to first establish the goals of the strategy. This collaboration requires the community to synchronize planning with relevant regional bodies and ensure its conformance with state and federal guidelines and regulations when they exist. A critical part of the goals of the strategy is to define the relative levels of protection both in front of and behind levees and then to lay out the appropriate combination of mitigation measures (structural, nonstructural, or insurance) to achieve those levels.

Today, there is no national standard for level or degree of protection for flood-prone communities, except for those choosing to participate in the NFIP. Under the NFIP, those in the SFHA must take action to mitigate against the one percent annual chance flood. Those behind levees must ensure that the levees protect against the one percent flood. However, this criterion is a standard for participation in the insurance program, not a life-safety or property protection standard. Since the decision was made in the 1970s to use this one percent standard for NFIP accreditation, the predominance of the written reports concerning levees have argued for a minimum standard of 0.2 percent annual chance for levees protecting urban areas (see Chapter 2 and Appendix F).

As indicated in Chapter 2, when the federal government began to build levees under its flood control authorities, it began with relatively high standards. Urban area levees and those protecting the Lower Mississippi Valley were to be built to pass a standard project flood or higher level flood. Agricultural areas were typically protected to near the 2-3 percent annual chance level. Executive Order 11988, March 1977, directed federal agencies supporting activity in the floodplain to require critical facilities to be located outside the 0.2 percent floodplain or be elevated above that level, but few similar requirements were imposed by state governments.

In 1973, the U.S. Water Resources Council issued Principles and Standards (WRC, 1973), which established national and regional economic development, environmental quality, and other social effects as criteria for evaluation of federal water resource projects. These standards permitted a variety of justifications for level of protection to be considered. In 1983, the WRC established, in Principles and Guidelines, national economic development as the single objective (WRC 1983) and it quickly became apparent that the optimum level of protection for a levee given the economic benefits was at or near the one percent level because reaching that level created the "benefit" of not having to buy insurance when located behind such levees. Life safety was not a consideration in the calculus.[5]

In 2007, the State of California legislated that new urban levees would have to be constructed at the 0.5 percent (200-year) level as a result of evaluation of the flood risks facing the state and the lessons learned from Hurricane Katrina. This represents a critical evolution in thinking—that it is appropriate to tailor flood risk management strategies to individual communities, geographic areas, or watersheds by carefully selecting a level of protection that reflects the life safety and economic assets in a given area. This includes recognition that the one percent annual chance flood standard of the NFIP might not serve as an appropriate life safety or property protection standard. Also critical to evaluating flood risk management strategies at the local level is the ability to acquire the resources it will need to carry out the floodplain management strategy.

Development of a flood risk management strategy focused solely on protection of areas behind levees would be neglecting the comprehensive nature of the flood risk challenge. It is critical to coordinate efforts to deal with one part of the floodplain with planning and other areas. Also critical is the consideration of all types of development and activity in local land-use plans, including residential, commercial, industrial, and agricultural needs of the community, placement of these components in a community, and an economic analysis of the costs and benefits of the flood risk management strategies being examined. Do the benefits to be gained merit the commitment of the

[5] In the 2007 Water Resources Development Act, Congress directed the revision of Principles and Guidelines to include "(1) seeking to maximize sustainable economic development; (2) seeking to avoid the unwise use of floodplains and flood-prone areas and minimizing adverse impacts and vulnerabilities in any case in which a floodplain or flood-prone area must be used; and (3) protecting and restoring the functions of natural systems and mitigating any unavoidable damage to natural systems." To date the revision has not occurred.

resources needed to carry out the floodplain management strategy? However, approval of community participation in the NFIP and accreditation of levees in these communities is not dependent on the cost-effectiveness of the levee or other mitigation measures incorporated in the community flood risk management strategy. Economic decisions are more closely tied to the analyses conducted by federal or state agencies to meet the criteria for participation in or construction of flood-related infrastructure under other government programs.

Future Conditions and Climate Change

Actions to shape flooding in one location can have impacts on flood stages (i.e., flood risk scenarios) at other locations, a realization that first surfaced in the late 1800s and early 1900s in the Sacramento River Basin, California. At the time, ad hoc levee construction generated significant, unintended hydrologic impacts at both the local and regional levels, driving the state to eventually implement statewide flood control measures through a State Reclamation Board. Today, this issue still plagues the Sacramento–San Joaquin Delta area, and a Central Valley Flood Protection Plan was adopted in June 2012. This plan promotes integrated, systemwide wise flood management, a key implication of which is that flood management investments (e.g., levees) will not result in an increased risk of flooding in other areas (CVFPP, 2012).

Community flood risk scenarios will continue to evolve as change occurs. Climate change will have a variety of regional impacts, and the geographic location of a community will affect how changing conditions affect risk. Some areas will have more droughts, some will have more frequent floods, and others will have more intense floods. Research to understand these hydrologic changes is ongoing (NRC, 2011, 2012a). A recent special report of the Intergovernmental Panel on Climate Change (IPCC, 2012) indicates a likely increase in many regions of the frequency of heavy precipitation events, and when coupled with increasing vulnerability presents a myriad of challenges for coping with climate-related disastersIPCC. Galloway (2009) cites 11 major international studies conducted from 1987 to 2002 that all predict significant climate change–induced hazards, including increased flooding, higher mean atmospheric temperatures, higher global mean sea levels, increased precipitation, increased strength of storms, more energetic waves, storm surges that reach further inland, undercapacity of urban sewerage and drainage systems, increased vulnerability of port cities, and disproportionate impacts on disadvantaged population segments (Galloway, 2009). The rise in sea level and the increase in storm surge due to climate change puts many coastal areas at risk from intensified flooding (NRC, 2010).

Hirsch and Ryberg (2012), in examining trends in annual floods at 200 stream-gauge sites in the United States, found that , while there appeared to be no strong statistical evidence for flood magnitudes increasing with increasing global mean carbon dioxide concentration, there were differences in flood magnitudes among the four quadrants of the conterminous United States (Figure 6-8). They indicate that the attention should be paid to the effects of changes in the relative "importance of the role of snowpack and rain on snow events." Raff (2013) suggests that the increase in magnitude of floods in the northeastern and midwestern United States (Figure 6-9, *Upper Right*), may have consequences in the Upper Mississippi, Ohio, and Missouri watersheds (Hirsch and Ryberg, 2012; Raff, 2013).

The Draft National Climate Assessment, issued in January 2013 by the National Climate Assessment Development Advisory Committee, begins with the statement:

> Climate change is already affecting the American people. Certain types of weather events have become more frequent and/or intense, including heat waves, heavy downpours, and, in some regions, floods and droughts. . . . The largest increases have occurred in the Northeast, Midwest, and Great Plains, where heavy downpours have exceeded the capacity of infrastructure such as storm drains, and have led to flooding events and accelerated erosion.

The report goes on to point out the increasing vulnerability to flooding of those in floodplains and coastal areas (Figure 6-10; NOAA, 2013). For example, prior to Hurricane Sandy, FEMA, as part of its continuous updating of FIRMs had begun revision of FIRMs for the New York Metropolitan Area and had planned to deliver them to state and local officials in mid-2013 for comment. When Hurricane Sandy hit New York, FEMA accelerated the release of information about the BFEs that would be seen on the new maps so that individuals and community officials could plan the recovery and reconstruction accordingly. These advisory BFEs for A and V zones, now in the hands of officials and the public, reflect the anthropogenic and physical changes that have taken place in the

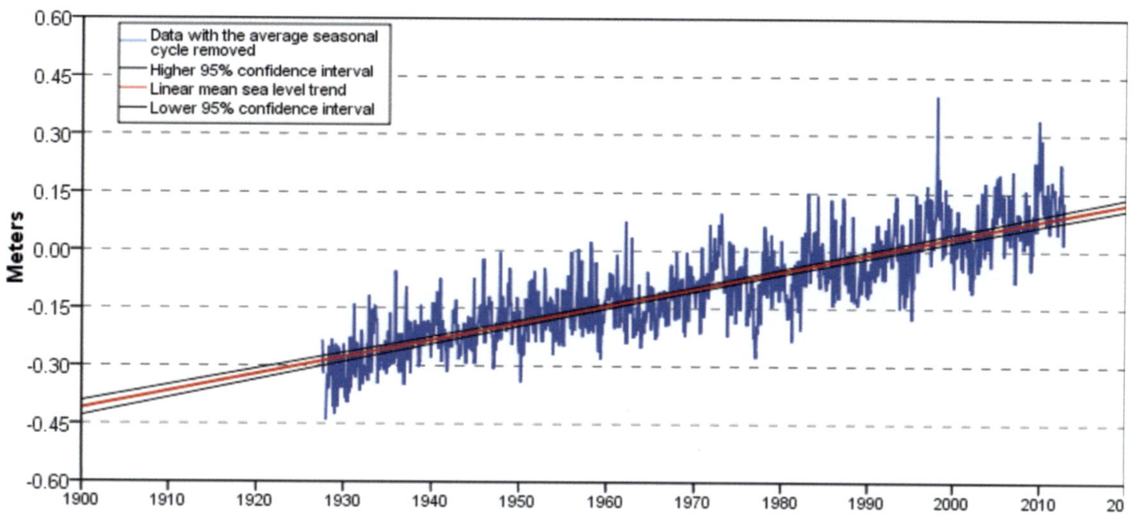

FIGURE 6-8 Average sea level trend at Seawells Point, Virginia, recording a 4.44-mm/year rise with a 95 percent confidence interval of ±0.27 mm/year based on data from 1927 through 2006. This is equivalent to 1.46 feet in 100 years. Data from NOAA's Seawells Point monitoring station.
SOURCE: http://tidesandcurrents.noaa.gov/sltrends/sltrends_station.shtml?stnid=8638610 Sewells Point, VA.

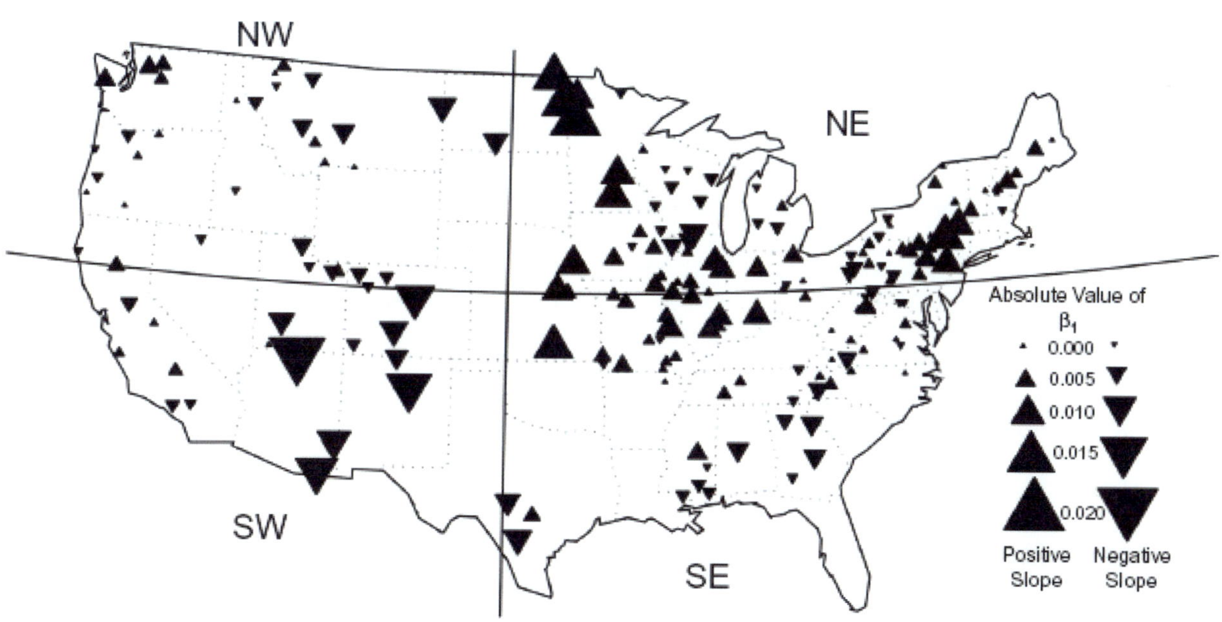

FIGURE 6-9 Areas of increases and decreases in flood magnitude trends at 200 U.S. Geological Survey stream gauge stations. Upward pointing triangles represent increases in the magnitude; downward pointing triangles, decreases.
SOURCE: Hirsch and Ryberg (2012).

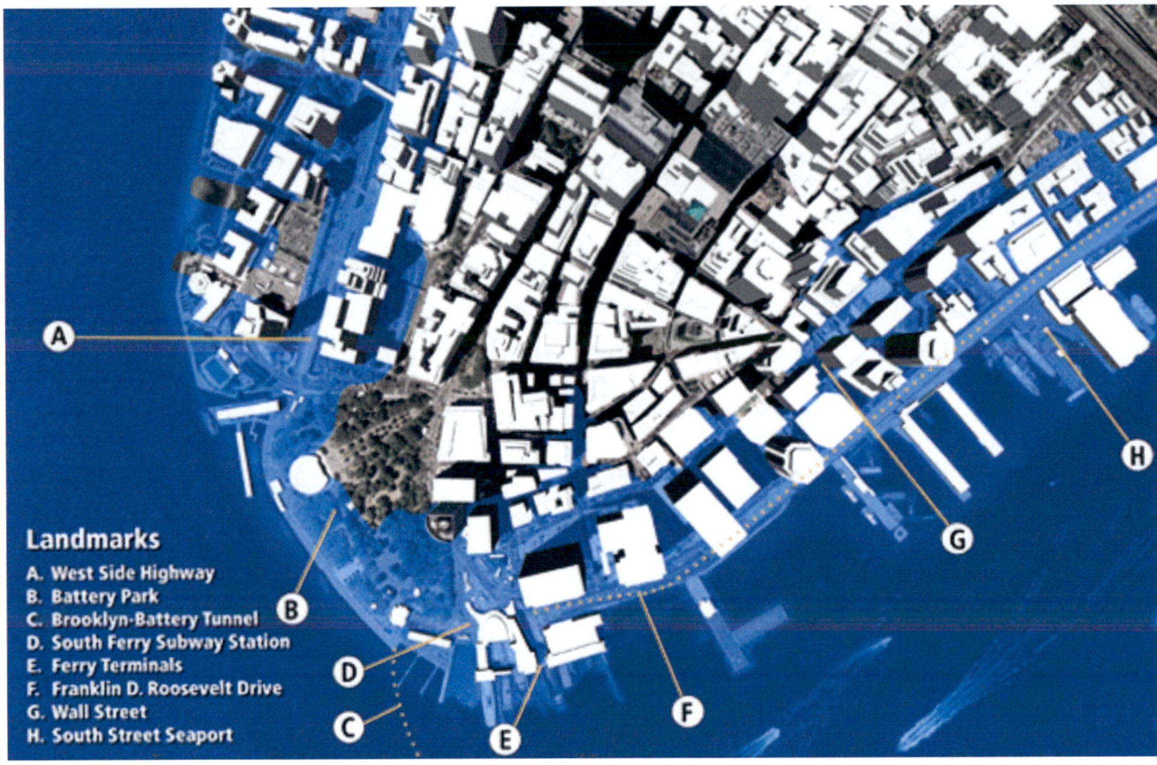

FIGURE 6-10 The challenge of rising sea level and the expansion of the one percent annual chance floodplain to New York City. The light blue area represents a 2012 depiction of the one percent annual chance floodplain of Lower Manhattan, New York City. With sea-level rise, a far larger area would be flooded in the event of a one percent annual chance flood event. (Note that flooding during Hurricane Sandy (2012) exceeded the one percent annual chance flood.) With sea-level rise, the one percent annual chance flood will inundate a larger area of New York City (not pictured here).
SOURCE: USGCRP (2009).

over 25 years since many of the maps had been developed; the new BFEs range from 2 to more than10 feet higher than previously depicted on FIRMs. Attention to these new elevations will ensure that those who were damaged by Sandy will be able to rebuild or relocate knowing more about the 2013 hazard they face (FEMA, 2013; Jeffrey Woodward, FEMA, personal communication, February 27, 2013).

A soon-to-be-released FEMA study of the impacts of climate change and population growth on the NFIP indicates that the size of the nation's floodplains will grow between 40 and 45 percent over the next 90 years and double the number of needed flood policies by the end of the century (Figure 6-11). It is speculated that this will require FEMA to incorporate the effects of climate change in the operation of the NFIP (Lehmann, 2011). Under current regulations, FEMA does not include consideration of future conditions in development of FIRMs.

The 2012 Biggert-Waters Act established a Technical Mapping Advisory Council to serve as an advisory body dealing with map modernization issues.[6] The legislation directs the Council to:

> consult with scientists and technical experts, other Federal agencies, States, and local communities to (A) develop recommendations on how to (i) ensure that flood insurance rate maps incorporate the best available climate science to assess flood risks; and (ii) ensure that the Federal Emergency Management Agency uses the best available methodology to consider the impact of (I) the rise in the sea level; and (II) future development on flood risk; and (B) not

[6] Section 100215.

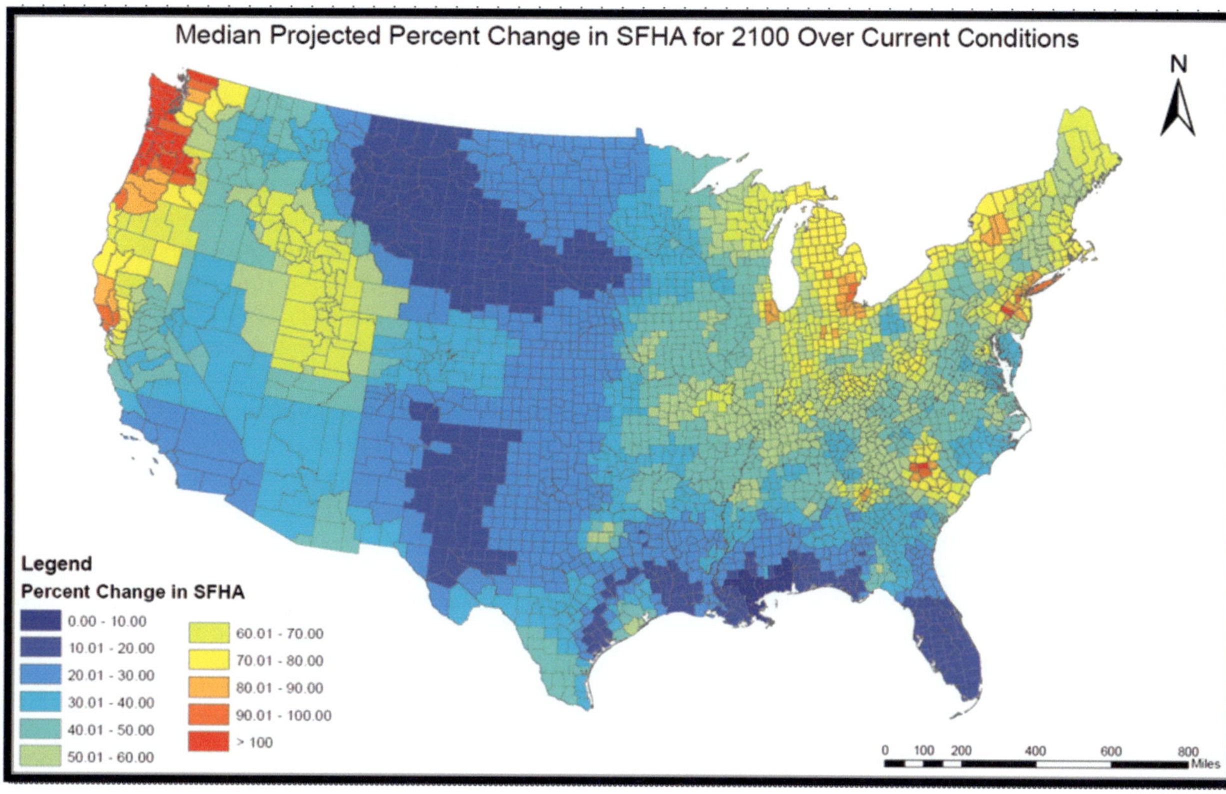

FIGURE 6-11 The average projected percent change in the size of the Special Flood Hazard Area by the year 2100 as a result of climate change and demographic change, in counties across the United States.
SOURCE: Crowell (2010).

> later than 1 year after the date of enactment of this Act, prepare written recommendations in a future conditions risk assessment and modeling report and to submit such recommendations to the Administrator.

The Biggert-Waters Act further charges the Administrator to review and update NFIP rate maps and to incorporate any future risk assessment submitted by the Technical Mapping Advisory Council in any such revision or update.[7]

Implementing Flood Risk Management

Flood risk management begins with goal and policy development, identification of the hazard, analysis that includes assessment of the hazard's potential impact, and development of flood risk management strategies, representing the amalgamation of the various measures discussed in this chapter and Chapter 7 available to deal with the flood challenge (Figure 3-3). Flood risk management strategies are represented by portfolios of mitigation measures. A structural approach—that is, construction of a levee—is immediately visible and for many flood events provides adequate mitigation of the effects of rising waters. Implementation of nonstructural mitigation approaches is not always well received. Efforts to promote wise development or enhance building codes are sometimes seen as efforts to limit growth and reduce economic vitality in the affected area. Requiring communities to elevate homes can be seen as creating visual blight and similarly reducing the economic utility of the property. Less thought is given to the potential economic consequences should a flood actually occur. Federal project development guidance makes structural projects more feasible and requires less contribution from local governments for their execution

[7] Section 100216.

than would be expected of nonstructural efforts. Simply looking at an economic balance sheet tilts decision making toward structural approaches.

Successful flood risk management strategies are developed through a bottom-up process that is supported from the top down. Development of flood risk strategies by the community is critical, as is support from and collaboration with FEMA (Box 6-3) and other federal and state agencies. FEMA supports mitigation projects through a variety of hazard mitigation assistance (HMA) grants, administered through several programs that provide funding for eligible activities that reduce disaster loss and protect life and property from future damages. For example, FEMA supports mitigation projects through the Hazard Mitigation Grant Program (HMGP), a program that provides state and local government with grants for projects that reduce or eliminate losses of life and property from future disasters. Projects include elevation of a home, purchase of property subject to repetitive damage, and retrofitting, etc. HMGP grants may address a variety of risks and are not limited to flood risk; grants are only available to those in a presidentially declared disaster area and can be significant. FEMA also issues flood mitigation assistance (FMA) grants, severe repetitive loss (SRL) grants, and repetitive flood claim mitigation (RFC) grants, and pre-disaster mitigation (PDM) grants. FMA, SRL, and RFC grants are designed to, over time, reduce flood losses to the NFIP by mitigating insured properties with particular attention to repetitive loss and severe repetitive loss properties. PDM grants provide support for multihazard mitigation (planning and project implementation) prior to a disaster event (FEMA, n.d.b).

Mitigation projects that are eligible for HMA assistance include, for example, property acquisition, structure elevation, dry floodproofing, and minor flood control projects, such as the installation or modification of culverts, and stormwater management activities such as creating retention and detention basins. Ineligible projects generally include major flood control projects such as "construction, demolition, or repair of dams, dikes, levees, floodwalls, seawalls, groins, jetties, breakwaters, and erosion projects related to beach nourishment or re-nourishment," that is, projects where catastrophic failure is a possibility and the potential for loss of life and property is too great (FEMA, 2010a; Ryan Janda, FEMA, personal communication, January 29, 2013). Collectively, these grant programs become important parts of the risk management portfolio.

Communities behind levees are diverse in their physical and social characteristics. They have different types of flood risk that require a combination of risk management measures targeted at households, neighborhoods, and regions. They face a rapidly changing future, and a one-size-fits-all solution is not sufficient to adequately manage risk for all communities across the nation. The National Research Council report, *Disaster Resilience: A National Imperative* (NRC, 2012b), points out that informed decision making is key to managing risk. The development of information using a modern risk analysis tools will allow communities to utilize specific information in designing risk reduction strategies specific to that community and to visualize the risk throughout the communities over time.

BOX 6-3
Project Impact

Initiated in 1997, FEMA's Project Impact provided training and funding for disaster mitigation activities (flood, earthquake, fire, etc.) and facilitated the establishment of local partnerships in the community to mitigate the impacts of these disasters. The program established responsibility and authority for disaster mitigation at the community level. The program goals were accomplished by establishing responsibility and authority for mitigation at the community level. The actions were organized into four phases: (1) building partnerships, (2) identifying hazards and community vulnerability, (3) identifying and prioritizing risk reduction actions for the community, and (4) communicating success (FEMA, 1997).

Project Impact was discontinued at the federal level in 2001. However, some communities continued partnerships and mitigation activities where Project Impact was the catalyst for a long-term disaster management approach. Project Impact remains today a good model for encouraging and implementing community resilience.

Sayers et al. (in press) point out that risk-based methods enable uncertainty to be explicitly recognized and data collection prioritized to address areas where the lack of uncertainty is material to the choices being made. With science as the foundation of flood management measures, "better information leads to better decision-making and better flood management" (Schropp and Soong, 2006). With better knowledge, decisions makers can make land-use and other decisions that lead to reductions in loss of life and property damages (Brody et al., 2011).

Most communities participating in the NFIP focus their attention on those areas within the SFHA and overlook the risk to those outside the SFHA, especially those areas behind levees. **There is a clear need for a comprehensive, tailored approach to risk management behind levees that (1) is designed and implemented at the local level; (2) involves federal and state agencies, communities, and households; (3) takes into account possible future conditions; and (4) relies on an effective portfolio of structural measures, nonstructural measures, and insurance to reduce the risk to those behind levees.**

To reduce the flood risk to those behind levees, FEMA should encourage communities to develop and implement multimeasure risk management strategies for areas behind accredited levees. This strategy would contemplate the full range of foreseeable flood scenarios including levee overtopping or failure. The development of new risk management analysis tools will facilitate the identification of risks and suggest feasible approaches to dealing with the risks to those behind levees.

EFFECTIVE FLOOD RISK MANAGEMENT: ADMINISTRATION OF LEVEES

Development and maintenance of appropriate supporting policies and data affecting the portfolio of mitigation measures are both critical effective floodplain management strategies. For example, managing the inclusion of levees in the NFIP is an important mission of the Federal Insurance and Mitigation Administration and requires the development of policies and procedures as well as regulatory instruments. Furthermore, data on the extent and status of the nation's levees are critical to informing all stakeholders of the extent of flood risk and developing appropriate floodplain management strategies. This section addresses both the current status of and the need for revision to the principal regulatory document establishing procedures for inclusion of levees in the NFIP and efforts to develop and maintain a national inventory of levees.

Levee-Related Federal Regulations

Title 44 of the *Code of Federal Regulations* (CFR) contains several sections pertaining to levees in the NFIP including:

- 44 CFR §59.1, Definitions;
- 44 CFR §64.3, Flood Insurance Maps; and
- 44 CFR §65.10, Mapping of Areas Protected by Levee Systems.

Section 65.10 describes the "types of information FEMA needs to recognize, on NFIP maps, that a levee system provides protection from the base flood," and serves as the principal guidance document for levee development and certification.

The committee considered the three aforementioned sections of the CFR during the course of deliberations. The committee was asked specifically for advice on "Existing requirements for levee accreditation under 44 CFR §65.10" (Box 1-4). All three sections are relevant to the charge of this committee in the sense that Section 65.10 relies on the information contained in Sections 64.3 and 59.1. As indicated in Chapter 2 and Appendix F, the 2006 Interagency Levee Policy Review Committee (ILPRC) report to FEMA also addressed these three sections of Title 44, and the committee's advice generally supports the recommendations of the ILPRC.[8]

[8] The changes to Title 44 recommended by the ILPRC have yet to be acted upon by FEMA.

44 CFR §59.1, Definitions

In the past, existing roads and railroads, that is, embankments, were assumed to be included in the definition of a levee system and its accreditation even though the information concerning the integrity of the structures was unknown. As defined in 44 CFR §59.1, a levee system is: "a flood protection system which consists of a levee, or levees, and *associated structures*, such as closure and drainage devices, which are constructed and operated in accordance with sound engineering practices" (emphasis added). Although such embankments can provide a degree of flood protection, they can also be the source of catastrophic failures. If embankments are to be included in a levee system, they need to be subject to the same examination as levees designed specifically for flood protection purposes or the definition modified to exclude them because they were not designed to provide risk reduction from flood damages (ILPRC, 2006).

44 CFR §64.3 Flood Insurance Maps

Under current procedures, areas behind accredited levees are normally shown as Zone X or a shaded Zone X with the latter frequently indicating the presence of a levee. Section 64.3 identifies Zone X as either an area "of moderate flood hazards or areas of future-conditions flood hazard" or an area "of minimal hazards." FEMA's Map Service Center indicates that a shaded Zone X is an "Area of moderate flood hazard, usually the area between the limits of the 100-year and 500-year floods. Are also used to designate base floodplains of lesser hazards, such as areas protected by levees from 100-year flood, or shallow flooding areas with average depths of less than one foot or drainage areas less than 1 square mile" (Appendix E). Neither of these definitions adequately identifies areas behind accredited levees. Establishing a unique zone designation for areas behind levees would more clearly distinguish areas behind accredited levees from the 0.2 percent annual chance floodplain and the one percent annual chance floodplain and would provide clarity and assist in risk communication (ILPRC, 2006).

44 CFR §65.10, Mapping of Areas Protected by Levee Systems

If the definition of a levee system in 44 CFR §59.1 is modified to specifically exclude embankments, such as roads and railroads that were not engineered to provide risk reduction from flooding, similar modification of 44 CFR §65.10 is necessary (ILPRC, 2006). This would ensure that nonengineered levee systems are not certified as providing protection from the base flood.

As part of periodic update of FRIMs (map modernization), all levees are being subjected to analysis as they are mapped. Even though the map modernization program has a high priority, at current resource levels, the period between updates may be lengthy. It is important that, independent of the mapping of levees, FEMA maintain some degree of surveillance over the integrity of levees in the NFIP (ILPRC, 2006). At present, once accredited, a levee maintains this status until the next update. Rather than allowing accreditation to remain in effect until the FIRM is revised, periodic submission of operation and maintenance records as well as a written assessment of the levee systems performance in any flood events that have occurred to FEMA is needed. Furthermore, periodic recertification of the engineering and geotechnical aspects of the levee is critical. These continued actions verifying to FEMA that an accredited levee maintains its integrity would require modification of 44 CFR §65.10.

The design of levee systems is based on conditions at the time of the design including the flow and channel carrying capacity. As a result of upstream activity or normal geomorphic processes, the flow can be increased and the carrying capacity of these channels can be significantly reduced, thereby increasing the elevation of the base flood and reducing or eliminating the capability of an accredited levee to pass the base flood. Attention to the maintenance of these channels is an important part of maintaining the integrity of the levee system. Assurance that the flood carrying capacity of the main river channel is maintained is critical. Channel maintenance preserves the required level of protection, ensures that encroachments are controlled, and that an increased flow does not affect the integrity of the levee or the ability of the levee sponsor to maintain the levee and fight floods, when necessary (ILPRC, 2006).

To ensure that levees are designed to meet widely recognized standards, FEMA, in the past, has included references to the appropriate USACE manuals.[9] During the establishment of the NFIP and the inclusion of levees in the program, FEMA worked in close coordination with USACE to determine the appropriate design guidance to be used in construction of levees and in their evaluation and ceded to USACE the lead in several areas. The above recommendations reflect the 2006 documents available but need to be adjusted to reflect more recent publications.

An Inventory of the Nation's Levees

The first step in properly dealing with levee risk and developing flood risk management strategies is to identify the location, condition, and ownership of levees, as well as the extent of the areas protected by the levee and the consequence of failure or overtopping (i.e., assets and lives), if it should occur. Compiling this information and making it readily available is fundamental to ensuring public safety and reducing flood losses. Because levees can alter natural hydrologic flows, knowledge of their location, properties, and conditions is essential to understanding, quantifying, and mitigating flood hazards, including working to increase public awareness of vulnerability during a flood event. Although some responsibilities have been assigned by Congress to FEMA and USACE for developing and maintaining levee information, several states and communities also are working to develop and maintain this information.

The U.S. Army Corps of Engineer's National Levee Database

The biggest single resource for levee data in the United States is the National Levee Database (NLD). Administered and populated by the USACE, the NLD was authorized by Congress in the 2007 Water Resources Development Act. Congress directed the Secretary of the Army to establish and maintain a database with an inventory of the nation's levees; the database was to include:

> (A) location information of all Federal levees in the Nation (including global information system information) and, for non-Federal levees, such information on levee location as is provided to the Secretary by State and local governmental agencies; (B) utilizing such information as is available, the general condition of each levee; and (C) an estimate of the number of structures and population at risk and protected by each levee that would be adversely impacted if the levee fails or water levels exceed the height of the levee. (WRDA, 2007)

Appropriations for the database since 2007 have been insufficient to carry out the charge (Eric Halpin, USACE, personal communication, November 7, 2011). The NLD currently includes the more than 2,500 levees and associated structures (together known as "levee systems") that total more than 14,500 miles in length, cataloged in digital format. This includes levees that were built by USACE and are still operated and maintained by USACE (2,800 miles), built by USACE and turned over to local sponsors to operate and maintain (9,500 miles), or are under USACE oversight under Public Law 84-99[10] program that permits USACE to provide pre-and post-disaster assistance to nonfederal levees that meet certain standards of performance (2,200 miles) (Bryan Baker, USACE, personal communication, January 13, 2013; USACE, n.d.b). All data except information that could be considered to affect national security are accessible to both official data users and partners such as federal, state, and local governments, sponsors, and the general public (USACE, 2012c).[11] It should be emphasized that levee data are

---

[9] Reference USACE Engineer Manual EM 1110-2-1413, Engineering and Design, Hydrologic Analysis of Interior Areas as the acceptable/preferred methodology for interior drainage analysis. [65.10(b) (6]. Reference the USACE Levee Owners Manual for Non-Federal Flood Control Works as required for operation, maintenance, and inspection guidance and procedures for non-Federal levees and floodwalls [65.10(c) and 65.10(d)].

[10] Under Public Law 84-99, Flood Control and Coastal Emergency Act, USACE is authorized to undertake activities including disaster preparedness, advance measures, emergency operations (flood response and post-flood response), rehabilitation of flood control works threatened or destroyed by flood, protection or repair of federally authorized shore protective works threatened or damaged by coastal storm, and provisions of emergency water due to drought or contaminated source.

[11] See http://nld.usace.army.mil/egis/f?p=471:1:0::NO for additional information.

currently being collected from a variety of sources. Based on discussions with FEMA and USACE staff, the committee expects that the accuracy of the NLD and FEMA levee data will improve over time.

Currently, any data provided to USACE for inclusion in the NLD on levees that do not fall under the auspices of USACE and its national portfolio are collected on a voluntary basis. Efforts are under way to integrate the levee data collected on the communities that participate in the NFIP into the NLD, which will improve the information in the levee database (NRC, 2012b). Though USACE is working closely with other federal agencies and nonfederal levee owners to incorporate their data into the NLD on a voluntary basis, this current effort is not likely to yield the desired results (e.g., the building of a comprehensive national levee inventory) because of its voluntary nature (Eric Halpin, USACE, personal communication, 2011). Additionally, information collected and information voluntarily supplied are not likely to be in a format consistent across states, tribes, and local and regional entities, challenging the development of a comprehensive understanding.

FEMA Mid-Term Levee Inventory

FEMA has long maintained an inventory of NFIP-accredited levees; however, in 2006 the ILPRC reported that "various levee databases have been developed, but for the most part they are not geospatially referenced, are incomplete, and lack usable information concerning the condition of the levees and attendant supporting structures" (ILPRC, 2006). As a result, in FY 2007, FEMA funded and developed the FEMA Mid-Term Levee Inventory (MLI) database, which was designed to complement the NLD and uses a data model extracted from the NLD's database. Whereas the NLD inventories levees of all protection levels, the MLI gathered levee data for structures designed to provide protection from a base (one percent annual chance) flood, to better to support the production of countywide flood insurance studies and Flood Insurance Rate Maps (FIRMs) and any other levees that were located in the area covered by FIRMs (FEMA, 2012b; Figure 6-12).

Maintenance of the MLI data is performed at FEMA's regional level and additional levee data is added as it becomes available (FEMA, 2012b). As of November 2012, the inventory identified approximately 29,800 miles of levees. Of those, approximately 5,100 miles have been or are accredited or in Provisionally Accredited Levee (PAL) status (i.e., NFIP levees) and 22,000 miles are not part of the NFIP but are located on FIRMs and could affect hydraulics in those areas or provide protection at less than the one percent annual chance flood level (FEMA, 2012b). These latter levees were identified by FEMA's Production and Technical Services contractors in counties covered by FIRMs.

Overlaps

There is currently no method for determining whether a levee in the NLD is also in the FEMA MLI except through a one-by-one comparison of certain levee characteristics. Because some data elements in the two inventories are different, USACE and FEMA have been attempting to manually look for overlaps, but have yet to complete the effort. A preliminary analysis by both agencies indicates that approximately 3,400 miles of accredited or PAL levees (66 percent of NFIP levee miles) may be operating under oversight of both agencies (Figure 6-13). Because the 2012 Flood Insurance Reform and Modernization Act requires the two agencies to synchronize their analytical methodologies where both agencies inspect the same levee, FEMA and USACE have accelerated their efforts to define overlaps (David Bascom, FEMA, personal communication, November 1, 2012; Bryan Baker, USACE, personal communication, December 9, 2012).

Total Miles of Levees in the United States

The National Committee on Levee Safety (NCLS), a group of federal, state, local, and private-sector members with expertise in representing national interests in levee safety, was convened in 2008 to prepare recommendations and a strategic implementation plan on a National Levee Safety Program with direction from Congress under the Water Resources Development Act of 2007 (NCLS, 2009). The draft NCLS report was issued in 2009 (ASFPM, 2012).

Acknowledging the fact that little was currently known about the specific size of the national levee portfolio outside efforts at the federal level, the NCLS formulated an estimate of the number of privately owned and state levees by extrapolating data entered by the state of California into their levee inventory database and then applying that ratio of USACE program and other federal levees to private and nonfederally owned levees across the nation. The NCLS estimated that there may be 100,000 or more miles of levees (including thousands of miles of irrigation and water-supply canal embankments) that were built by federal, public, or private entities and operated by nonfederal, public, or private organizations. The current inventories in the NLD and MLI have identified less than 32,000 miles of levees. Mapping under the NFIP covers areas containing 92 percent of the U.S. population and more than 21,000 communities with flood-prone areas, logically the areas in need of levees. Analysis, discussions with members of the NCLS, and site visits by this committee indicate that levees not contained in the NLD and MLI probably represent small individually owned levees, private levees used to protect industrial assets, and roads and railroads that act as levees. It is also likely that the NCLS estimate, based on the California inventory, overestimated the number of levees, given that the California levee population is one of the densest in the nation and is not reflective of the nation as a whole (Figure 6-10). Based on the judgment of the committee, the total mileage of significant levees in the United States, excluding embankments, is estimated to be under 50,000 if not under 40,000 miles.

The NCLS report recommended that Congress grant USACE the authority and appropriation to expand and maintain the National Levee Database by conducting a one-time inventory and assessment of all nonfederal levees and canal embankments[12] in the United States (NCLS, 2009). Such an action would allow for baseline information on levee performance to be collected for the entirety of the nation's levee systems, resulting in an accurate, cost-effective, and efficient decision-making process when it comes to managing the nation's levee portfolio. Upon completion of this one-time inventory, the responsibility for periodic inventory and inspection updates to the NLD would fall to the states where these levees are located. Congressional action has yet to be taken on this recommendation.

State Levee Inventory Efforts

As of 2006, only Wisconsin and North Dakota had geospatial inventories of levees; Illinois, Mississippi, North Dakota, Ohio, and California had some levee database efforts under way (ILPRC, 2006). Today, the majority of states still do not have levee inventories (Larry Larson, ASFPM, personal communication, September 2012). As an example of what can be accomplished, the California Department of Water Resources (CA DWR) has created the California Levee Database (CLD), a "statewide central repository for levee and flood control structures for use by public agencies for flood risk assessments and emergency response" (Rod Mayer, CA DWR, personal communication, 2012). Using data from sources such as the NLD, FEMA's Mid-Term Levee Inventory, CA DWR, and the California Emergency Management Agency, as well as local agencies and publicly available maps, the CLD includes more than 13,725 miles of levees and levee-like structures within the state. Currently, the inventory includes 92 percent of all state–federal State Plan of Flood Control (SPFC) levees[13] in the state, 85 percent of all non-SPFC levees, and 71 percent of all local levees.

Washington State has also taken steps to create a levee inventory. In the absence of a comprehensive national levee inventory, the state, well aware of its status as being one of the more flood-prone states in the nation, commissioned and conducted the Statewide Levee Inventory and Flood Protection Study to better understand the current status of accredited levees within its borders. This study was divided into two parts: the Washington State Final Report on Levee Certification and Accreditation, which summarized current levee policies and practices, and the development of a statewide inventory that includes the location, level of protection, and certification/

[12] Canal embankments, although not designed to protect areas from flooding, can pose threats to some communities. When embankment walls are aboveground, the water they contain can break through and flood nearby areas. In 2008, over 300 homes in Fernley, Nevada, flooded when the walls of an irrigation canal collapsed. In 2012, the first court settlement of numerous lawsuits awarded $10 million to 600 local residents for damages.

[13] SPFC levees are federally authorized levees located in California's Central Valley for which the state has provided assurances of cooperation with the federal government (CA DWR, 2010).

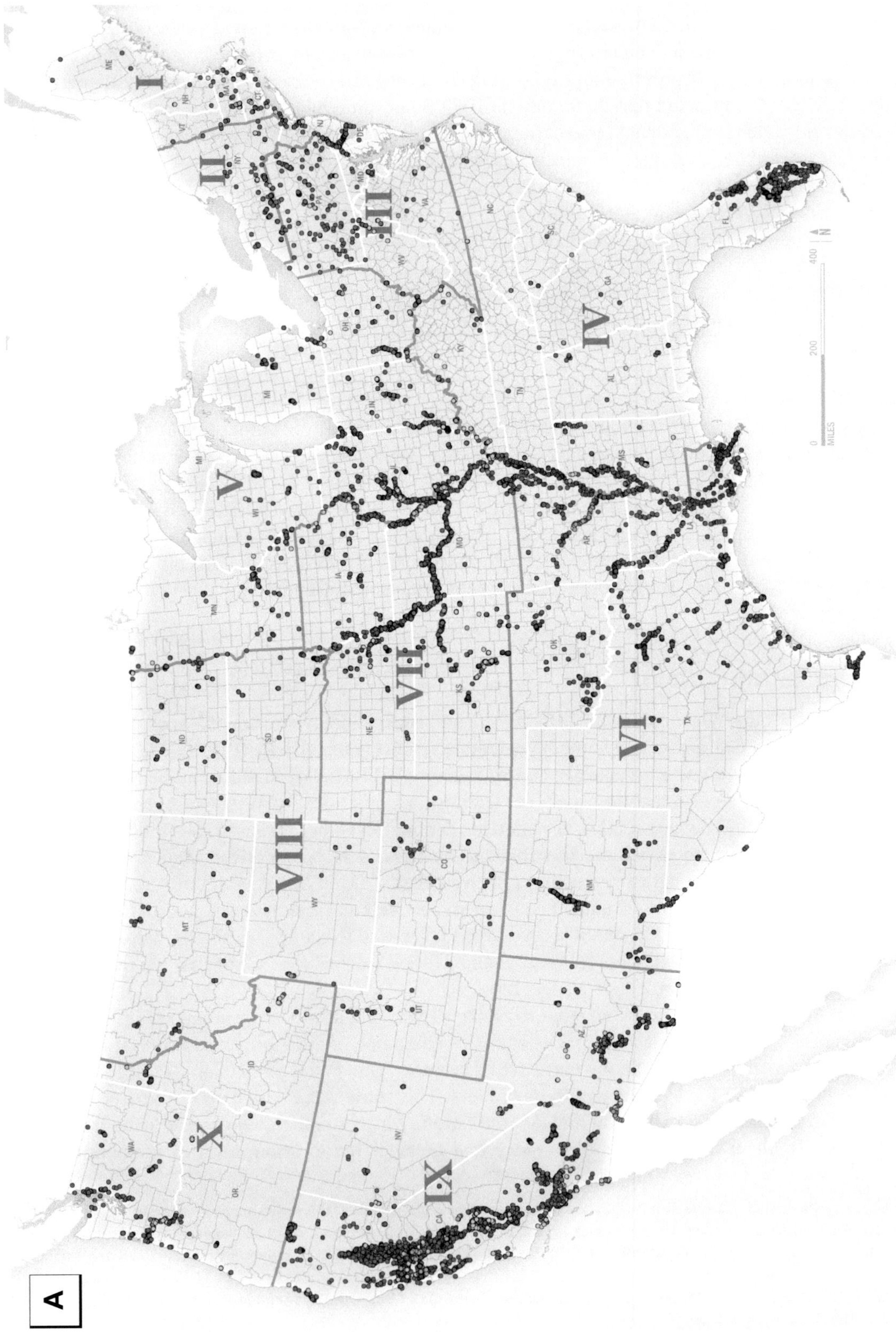
A
I
II
III
IV
V
VI
VII
VIII
IX
X
0
200
400
MILES
N

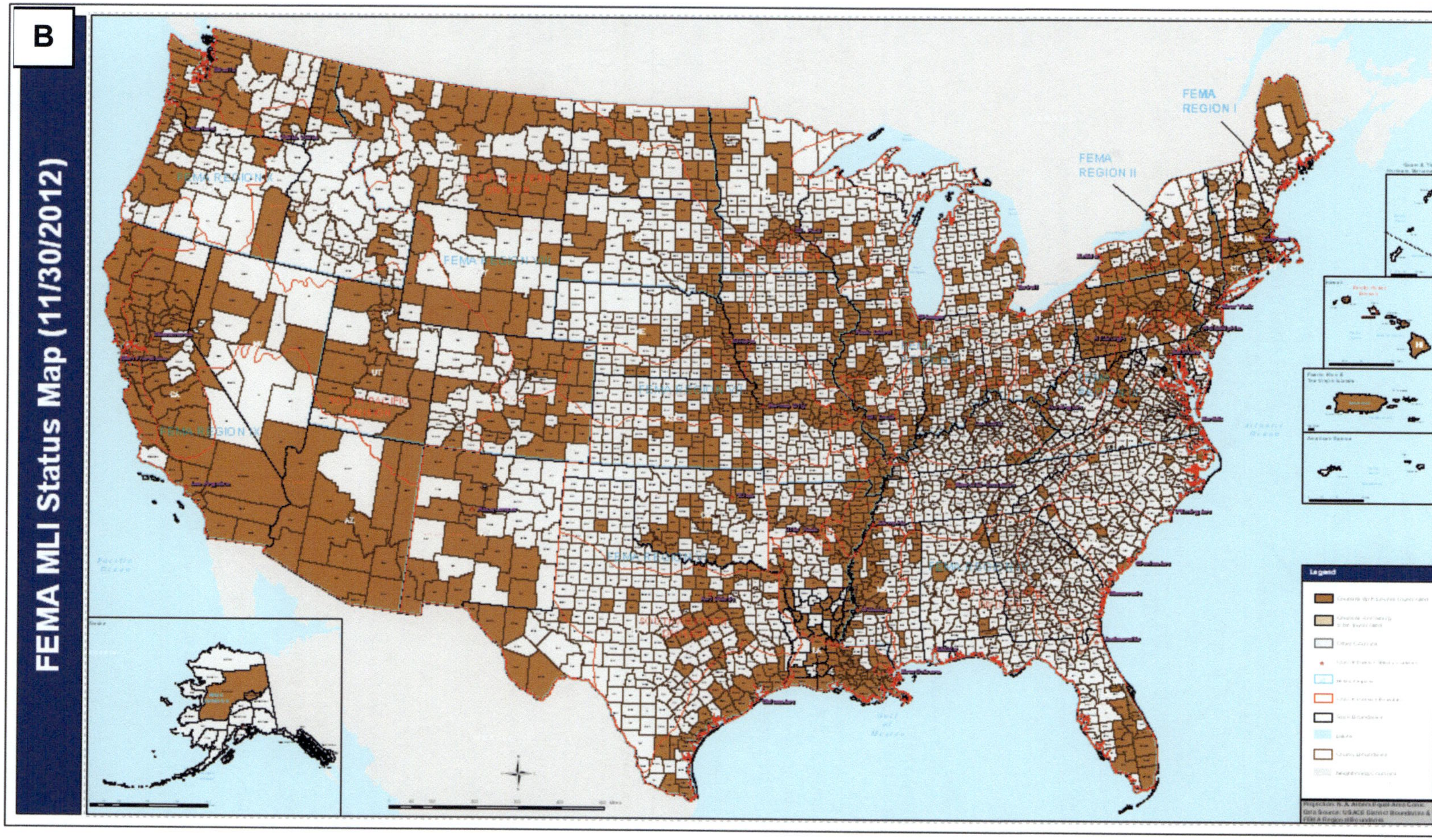

FIGURE 6-12 (*A*) National Levee Status Map capturing levees inventoried in the MLI. Roman numerals denote the 10 individual FEMA regions. (*B*) Counties across the nation containing levees captured in the MLI. SOURCE: (A) Modified from FEMA (2011); (B) FEMA (2012b).

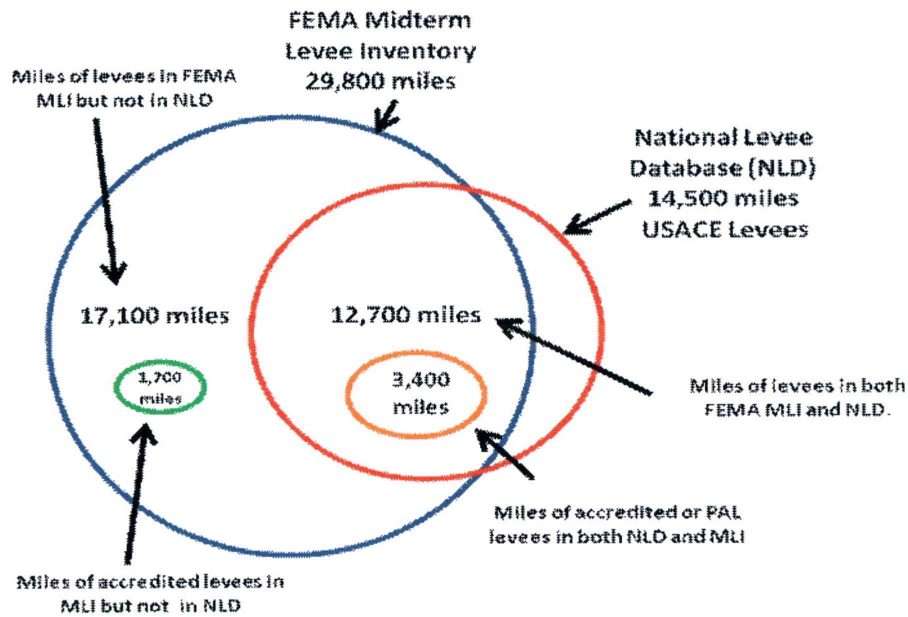

FIGURE 6-13 Overlap between USACE NLD inventory of USACE levees and levees identified in FEMA's MLI. FEMA's MLI includes 29,800 miles of levees, and USACE's NLD identifies 14,500 miles of levees; 17,100 miles of levees in FEMA's MLI are not included in USACE's NLD, approximately 1,700 of which are accredited; 12,700 miles of levees in FEMA's MLI are included in the USACE's NLD, approximately 3,400 of which are accredited or PAL. This represents a 66 percent overlap. All values are approximate.

accreditation status of all currently known levees[14] at the statewide level (Resilience Institute, 2010). Of the 697 miles of identified levees found within the state, 18 percent (approximately 125 miles) were found to currently be accredited, provisionally accredited, or de-accredited. Only 9 percent (approximately 62 miles) of all levee miles were found to be accredited. The long-term goal is that the Washington inventory will be strengthened to the point where it will be a useful tool in developing a prioritized list of levees needing site-specific attention (such as repair, certification and accreditation, and/or removal) and in the prioritization of state support for such levees (Resilience Institute, 2010).

REFERENCES

ASFPM (Association of State Floodplain Managers). 2012. National Committee on Levee Safety (NCLS). Available online at http://www.floods.org/index.asp?menuID=338. Accessed August 7, 2012.

Boon, J. D., J. M. Brubaker, and D. R. Forrest. 2010. Chesapeake Bay Land Subsidence and Sea Level Change: An Evaluation of Past and Present Trends and Future Outlook. Available online at http://web.vims.edu/GreyLit/VIMS/sramsoe425.pdf?svr=www. Accessed February 21, 2013.

Briaud, J. L., H. C. Chen, A. V. Govindasamy, and R. Storesund. 2008. Levee erosion by overtopping in New Orleans during the Katrina hurricane. Journal of Geotechnical and Geoenvironmental Engineering 134(5): 618-632.

Brody, S. D., S. Zahran, P. Maghelal, H. Grover, and W. E. Highfield. 2007. The rising costs of floods: Examining the impact of planning and development decisions on property damage in Florida. Journal of the American Planning Association 73(5): 330-345.

[14] The State of Washington defines a levee as a "fabricated barrier along rivers, streams, channels, sloughs, and other water courses" (Resilience Institute, 2010).

Brody, S. D., W. E. Highfield, and J. E. Kang. 2011. Rising Waters: Causes and Consequences of Flooding in the United States. Cambridge, UK: Cambridge University Press.
Burby, R. J., and L. C. Dalton.1994. Plans can matter: The role of land use plans and state planning mandates in limiting the development of hazardous areas. Public Administration Review 54(3): 229-238.
CA DWR (California Department of Water Resources). 2010. Draft State Plan of Flood Control Descriptive Document. Available online at: http://www.water.ca.gov/cvfmp/docs/DRAFT_SPFC_Descriptive_Doc_20100115.pdf. Accessed July 31, 2012.
Crowell, M. 2010. Impact of Climate Change on the NFIP. Presentation to the Coastal Engineering Research Board Meeting, June 22. Vicksburg, MS: USACE. Available online at http://chl.erdc.usace.army.mil/dirs/events/319/08%2087th%20 CERB%20Crowell.pdf. Accessed March 3, 2013.
CVFPP (Central Valley Flood Protection Plan). 2012. 2012 Central Valley Flood Protection Plan, a Path for Improving Public Safety, Environmental Stewardship, and Long-Term Economic Stability. Available online at http://www.cvfpb.ca.gov/CVFPP/index.cfm. Accessed January 28, 2013.
FEMA (Federal Emergency Management Agency). 1997. Building a disaster-resistant community: Project Impact. The Project Impact Guidebook. Washington, DC: FEMA.
FEMA. 2000. Above the Flood: Elevating Your Floodprone House. Available online at http://www.fema.gov/library/viewRecord.do?id=1424. Accessed November 28, 2012.
FEMA. 2006. The National Levee Challenge: Levees and the FEMA Flood Map Modernization Initiative. Washington, DC: FEMA.
FEMA. 2007. Selecting Appropriate Mitigation Measures for Floodprone Structures. FEMA 551. Available online at http://www.fema.gov/library/viewRecord.do?id=2737. Accessed January 22, 2013.
FEMA. 2010a. Hazard Mitigation Assistance Unified Guidance on the Hazard Mitigation Program, Pre-Disaster Mitigation Program, Flood Mitigation Assistance Program, Repetitive Flood Claims Program, and Severe Repetitive Loss Program. Available online at http://www.fema.gov/library/viewRecord.do?id=4225. Accessed January 29, 2013.
FEMA. 2010b. Mitigation Planning Fact Sheet on Risk MAP. Available online at http://www.fema.gov/library/viewRecord.do?id=2066. Accessed November 14, 2012.
FEMA. 2011. Midterm Levee Inventory Project Summary Report. Available online at: http://www.r3levees.org/wiki/images/0/03/MLI_Summary_Report_20111130_Final.pdf. Accessed August 8, 2012.
FEMA. 2012a. Analysis and Update of National Flood Hazard Layer Demographics and NFIP Policy and Claims Data. Prepared by Risk Assessment, Mapping, and Planning Partners, project report for task order #HSFEHQ-11-J-002. Washington, DC: FEMA.
FEMA. 2012b. Midterm Levee Inventory Project Summary Report: Standard Operations Task Order 4, November. Washington, DC: FEMA.
FEMA. n.d.a. Non-residential Floodproofing Requirements and Certification for Buildings Located in the Special Flood Hazard Areas in Accordance with the National Flood Insurance Program. Technical Bulletin 3-93. Available online at http://www.fema.gov/library/viewRecord.do?id=1716. Accessed February 15, 2013.
FEMA. n.d.b. The Unified Hazard Mitigation Assistance Grant Programs Factsheet. Available online at http://www.fema.gov/library/viewRecord.do?id=3648. Accessed January 29, 2013.
FEMA. 2013. Hurricane Sandy Advisory Base Flood Elevations (ABFEs) in New Jersey and New York. FEMA Region 2. Available online at http://www.region2coastal.com/sandy/abfe, accessed February 27, 2013.
FM Global. 2006. Approval Standard for Flood Abatement Equipment, Class Number 2510. Available online at http://www.fmglobal.com/assets/pdf/fmapprovals/2510.pdf. Accessed December 6, 2012.
FM Global. n.d. Understanding the Benefit, FM Approved Flood Abatement. Available online at http://www.fmglobal.com/assets/pdf/P11204.pdf. Accessed December 6, 2012.
FWS (U.S. Fish and Wildlife Service). 2002. The Coastal Barrier Resources Act: Harnessing the Power of Market Forces to Conserve America's Coasts and Save Taxpayers Money. Available online at http://www.fws.gov/CBRA/Docs/Taxpayer-SavingsfromCBRA.pdf. Accessed November 28, 2012.
FWS. 2012. Background information on the Coastal Barrier Resource Act. Available online at http://www.fws.gov/CBRA/Act/index.html. Accessed November 28, 2012.
Galloway, G. 2009. Reacting to Climate Change, Floods, and Uncertainty. Available online at http://www.colorado.edu/hazards/. Accessed October 31, 2012.
Galloway, G. E, D. F. Boesch, and R. R. Twilley. 2009. Restoring and protecting coastal Louisiana. Issues in Science and Technology 25(2): 29-38.
GAO (Government Accountability Office). 2007. IRS Emergency Planning, Headquarters Plans Supported Response to 2006 Flooding, but Additional Guidance Could Improve All Hazard Preparedness. GAO-07-579. Washington, DC: GAO.

Hirsch R. M., and K. R. Ryberg. 2012. Has the magnitude of floods across the USA changed with global CO2 levels? Hydrological Sciences Journal 57(1): 1-9.

ILPRC (Interagency Levee Policy Review Committee). 2006. Sharing the Challenge: Floodplain Management into the 21st Century. Washington, DC: ILPRC.

IPCC (Intergovernmental Panel on Climate Change). 2012. Managing the Risks of Extreme Events and Disasters to Advance Climate Change Adaptation. A Special Report of Working Groups I and II, C. B. Field, V. Barros, T. F. Stocker, D. Qin, D. J. Dokken, K. L. Ebi, M. D. Mastrandrea, K. J. Mach, G.-K. Plattner, S. K. Allen, M. Tignor, and P. M. Midgley, eds. Cambridge, UK, and New York: Cambridge University Press.

Jones, C. P., W. L. Coulbourne, J. Marshall, and S. M. Rogers, Jr. 2006. Evaluation of the National Flood Insurance Program's Building Standards. Washington, DC: American Institutes for Research.

Kousky, C., and L. Shabman. 2012. The Realities of Federal Disaster Aid: The Case of Floods. Issue Brief 12-02. Washington, DC: Resources for the Future.

Kunreuther, H. 1996. Mitigating disaster losses through insurance. Journal of Risk and Uncertainty 12: 171-187.

Lehmann, E. 2011. Flood-Prone Land Likely to Increase by 45%—A Major Challenge to Federal Insurance Program. Available online at http://www.nytimes.com/cwire/2011/07/22/22climatewire-flood-prone-land-likely-to-increase-by-45-a-19117.html?pagewanted=all. Accessed January 28, 2013.

National Capital Planning Commission. 2008. Report of Flooding and Stormwater in Washington, DC. Washington, DC: National Capital Planning Commission.

NCLS (National Committee on Levee Safety). 2009. Draft Report to Congress on Recommendations for a National Levee Safety Program. Available online at http://www.iwr.usace.army.mil/ncls/docs/NCLS-Recommendation-Report_012009_DRAFT.pdf. Accessed August 8, 2012.

NOAA (National Oceanographic and Atmospheric Administration). 2013. 2013 Draft National Climate Assessment Report. Available online at http://ncadac.globalchange.gov/. Accessed January 15, 2013.

NPR (National Public Radio). 2012. Norfolk, Va., Puts Flooding Survival Plan to the Test. November 6, 2012. Available online at: http://www.npr.org/2012/11/06/164362276/norfolk-va-puts-flooding-survival-plan-to-the-test. Accessed November 28, 2012.

NRC (National Research Council). 2010. Adapting to the Impacts of Climate Change. Washington, DC: The National Academies Press.

NRC. 2011. Building Community Disaster Resilience Through Private–Public Collaboration. Washington, DC: The National Academies Press.

NRC. 2012a. Dam and Levee Safety and Community Resilience. Washington, DC: The National Academies Press.

NRC. 2012b. Disaster Resilience: A National Imperative. Washington, DC: The National Academies Press.

Opperman, J. J., G. E. Galloway, J. Fargione, J. F. Mount, B. D. Richter, and S. Secchi. 2009. Sustainable Floodplains Through Large-Scale Reconnection to Rivers. Science 326: 1487-1488.

Raff, D. 2013. The 21st Century River: Changing Hydrology. Presentation to the Lower Mississippi Vision Summit, New Orleans, LA, January 9. Washington, DC: Meridian Institute.

Resilience Institute. 2010. Washington State Statewide Levee Inventory and Flood Protection Study: Report on Certification and Accreditation. Bellingham, WA: Western Washington University.

Rogers, J. D. 2009. We Have to Employ Systems Engineering Concepts in the 21st Century. Presentation at Flood Protection and Ecosystem Restoration Flood Conference, Louisiana State University, January 23. Available online at http://www.laseagrant.org/floodprotection/ppts/3b-Rogers4.pdf. Accessed November 15, 2012.

Sayers, P., G. Galloway, E. Penning-Rowsell, F. Shen, K. Wang, Y. Chen, and T. Le Quesne. In press. Flood Risk Management: A Strategic Approach—Consultation Draft. UNESCO on behalf of World Wildlife Fund—UK/China and the General Institute of Water Design and Planning, China.

Schropp, M. H. E., and T. W. Soong. 2006. Scientific Developments ISFD3. Pp 247-250 in Floods, From Defence to Management, J. van Alphen, E. Van Beek, and M. Taal, eds. London, UK: Taylor and Francis Group.

Storesund, R., R. G. Bea, M. Bernhardt, J.-L. Briaud, D. Franken, A. Govindasamy, S. Gyu, D. Kim, M. Leclair, G. Hempen, N. R. Maerz, J. D. Rogers, and C. M. Watkins. 2009. 2008 Midwest Levee Performance Investigation: Summary of Field Reconnaissance. Arlington, VA: National Science Foundation.

USACE (U.S. Army Corps of Engineers). 1986. Overtopping of Flood Control Levees and Floodwalls. Technical Letter No. 1110-2-299. Washington, DC: USACE.

USACE. 2006. Risk Analysis for Flood Damage Reduction Studies. ER 1105-2-101. Washington, DC: USACE.

USACE. 2012a. Inspection of Completed Works [Levee Safety Program] Factsheet as of February 2012. Baltimore District. Available online at http://www.nab.usace.army.mil/Factsheets/PDFs/Operations/MD-NY-PA-VA-WV-LeveeSafetyProgram-O&M.pdf. Accessed January 25, 2013.

USACE. 2012b. Northwest Division. Missouri River Flood Task Force (MRFTF). Available at http://www.nwd-mr.usace.army.mil/rcc/MRFTF/default.html. Accessed October 2012.

USACE. 2012c. What We Know About USACE Levees. Washington, DC: USACE.

USACE. n.d.a. National Nonstructural Flood Proofing Committee. Available online at http://www.usace.army.mil/Missions/CivilWorks/ProjectPlanning/nfpc.aspx. Accessed November 15, 2012.

USACE. n.d.b. USACE Disaster Operations Public Law 84-99. Available online at http://www.nfrmp.us/docs/PL84-99factsheet.pdf. Accessed February 18, 2013.

USGCRP (U.S. Global Change Research Program). 2009. Global Climate Chance Impacts in the United States. Regional Climate Impacts, Northeast. Available online at http://nca2009.globalchange.gov/. Accessed February 19, 2013.

WRC (Water Resources Council). 1973. Water and Land Related Resources: Establishment of Principles and Standards for Planning. Washington, DC: WRC.

WRC. 1983. Economic and Environmental Principles and Guidelines for Water and Related Land Resources Implementation Studies. Washington, DC: WRC.

WRDA (Water Resources Development Act). 2007. Public Law 110-114. Available online at http://www.evergladesplan.org/wrda2000/wrda_docs/wrda_2007.pdf. Accessed February 27, 2013.

Xu, Y., L. Li, and F. Amini. 2012. Slope stability analysis of earthen levee strengthened by high performance turf reinforcement mat under hurricane overtopping flow conditions. Geotechnical and Geological Engineering 30(4): 893-905.

7

Understanding and Communicating Flood Risk Behind Levees

Inherent to the National Flood Insurance Program's (NFIP's) purpose of providing the opportunity for insurance purchase in flood-prone areas and the need to reduce flood damages across the nation is the need to develop public awareness of flood risk. Risk communication is the purposeful exchange of information about the types of risks and their likely impacts, and alternatives for managing the risks including benefits and costs. It is an iterative and interactive process involving a two-way mode of communication between sources and recipients of the information (Morgan et al., 2001). NRC (1989) defines risk communication as

> an interactive process of exchange of information and opinion among individuals, groups, and institutions. It involves multiple messages about the nature of risk and other messages, not strictly about risk, that express concerns, opinions, or reactions to risk messages or to legal and institutional arrangements for risk management.

Risk communication includes both the risk information itself (messages) and sources of such information (messenger), as well as the channels (media) used in the delivery of the communication (e.g., mass media, social media, text messaging, face-to-face).

The goals of risk communication are to foster a communications environment based on trust and credibility, to provide sufficient information to an audience so that they can make informed decisions about risk-reducing behaviors, and to involve stakeholders in the decision-making process (NRC, 1989). Thus the risk communication process begins with the development of the risk management strategy including the identification of risk, its assessment, and the management of alternatives. Risk can be viewed very differently by sources and recipients. For example, the scientific community prefers to present information in an objective manner generally using statistical information and probabilities. The general public generally looks at risk information from a more subjective perspective and how it relates to their individual experience and self-benefit. In the latter case, perception and trust in and perceived credibility of the messenger are key determinants of the positive receptivity of the message (Fischhoff and Kadvany, 2011).

The challenges associated with communicating levee-related flood risk begin with challenges associated with the public perception of flood risk. Communicating the concept of residual risk behind a levee presents an additional challenge beyond the public perception of flood risk. The variety of levee-related stakeholders, including several federal agencies with different roles and responsibilities who are sources and recipients of risk information, add additional complexity. So, rather than one entity confronted with the challenge of communicating risk to the public, levee-related flood risk communication efforts face the challenge of presenting a unified, effective communication among governments at all levels, as well as among agencies, the private sector, and the public.

Risk communication activities include both "regulatory" and "nonregulatory" communication products. Regulatory flood risk communication products are those centered on FEMA's Flood Insurance Rate Maps (FIRMs) and related documents (e.g., flood insurance studies), and nonregulatory products are all other FEMA products that communicate flood- and levee-related flood risk.

UNDERSTANDING FLOOD RISK PERCEPTION

Communication of flood risks to residents has a number of inherent challenges that affect stakeholders and begin with their perception of flood risk. Risk communication efforts can be enhanced by understanding risk perception.

First, people generally underestimate flood risk (Botzen et al., 2009; Kousky and Kunreuther, 2010; Kousky, 2011; Wood et al., 2012). This is due to misunderstanding about the frequency of flooding. Such underestimates are based on the tendency to recall or remember the most severe, or infrequent flood events, rather than the less severe but more frequent ones. It is also based on misunderstandings of the likely damages from various magnitudes of flooding.

Second, residents often misunderstand the term "100-year flood" and assume that if an event occurs, it will not happen for another 100 years (Pielke, 1999; Kousky and Kunreuther 2010; USGS, 2010; Ludy and Kondolf, 2012). The designation of the one percent annual chance flood is more effective in communicating uncertainty and is being promoted as such (Bell and Tobin, 2007; USGS, 2010). However, this does not necessarily translate into protective actions such as the adoption and/or implementation of risk reduction measures, as illustrated by low insurance penetration rates (Chapter 5).

Third, there is mixed empirical evidence to support the conclusion that increased risk perception translates into actions to reduce flood risk (e.g., mitigation) (Terpstra et al., 2009; Bubeck et al., 2012; Kellens et al., 2012; FEMA's annual national surveys, mentioned later in this chapter). Although there is a positive association between risk perception and protective actions, it is more likely that other factors are more important, such as personal experience, personal attributes, and evaluations of the protective actions themselves.

Fourth, undertaking protective actions against flood risk, such as purchasing insurance and elevating homes, is more likely to be based on their efficacy and cost of implementation (Soane et al., 2010; Harvatt et al., 2011). Residents will evaluate the likely outcomes of protective actions according to varying levels of desirability, and if the outcome is viewed favorably and one that is desirable, it is more likely that a household or resident undertake that protective action (Grothmann and Reusswig, 2006; Bourque et al., 2012; Keller et al., 2012; Lindell and Perry, 2012; Wood et al., 2012).

Fifth, there is considerable trust in flood control structures in preventing flood risks (Ludy and Kondolf, 2012). This is shown by lower insurance purchase rates in areas with levee protection compared with those without the same, for example, in the St. Louis region (Kousky, 2011). Tobin (1995) notes that levees can actually increase the potential for flood loss by generating a false sense of security, the "levee effect," due to perception that the possibility of flooding has been eliminated, the lack of preparation for a flood event, and greater development behind the levee adding to the amount of property placed at risk. Thus, a fallacy that "levees prevent damages" occurs in society (Pielke, 1999), confounding the effective communication of residual risk in leveed areas.

Sixth, people will take the opinions and actions of friends, family, and neighbors into account in behavioral responses, a social exchange that is often termed milling or information seeking. In this regard, social networks are often more important than official sources of information about risks and mitigation options (Harvatt et al., 2011; Bourque et al., 2012).

Seventh, risk perception and behavioral responses are often influenced by experience with flooding. If there is little or no experience with flooding, then the negative effects of floods tend to be underestimated (Siegrist and Gutscher, 2008) and this provides less motivation to residents or households to engage in protective responses (Harvatt et al., 2011).

Eighth, risk communication tends to focus on the technical or material costs/benefits to trigger mitigation, yet recent research suggests that the emotional and experiential aspects of flooding provide more impetus for undertaking behavioral responses (Siegrist and Gutscher, 2008; Harries, 2012).

Ninth, a number of studies have examined the adoption of flood insurance and a number of their findings are relevant. In particular, if homeowners received information about the risk they will face over a longer time period (e.g., based on a 30-year period or the lifetime of a mortgage), they perceived more potential damage to their house than if they received the same risk information for 1 year (Keller et al., 2006). Also, the median length of time for holding onto flood insurance policies in the United States is 2 to 4 years, with people who have experienced small flood claims holding onto their insurance longer than those who experienced larger flood claims (Michel-Kerjan et al., 2012).

Finally, a number of studies on St. Louis found that insurance purchase was rather low, even with mandatory requirements (Kousky and Kunreuther, 2010; Kousky, 2011). Flood risk was a key determinant of the purchasing decision. As expected, there was more insurance purchase in the one percent annual chance floodplain due to the increased flood risk and mandatory purchase requirement. On the other hand, insurance purchase behind levees was less as residents perceived the area as "safe" (Kousky and Kunreuther, 2010; Kousky, 2011).

Development of appropriate strategies for risk communication is part of the overarching formulation of flood risk management (Chapter 3, Figure 3-3; Chapter 6). In developing risk communication strategies, key challenges identified from a further understanding of risk perception need to be considered:

1. The meaning of the phrase "100-year" floodplain and, to a lesser degree, the one percent annual chance flood is often misunderstood.
2. Awareness of risk does not always lead to actions to reduce or mitigate risk.
3. People assume that flood control structures are "safe" with no need to take additional actions to mitigate risk.
4. Experience with floods matters and influences the actions that residents and communities take to mitigate the impacts.
5. People will seek advice from friends, family, and neighbors on the determination of risk and actions to reduce flood risk.
6. Personalized risk assessment, based on experience and feelings, is more likely to motivate behavioral action than simply providing technical information about risks and consequences.

Understanding the above factors ensures that risk communication messaging is able to overcome the barriers associated with communicating levee-related flood risk.

BRIEF HISTORY OF RISK COMMUNICATION IN THE NFIP

Early in the NFIP's history, it was thought that risk communication efforts would be largely undertaken by communities participating in the NFIP. FEMA provided guidance for local officials to use in efforts at the local or state level to reduce flood losses in high-risk areas and identified "areas behind unsafe or inadequate levees" as one of nine high-risk areas (FEMA, 1987). A "regular schedule for communication between engineering, planning, and emergency management personnel to provide consideration of levee hazards in land use decisions" was suggested for communities with levees (FEMA, 1987). For the most part, until 2005, FEMA used the term "risk" to describe a "hazard" as a consequence; probabilities other than flood frequencies were not considered.

Early outreach to communities revolved around adoption of the initial mapping products: Flood Hazard Boundary Maps (FHBMs) and Flood Insurance Rate Maps (FIRMs) (Figures 7-1 and 7-2). The FHBM, an official map of a community issued by FEMA, delineated Special Flood Hazard Areas (SFHAs) and was prepared using available data or approximate methods to identify the areas within a community subject to inundation by a one percent annual chance flood (FEMA, 1995). The FIRM, also an official map of the community issued by FEMA, delineates both the special hazard areas and the applicable risk premium zones.[1] The FIRM is developed using the

[1] Title 44, Chapter 1, Code of Federal Regulations (1990), National Flood Insurance Program (Regulations for Floodplain Management and Flood Hazard Identification).

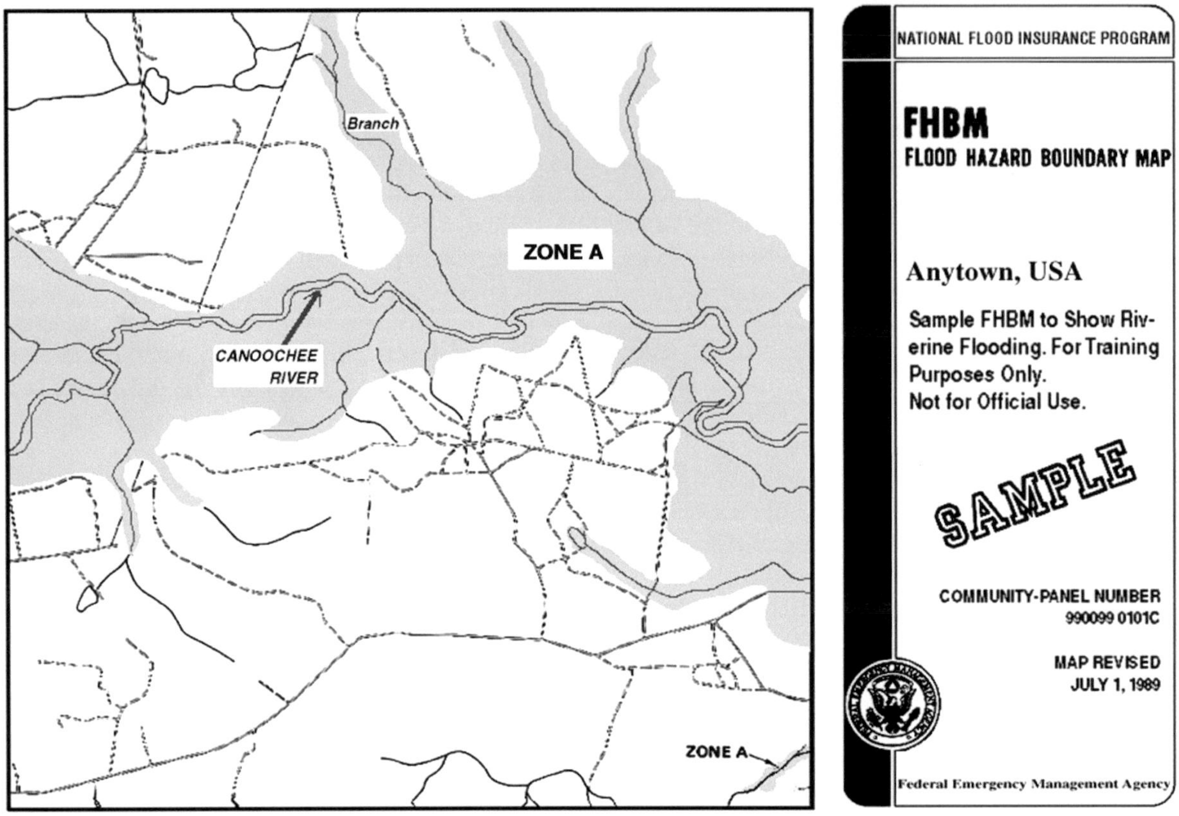

FIGURE 7-1 Example of a Flood Hazard Boundary Map.
SOURCE: FLOOD MAPS, May 1, 2008. Available online at http://www.fema.gov/pdf/nfip/manual200810/16map.pdf.

hydrologic and hydraulic analyses to develop base flood elevations and designate risk zones during flood insurance studies (FISs) (FEMA, 1995). The Digital Flood Insurance Rate Map (DFIRM) is a digital conversion of the paper FIRM that has been updated and georectified.

The 1968 National Flood Insurance Act intended that all SFHAs be identified and maps published within 5 years, and that these data be used to establish within 15 years the flood risk zones and associated insurance premium rates based on risks involved and accepted actuarial principles (FEMA, 1995). When it became evident that the needed FISs would not be completed in that time frame, Congress took action and authorized an Emergency Program (FEMA, 1995). Until the more detailed FIRMs were available, the FHBMs could be used to expand participation in the NFIP; this Emergency Program could offer insurance at nonactuarial, federally subsidized rates (FEMA, 1995). To further encourage participation in the NFIP, Congress required in the Flood Disaster Protection Act of 1973 that communities with flood-prone areas be notified of their flood hazards; this was also accomplished using the FHBM (FEMA, 1995). These initial risk (or hazard) communication efforts served to advise communities of their flood hazard as early as practicable, and sought to encourage action by communities to address and lessen their vulnerability. In the 1990s, the Mitigation Directorate was established within FEMA, Project Impact was launched (FEMA, 2011a), and the Community Rating System codified in the National Flood Insurance Reform Act of 1994 (H.R. 103-3191), all setting the stage for increased community involvement.

Historically there was little information about levee risk available to local communities other than FHBMs and FIRMs, which are focused on identifying the hazard, or areas flooded by the 0.2 percent and one percent annual chance floods for food insurance rating purposes. Today, the primary nationwide resource for information

FIGURE 7-2 A Flood Insurance Rate Map for a section of Greenville, Mississippi, along the Mississippi River. The blue areas are within the SFHA (Zone AE) and the gray area (Shaded Zone X) represents areas behind accredited levees.
SOURCE: FEMA Map Service Center. Available at http://map1.msc.fema.gov/idms/IntraView.cgi?KEY=93429510&IFIT=1

on flood-prone areas remains the FIRM. Private insurers assess exposure and offer flood insurance coverage based on FIRMs. To illustrate, preliminary estimates reported private-sector (re)insurance losses as a result of Hurricane Sandy in 2012 that range from $10 billion to $25 billion (BestWire, 2012; Insurance Insider Weekly Newsletter, 2012). However, this creates a challenge in leveed areas because older FIRMs do not explicitly identify areas as being behind accredited levees; thus, levee risk cannot be identified using the map alone.

Newer FIRMs, such as in Figure 7-2, indicate areas behind accredited levees as being Shaded Zone X; however, this zone designation is used to identify hazard areas of moderate to low risk. This lack of clear information promotes the misperception that areas behind levees are not subject to flooding and "safe." This misperception is also promoted by the mandatory purchase requirement (Chapter 5).

TABLE 7-1: Timeline of Establishment of Programs and Efforts That Communicate Flood Risk Information Behind Levees to and from the National Level

Program or Activity	Date of Establishment	Flood Risk Communication Efforts Specific to Levees
So You Live Behind a Levee (ASCE)[a]	2010	Public education booklet about risk living behind a levee
Silver Jackets (USACE, FEMA, and state agencies)[b]	2006	Flooding with some specifics on levees
National Flood Risk Management Program (USACE)[c]	2006	Reducing overall flood risk including appropriate use of levees
FloodSmart (FEMA)[d]	2003	Levee simulator Levee toolkit
Community Rating System (FEMA)[e]	1990	Credits for levee maintenance given to communities to reduce insurance rates
Stakeholder organizations (ASFPM and NAFSMA)[f]	1976, 1978	State and local flood and levee risk

[a]See http://www.asce.org/Content.aspx?id=2147488910.
[b]See http://www.nfrmp.us/state/.
[c]See http://www.nfrmp.us/.
[d]See http://www.floodsmart.gov/floodsmart/.
[e]See http://www.fema.gov/national-flood-insurance-program/community-rating-system.
[f]See http://www.floods.org/.

CURRENT FLOOD RISK COMMUNICATION EFFORTS

Throughout its duration the NFIP and its risk communication strategy continued to evolve toward one that has a more engaged agency working in a shared effort with the many stakeholders that make use of and are impacted by the NFIP. A number of federal agencies, states, local communities, and nongovernmental organizations are actively engaged in flood risk communication efforts that share information about levees directly or indirectly and support the mission of the NFIP in general and risk communication about levees in particular (Table 7-1). A number of the more significant efforts in communicating levee flood risk are briefly described below.

FloodSmart

FloodSmart began in 2004 as a marketing campaign to increase the number of policies in force (Vrem, 2010). It has evolved into a multifaceted communication tool. Presently, FloodSmart, its website,[2] and its associated activities are the primary communication avenues for information regarding the NFIP and the treatment of levees within the NFIP.[3]

FloodSmart provides Web-based information in five main areas related to flood risk generally and flood risk behind levees specifically. First, flooding and flood risk includes information on causes of flooding, defining flood risk, tutorials on understanding flood maps as well as information regarding map changes and updating flood maps. In addition to this, the website has three multimedia and interactive applications or tools: one on flood risk scenarios, one on the cost of flooding, and a third called the levee simulator, which illustrates how levees work and the causes of levee failures. The second main topical area is the detailed information regarding the NFIP itself. Here the site describes coverage within the NFIP and when insurance is required, and locator information for insurance agents and estimations of premiums. The third area focuses on residential coverage which describes policy options for homeowners, renters, and condo owners/renters, including basic information, rates, what is covered,

[2] See FloodSmart.gov.

[3] FEMA has other Web-based applications and publications for communicating flood risk, for example, the eWatermark publication available online at http://www.nfipiservice.com/watermark/index.html.

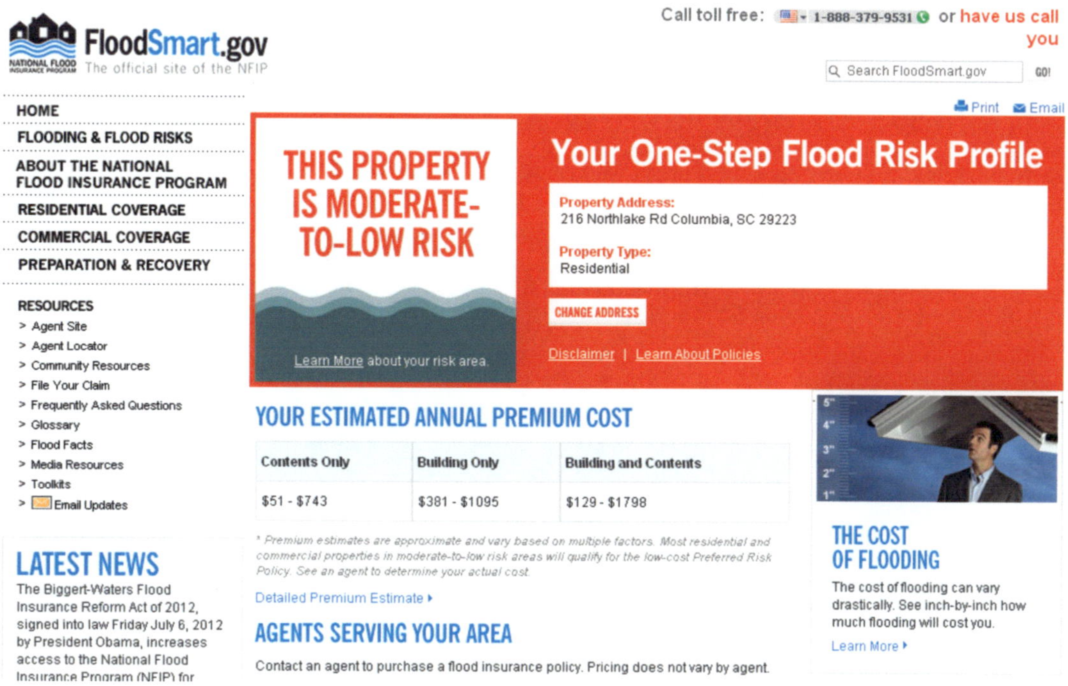

FIGURE 7-3 FEMA's one-step flood risk profile that provides quick information on flood risk, estimated flood insurance premiums, and insurance agents in the area.
SOURCE: www.floodsmart.gov/floodsmart.

questions for the agent, and locations of agents. The fourth area provides similar information for commercial properties. Finally, the last area communicates information on preparation and recovery, especially on what to do after the flood and how to file a claim.

On all of the FloodSmart webpages, there is an interactive one-step flood risk profile where one can insert the property address and the level of risk along with estimated annual premiums, and a list of agents serving the area appear (Figure 7-3). As an aid for communication, the site also has videos, media resources including public announcement scripts, and background information about the NFIP. A Levee Toolkit is available to educate citizens about flood risk associated with living behind a levee and information about flood insurance behind a levee.[4] The toolkit does so by describing three scenarios a community might face: (1) levees are no longer accredited, (2) levees are newly accredited, and (3) levees are under review or provisionally accredited. The website is easy to navigate and filled with informative content for interested parties.

Risk MAP

Begun in FY 2009, Risk Mapping, Assessing, and Planning (Risk MAP) is the extension of FEMA's map modernization efforts to improve flood hazard mapping through advanced technologies. According to FEMA, "The vision for Risk MAP is to deliver quality data that increases public awareness and leads to action that reduces risk to life and property"[5] (Figure 7-4).

---

[4] See http://www.floodsmart.gov/toolkits/.

[5] See http://www.fema.gov/rm-main#.

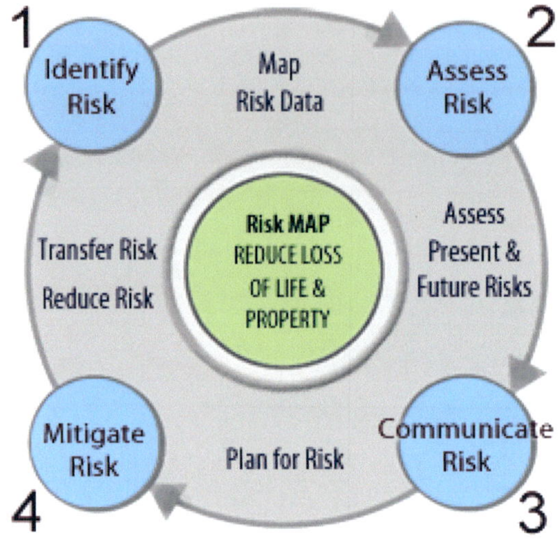

FIGURE 7-4 The vision for the Risk MAP life cycle, which begins with identifying risk, assessing risk, communicating risk, and mitigating risk.
SOURCE: http://www.fema.gov/rm-main.

The goals for Risk MAP, as identified in the Risk MAP multiyear plan (spanning FY 2010 through FY 2014), are to

> Address gaps in flood hazard data to form a solid foundation for flood risk assessments, floodplain management, and actuarial soundness of the NFIP,
>
> 1. Ensure that a measurable increase of the public's awareness and understanding of risk management results in a measurable reduction of current and future vulnerability to flooding,
> 2. Lead and support state, local, and tribal communities to effectively engage in risk-based mitigation planning resulting in sustainable actions that reduce or eliminate risk to life and property from natural hazards,
> 3. Provide an enhanced digital platform that improves management of limited Risk MAP resources, stewards information produced by Risk MAP, and improves communication and sharing of risk data and related products to all levels of government and the public,
> 4. Align Risk Analysis programs and develop synergies to enhance decision-making capabilities through effective risk communication and management." (FEMA, 2009)

Thus, Risk MAP is both a program for planning and assessment and a platform for delivering high-quality georeferenced spatial data, initially focused on flood risk but with capabilities for multihazard risk assessments. Moving to a GIS-based platform permits rapid updates of technical data.

Living with Levees is a specific FEMA website within the Risk MAP program, devoted specifically to levee risk rather that a discrete program of activities.[6] This website provides detailed information for a number of stakeholders including home and business owners; community officials; real estate, insurance, and lending professionals; technical partners (e.g. engineers, surveyors, contractors); and the media. Most of the links are to additional informational brochures or tools such as FloodSmart's levee simulator. The website contains general information that is useful, but detailed technical information that could help households understand or mitigate their flood risks is limited. Most of the website features links to FEMA program descriptions, documents, and online tutorials on NFIP, and the use of geospatial information systems (GIS) in flood mapping.

[6] See http://www.fema.gov/living-levees-its-shared-responsibility.

Community Rating System

The CRS is a part of the NFIP designed to provide incentives to communities to go beyond the minimum level of requirements of the NFIP. It rewards communities through discounts on flood insurance from 5 to 45 percent based on floodplain management and public information activities. Points are awarded in four broad categories (called a series): public information, mapping and regulations, flood damage reduction, and flood preparedness. Within each of these are a number of activities. For example, the public information category includes making information available on flood insurance rate maps and outreach projects (series 300) (FEMA, n.d.). Each activity has a maximum number of points allotted. In the public information series, for example, the maximum points are awarded for activities related to outreach projects (380). In the flood damage reduction activity series (500), the maximum points are awarded for activities related to acquisition and relocation of properties (3,200). The credit points are cumulative, and for each increment of 500 points earned, a reduction of 5 percent in the premiums for flood insurance is realized (FEMA, n.d.).

Out of the total number of communities in the NFIP (21,881[7]), only 1,211 are in the CRS program (approximately 5.5 percent). However, policies held within CRS communities represent 68 percent of the policies in force and 69 percent of the insurance in force[8] (Bill Blanton, FEMA, personal communication, July 10, 2012). Once points are earned, classifications are awarded and premium reductions are given to the community. The majority of participating communities are in the introductory class (56 percent) with premium reductions of 5-10 percent, followed by the intermediate class (43 percent) with percent premium reductions of 15-25 percent, and then the advanced class (less than 1 percent) with premium reductions of 30 percent or more (FEMA, 2012).

High Water Mark Initiative

The High Water Mark Initiative (also known as Know Your Line: Be Flood Aware) is a recent initiative involving nine federal agencies,[9] designed to raise flood awareness and motivate actions at the individual level to reduce risk.[10] The initiative began in 2012 by soliciting six pilot communities to showcase the most severe flood that has affected the community by posting high-water-mark signs on prominent buildings. As such, it relies on local implementation to offer a testimony that identifies how high floodwaters have risen in the past. It is unclear as to whether the pilot program will illustrate high water marks behind the levee, pre-levee. This would be an important way to communicate flood risk.

National Flood Risk Management Program

The National Flood Risk Management Program (NFRMP) established by USACE in 2006 is designed to integrate and synchronize flood risk management programs across USACE, with federal, state, and local partners. The risk communication efforts of this program include providing risk information to the public from flood damage reduction infrastructure, improving public awareness and understanding of flood hazards and risk, and providing accurate floodplain management information to the public and decision makers. These goals are achieved through an inventory and assessment of levees that are under USACE jurisdiction (the National Levee Database, discussed in Chapter 8), and working collaboratively with FEMA to jointly develop risk communication strategies.

An innovative, collaborative initiative, the Silver Jackets Program is designed to bring together state, federal, and local agencies to "learn from each other and apply this knowledge to reduce risk." Part of the larger USACE National Flood Risk Management Program, the goals of the Silver Jackets program are to

---

[7] As of June 2012.

[8] Insurance in force is the total of the limit of insurance purchased on each policy.

[9] FEMA, National Oceanographic and Atmospheric Administration (NOAA), National Park Service, U.S. Army Corps of Engineers (USACE), U.S. Department of Agriculture, U.S. Department of Housing and Urban Development, U.S. Geological Survey (USGS), and U.S. Small Business Administration.

[10] See http://onlinepubs.trb.org/onlinepubs/conferences/2012/securitysummit/presentations/brown.pdf.

- Create or supplement a mechanism to collaboratively address risk management issues, prioritize those issues, and implement solutions;
- Increase and improve risk communication through a unified interagency effort;
- Leverage information and resources, including providing access to such national programs as FEMA's Risk MAP program and USACE's Levee Inventory and Assessment Initiative;
- Provide focused, coordinated hazard mitigation assistance in implementing high-priority actions such as those identified by state mitigation plans;
- Identify gaps among the various agency programs and/or barriers to implementation, such as conflicting agency policies or authorities, and provide recommendations for addressing these issues.[11]

The Silver Jackets process is centered at the state level and supported at the federal level primarily by USACE and FEMA. State agencies form federal-state-stakeholder teams that meet regularly to discuss a state's flood risk management priorities among stakeholders. Currently, 33 states have interagency, Silver Jacket teams. With the state in the lead, each team is encouraged to include other federal agencies that could advance flood risk management goals in the Silver Jackets forum, for example, NOAA–National Weather Service, Department of the Interior, and the USGS.

Living Behind a Levee

The American Society of Civil Engineers (ASCE) is a nonprofit professional society. One of its many roles is to help protect public safety through improved learning within the profession, but also among the general public. One key focus of the ASCE is to addresses disaster preparedness and response. The flood risk communication aspect of the ASCE is best seen in the preparedness guide, *So, You Live Behind a Levee! What You Should Know to Protect Your Home and Loved Ones from Floods*. Written for both the engineering and nonengineering public, the brochure emphasizes four essential levee facts: (1) flooding will happen, (2) risks associated with flooding vary, (3) no levee is floodproof, (4) actions taken now will save lives and property.

CONTEMPORARY FLOOD RISK MAPPING

Twenty-first century geospatial information technology has opened the door for vastly improved risk identification and communication. The development of GIS has provided decision makers, land-use planners, floodplain managers, and numerous other parties connected with flood hazard to conduct detailed analyses of the multiple flood scenarios that bring together the built environment with the natural environment to create flood risk. Use of GIS in management of communities is ubiquitous, and few communities lack some form of GIS. FEMA's DFIRMs provide an example of the use of GIS to identify flood hazards.

Over the last decade, considerable effort, at the global scale, has been placed on research to improve the representation of flood risks on flood maps and to make the maps both dynamic and interactive. Today, maps are available that permit those living in the floodplain to determine the level of inundation of their property under various flood condition scenarios; to examine the consequences of a particular flood event, taking into account the risk-based analysis that produced the information; and to link real-time weather information with flood potential in a given region (Figure 7-5 and 7-6). New flood maps not only illustrate the floods and their consequences but also can dynamically portray such elements as availability of evacuation routes, shelter status, and threat levels. The future of such approaches is illustrated in the National Research Council report, *Mapping the Zone Improving Flood Map Accuracy* (NRC, 2009).

FEMA's Risk MAP program is carefully examining new tools and, where appropriate, using them to better communicate flood risk under a variety of circumstances (Figure 7-7). Tying risk-based analysis to GIS will enable those living behind levees to gain a true perspective of their flood risk.

[11] Both quotations are from http://www.nfrmp.us/state/.

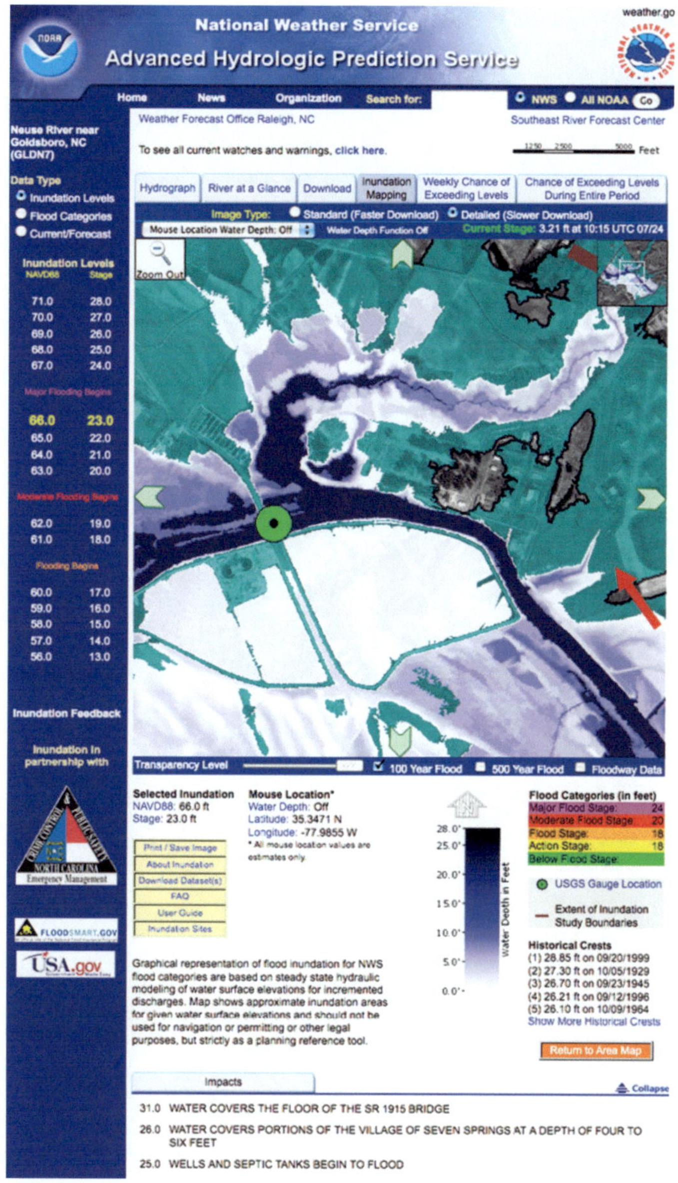

FIGURE 7-5 National Oceanic and Atmospheric Administration's National Weather Service (NWS) flood inundation map of a segment of the Neuse River near Goldsboro, North Carolina, showing the extent of flooding when water levels are forecast to rise to a stage of 23 feet (blue) and the location of the one percent annual chance floodplain (blue green) from a FEMA map. The darker the blue, the greater is the depth of inundation. The water depth is 0 to 2 feet near the edge of the Seymour Johnson Air Force Base runway (red arrow). The green circle shows the U.S. Geological Survey (USGS) stream gauge where the NWS provides the river forecast. The topographic data, digital elevation models, and hydraulic models underlying the map were produced by the USGS office in Raleigh and the North Carolina Floodplain Mapping Program.
SOURCE: http://water.weather.gov/ahps/inundation.php. Accessed January 28, 2013.

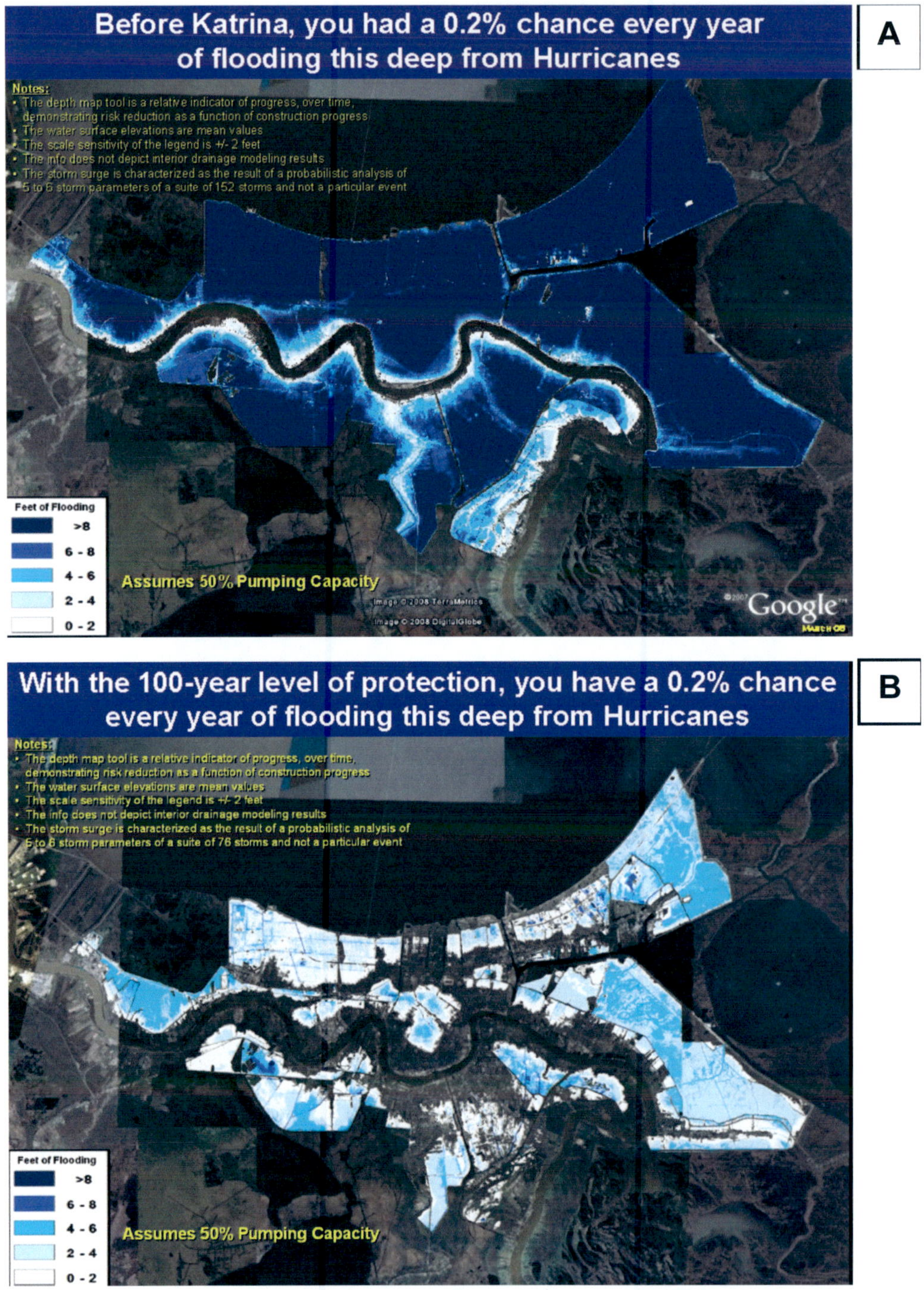

FIGURE 7-6 Flood risk maps for New Orleans, illustrating differences in levels of inundation for a 0.2 percent flood event (*A*) before and (*B*) after the completion of the post-Katrina one percent hurricane risk reduction levees. Water surface elevations are mean values, with a sensitivity of ±2 feet. The maps assume that the pumps inside the levees that evacuate interior floodwaters are operating at 50 percent capacity.

SOURCE: USACE, New Orleans District, http://www.mvn.usace.army.mil/hps2/hps_risk_depth_map.asp. Accessed February 17, 2013.

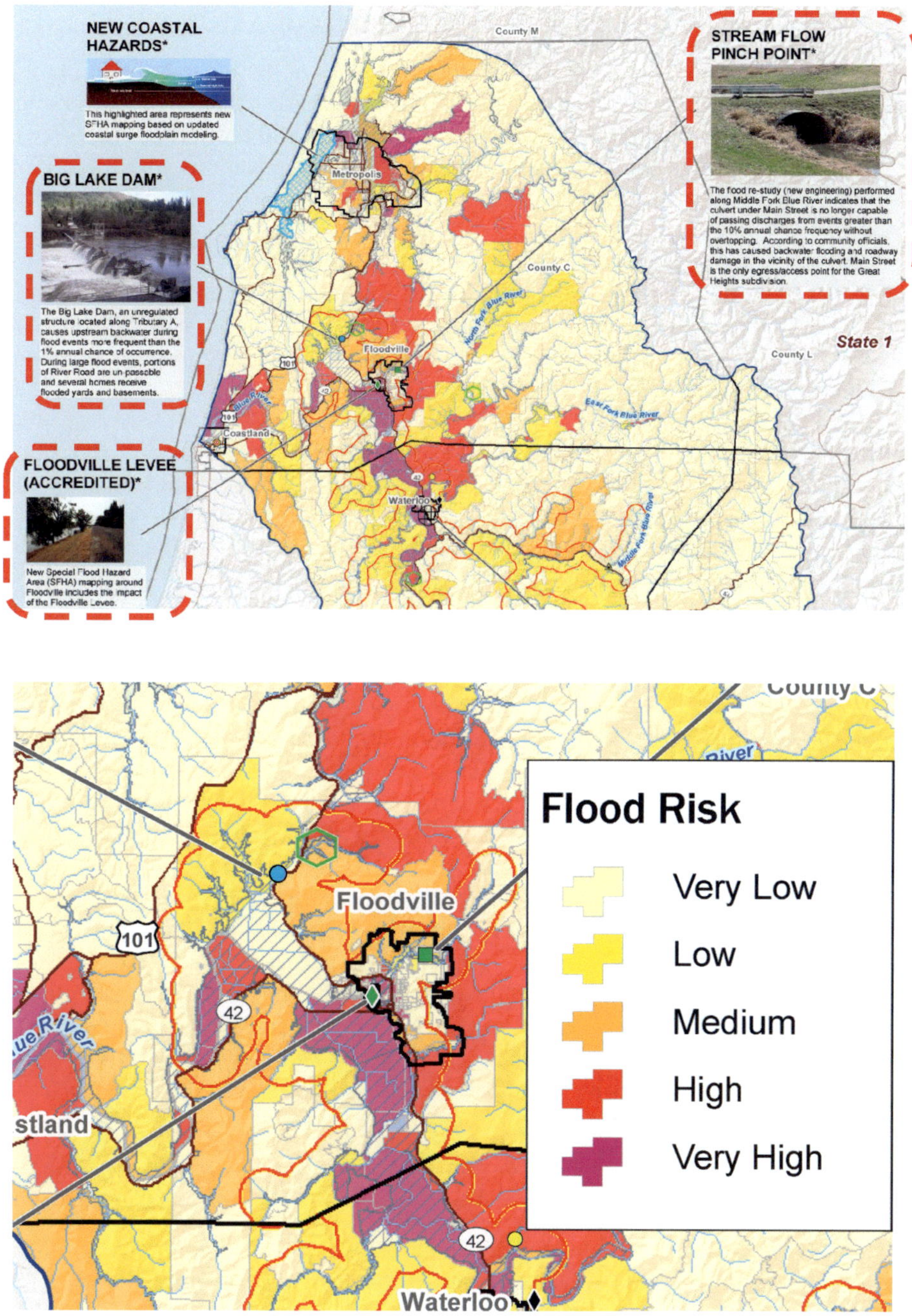

FIGURE 7-7 A Risk MAP pilot watershed-based flood risk map of "Watershed USA" containing nonregulatory data to aid in risk communication. At the top, the upper portion of Watershed USA is shown, containing detailed information about a variety of items influencing flood risk in the watershed. At the bottom, a closer view of the town of Floodville shows the different levels of flood risk by area.
SOURCE: Maps courtesy of FEMA Risk MAP.

EFFECTIVE FLOOD RISK COMMUNICATION

While FEMA, the USACE, and other groups have important roles at the federal level in providing information on flood risk, it is local and state governments that drive behavioral changes to mitigate the risk. This is recognized and built into communication efforts implemented at the national level, described above; these federal agencies maintain a focus on communicating at the local level. Also, organizations with a focus on the state and local levels that facilitate communication of the state and local perspective to the national scale play a particularly important role (Box 7-1).

Some flood-related risk communication efforts are very successful (Box 7-2), but metrics and evaluations of communication efforts are difficult to find. FEMA's annual national surveys are an exception. These surveys were conducted in 2010 and 2011 to evaluate progress toward achieving goal 2 of Risk MAP, above, to "ensure that a measurable increase of the public's awareness and understanding of risk management results in a measurable reduction of current and future vulnerability to flooding."[12] The surveys found that:

- the percentage of community members who believed that their community was at risk from flooding was 31 percent in 2012 and 41 percent in 2011, while fewer than 23 percent of those believing their community to be at risk believed that their home was at risk in 2011;
- 68 percent of local public officials thought their community was at risk for flooding;
- 47 percent of respondents expect to hear about flood risk from the mayor;
- local news, phone calls, and mailings were also preferred methods of communication at 87 percent, 25 percent, and 24 percent, respectively;
- when asked about how frequently people heard from local officials about flood risk, the most common response was "never" in 2010 (41 percent) and in 2011 (45 percent); and
- those in coastal areas were slightly more aware of flood risk to their community (54 percent) or home (19 percent) (FEMA, 2011b).

California, in partnership with FEMA and USACE, recently sent mailed notices to property owners located behind a state-federal project levee as part of their FloodSAFE California Flood Risk Notification Program.[13] The Living with Levees: Know Your Flood Risk mailing campaign was evaluated through a readership survey to determine and improve effectiveness in 2010 and 2011.[14] In 2010, 77 percent of survey respondents said that the mailing raised their awareness of flood risk "a lot" or "somewhat" (31 percent and 46 percent, respectively). This improved in 2011, with 85 percent noting that the mailing raised awareness of flood risk "a lot" or "somewhat" (28 percent and 57 percent, respectively). The mailing encouraged 12 percent and 14 percent of survey respondents to purchase flood insurance in 2010 and 2011, respectively. This corroborates risk awareness key principle 2, above, that although the mailing was effective at raising flood risk awareness, such awareness doesn't necessarily translate into action, that is, buying an insurance policy.

Information Access and Security: The Impact on Risk Communication

As discussed throughout this chapter, information regarding levee-related flood risk is becoming more sophisticated and accessible with, for example, actions taken to generate levee inventories and improved mapping. Furthermore, a modern risk analysis and better probability models and damage estimates (resulting in risk-based premiums at the structure level) will greatly increase communication of true risk to NFIP policyholders. New and improved information can be paired with development of community flood risk management strategies, including mitigation measures to reduce premiums. Improved mapping can be used to engage the private sector, so that insurance policies that are complimentary to the NFIP are offered to the public. Furthermore, the insurance industry

---

[12] One thousand telephone interviews were conducted at random to households in FEMA's 10 regions; 36 percent of respondents were from Risk MAP project areas, 87 percent of whom were homeowners and 12 percent renters.

[13] See http://www.water.ca.gov/floodmgmt/lrafmo/fmb/fas/risknotification/.

[14] Information included in this discussion was provided by FloodSmart California staff at the second meeting of this committee.

BOX 7-1
Stakeholder Organizations

The Association of State Floodplain Managers (ASFPM) is an organization of professionals involved in floodplain management, flood hazard mitigation, and flood preparedness and response. The mission of the organization is to "promote education, policies, and activities that mitigate current and future losses, costs, and human suffering caused by flooding, and to protect the natural and beneficial functions of floodplains—all without causing adverse impacts."[a] The ASFPM is organized into state chapters and, as such, ASFPM supports an avenue of communication between the states and the federal and local levels. ASFPM activities that promote communication include education of floodplain managers, technical conferences to bring together practitioners, and involvement in the establishment of FEMA's Community Rating System to reward communities for floodplain management practices. [b]

Similarly, the National Association of Flood & Stormwater Management Agencies (NAFSMA) "is an organization of public agencies whose function is the protection of lives, property and economic activity from the adverse impacts of storm and flood waters. The mission of the Association is to advocate public policy, encourage technologies and conduct education programs which facilitate and enhance the achievement of the public service function of its members."[c] Self-described as its members' voice in Washington, D.C., NAFSMA represents and communicates the perspective of state and local public agencies at the national level.

[a]See http://www.floods.org/.
[b]See http://www.floods.org/index.asp?menuID=425&firstlevelmenuID=179&siteID=1.
[c]See http://nafsma.org/about.php.

is in a position to support effective risk communication through its policy marketing, when provided with access to flood risk information.

Availability of information regarding the location and status of the nation's levees has a direct impact on successful flood risk communication efforts. Some data elements in the USACE's National Levee Database contain information about the integrity of the levees contained therein. Access to this information is restricted to government officials and is not available to the public. Restricting access to information regarding levees limits the ability of businesses, individuals, and in many cases, first responders to become aware of and properly prepare for possible emergency situations. Three recent reports (NRC, 2012a,b; Galloway et al., 2011) have addressed the need for full access to such information and recommended that this needed information be routinely released.

A WAY FORWARD

Understanding risk perception and evaluations of risk communication efforts sheds light on the importance of local communication and clear messaging, as well as challenges such as the assumption that flood control structures are "safe." Insight gathered at public meetings of this committee indicates that risk communication efforts surrounding NFIP levees have an advantageous window of attention during three situations: (1) after a flood disaster, (2) through the mapping programs, and (3) when a community participates in FEMA's CRS program. It also appears that during these three situations, particularly situations 1 and 2, considerable confusion and controversy or considerable success in communicating risk can occur.[15]

[15] This assessment is based on information gathering efforts during committee meetings in St. Louis, Missouri, Dallas, Texas, and Sacramento, California.

BOX 7-2
Flood Insurance Retention in Sacramento, California: A Transitional Community

The city of Sacramento sits at the confluence of the Sacramento and American rivers. Most of the city lies within the floodplains of these two rivers, around 55,000 acres containing approximately 110,000 damageable structures valued at over 30 billion dollars. More than 300,000 people live in these floodplains. The Sacramento levee system extends for approximately 100 miles around the city.

In 1978, Sacramento joined the NFIP and was mapped out of the one percent annual chance regulatory floodplain because of the levee system. Following a record flood in 1986, the city was mapped back into the SFHA (Zone A-99) in 1989. Improvements to the levee systems resulted in a revised Flood Insurance Rate Map for the American River area in 2004 and removal from the one percent annual chance regulatory floodplain.

Corresponding to the release of the 2004 map for the American River area, the Sacramento Area Flood Control Authority (SAFCA) commenced a 6-month flood insurance outreach campaign targeting approximately 46,000 property owners who were to be released from the mandatory purchase requirement.[a] The outreach campaign had several goals: (1) alert property owners to the risk of levee failure in their community, (2) urge property owners to take responsibility for this risk by maintaining a flood insurance policy, and (3) encourage property owners who are no longer subject to the mandatory purchase requirement to continue to carry a flood insurance policy (a preferred risk policy or PRP), and (4) urge stakeholders to engage in Sacramento's policy retention efforts. The campaign was implemented in five phases targeting the insurance agent, elected officials, lenders, property owners, and the media.

The choice to center part of the outreach plan around PRPs was based on a survey, conducted in March 2003, targeting homeowners in the American River area. Respondents said they were willing to maintain their insurance policies based on two conditions: (1) once the MPR was lifted, they could purchase flood insurance for approximately half what they currently paid;, and (2) if local flood officials communicated flood risk.

Seventy percent of the property owners released from the 2005 mandatory purchase requirement maintained flood insurance. Lessons learned include the effectiveness of an outreach campaign coinciding with a community's release from the flood insurance program and the importance of the more affordable PRP in encouraging policy retention (SAFCA, 2012).

---

[a]The campaign was jointly funded by NFIP and SAFCA (75 percent and 25 percent, respectively).

To be effective, levee risk communication efforts need to be anchored in several principles:

- delivery at the local level;
- products tailored to individual households, communities, and other stakeholders;
- delivery from a credible and trusted source;
- long-term (meaning efforts should extend over a period of time);
- contain consistent, correct, and nonconflicting content;
- encourage or motivate some behavior change;
- account for the values of target audiences or communities;
- employ multimodal networks; and
- provide repeat messaging (Peters et al., 1997; Mileti and Peek, 2002; Tinker and Galloway, 2008).

FEMA and others involved in risk communication at all levels should incorporate contemporary risk communication principles in the development of flood risk communication strategies and implementation

efforts. Seizing key windows of opportunity to communicate levee-related flood risk at the local level is critical to success. Correcting misperceptions, for example, that if you live behind a levee you are "safe," is also critical.

A systematic evaluation of risk communication efforts against measurable outcomes such as increases in insurance purchase, increases in mitigation activities, or a downward trend in damages could yield significant insights regarding levee-related risk communication. Appropriate assessment tools for monitoring the success of risk communication programs are critical, including (1) baseline information on the existing status of flood losses prior to program implementation, if available; and (2) predefined metrics for assessing the effectiveness in reducing risk as measured by increases in flood insurance policies, increases in the number and type of mitigation activities, improvements in the percentage reductions in NFIP premiums, and reductions in overall losses, among others. **FEMA should support evaluation of the success of risk communication efforts, including at the community level when appropriate, that is informed by appropriate assessment tools such as baseline information and predefined metrics.** Agencies and organizations outside of FEMA involved in communicating risk behind levees might find value in evaluating success of efforts as well.

REFERENCES

Bell, H. M., and G. A. Tobin. 2007. Efficient and effective? The 100-year flood in the communication and perception of flood risk. Environmental Hazards: Human and Policy Dimensions 7(4): 302-311.

BestWire. 2012. Update: Sandy losses by company: Insurance total could reach $25 billion. Available online at http://www.programbusiness.com/News/Sandy-Losses-by-Company-Insurance-Total-Could-Reach-25-Billion. Accessed February 21, 2013.

Botzen, W. J. W., J. Aerts, and J. C. J. M. van den Bergh. 2009. Willingness of homeowners to mitigate climate risk through insurance. Ecological Economics 68(8-9): 2265-2277.

Bourque, L. B., R. Regan, M. M. Kelley, M. M. Wood, M. Kano, and D. S. Mileti. 2012. An examination of the effect of perceived risk on preparedness behavior. Environment and Behavior, doi:10.1177/0013916512437596.

Bubeck, P., W. J. W. Botzen, and J. C. J. H. Aerts. 2012. A review of risk perceptions and other factors that influence flood mitigation behavior. Risk Analysis 32(9): 1481-1495.

FEMA (Federal Emergency Management Agency). 1987. Reducing Losses in High Risk Flood Hazard Areas: A Guidebook for Local Officials, FEMA 116. Available online at http://www.fema.gov/library/viewRecord.do?id=1508. Accessed November 14, 2012.

FEMA. 1995. Managing Floodplain Development in Approximate Zone A Areas: A Guide for Obtaining and Developing Base (100-Year) Flood Elevations, FEMA 265. Available online at http://www.fema.gov/library/viewRecord.do?id=1526. Accessed November 12, 2012.

FEMA. 2009. Risk Mapping, Assessment, and Planning (Risk MAP) Multi-Year Plan: Fiscal Years 2010-2014. Available online at http://www.fema.gov/library/viewRecord.do?id=3587. Accessed December 6, 2012.

FEMA. 2011a. FEMA Mitigation and Insurance Strategic Plan 2012-2014. FEMA P-857. Available online at http://www.fema.gov/library/viewRecord.do?id=4903.

FEMA. 2011b. 2011 Public Survey Findings on Flood Risk. Available online at http://www.fema.gov/library/viewRecord.do?id=6259. Accessed November 20, 2012.

FEMA. 2012. Community Rating System Participation National Map. Available online at http://www.fema.gov/library/viewRecord.do?id=6200. Accessed December 8, 2012.

FEMA. n.d. NFIP: Community Rating System: A Local official's guide to saving lives preventing property damage reducing the cost of flood insurance. Available online at http://www.fema.gov/library/viewRecord.do?id=3655. Accessed November 29, 2012.

Fischhoff, B., and J. Kadvany. 2011. Risk: A Very Short Introduction. Oxford, UK: Oxford University Press.

Galloway, G. E., G. B. Baecher, K. Brubaker, and L. E. Link. 2011. Review and Evaluation of the National Dam Safety Program. College Park, MD: Water Policy Collaborative. Available on line at www.fema.gov/library/viewRecord.do?id=5794. Accessed February 21, 2013.

Grothmann, T., and F. Reusswig. 2006. People at risk of flooding: Why some residents take precautionary action while others do not. Natural Hazards 38(1-2): 101-120.

Harries, T. 2012. The anticipated emotional consequences of adaptive behaviour-impacts on the take-up of household flood-protection measures. Environment and Planning A 44(3): 649-668.

Harvatt, J., J. Petts, and J. Chilvers. 2011. Understanding householder responses to natural hazards: Flooding and sea-level rise comparisons. Journal of Risk Research 14(1): 63-83.

Insurance Insider Weekly Newsletter. 2012. Reported Sandy losses to $10.4bn. December 17, Issue 506.

Kellens, W., T. Terpstra, and P. De Mayer. 2012. Perception and communication of flood risks: A systematic review of empirical research. Risk Analysis doi:10.1111/j.1539-6924.2012.01844.x.

Keller, C., M. Siegrist, and H. Gutscher. 2006. The role of the affect and availability heuristics in risk communication. Risk Analysis 26(3): 631-639.

Keller, C., A. Bostrom, M. Kuttschreuter, L. Savadori, A. Spence, and M. White. 2012. Bringing appraisal theory to environmental risk perception: A review of conceptual approaches of the past 40 years and suggestions for future research. Journal of Risk Research 15(3): 237-256.

Kousky, C. 2011. Understanding the demand for flood insurance. Natural Hazards Review 12(2): 96-110.

Kousky, C., and H. Kunreuther. 2010. Improving flood insurance and flood-risk management: Insights from St. Louis, Missouri. Natural Hazards Review 11(4): 162-172.

Lindell, M. K., and R. W. Perry. 2012. The protective action decision model: Theoretical modifications and additional evidence. Risk Analysis 32(4): 616-632.

Ludy, J., and G. M. Kondolf. 2012. Flood risk perception in lands "protected" by 100-year levees. Natural Hazards 61(2): 829-842.

Michel-Kerjan, E., S. Lemoyne de Forges, and H. Kunreuther. 2012. Policy tenure under the U.S. National Flood Insurance Program (NFIP). Risk Analysis 32(4): 644-658.

Mileti, D. S., and L. A. Peek, 2002. Understanding individual and social characteristics in the promotion of household disaster preparedness. In New Tools for Environmental Protection: Education, Information, and Voluntary Measures. Washington, DC: The National Academies Press.

Morgan, M. G., B. Fischhoff, A. Bostrom, and C. J. Atman. 2001. Risk Communication: A Mental Models Approach. Cambridge, UK: Cambridge University Press.

NRC (National Research Council). 1989. Improving Risk Communication. Washington, DC: National Academy Press.

NRC. 2009. Mapping the Zone, Improving Flood Map Accuracy. Washington, DC: The National Academies Press.

NRC. 2012a. Dam and Levee Safety and Community Resilience: A Vision for Future Practice. Washington, DC: The National Academies Press.

NRC. 2012b. Disaster Resilience: A National Imperative. Washington, DC: The National Academies Press.

Peters, R. G., V. T. Covello, and D. B. McCallum. 1997. The determinants of trust and credibility in environmental risk communication: An empirical study. Risk Analysis 17(1): 43-54.

Pielke, R. A. 1999. Nine fallacies of floods. Climatic Change 42(2): 413-438.

SAFCA (Sacramento Area Flood Control Agency). 2012. Data on Flood Insurance Purchase and Retention in the City of Sacramento, Memo to the Committee on Levees and the National Flood Insurance Program: Improving Policies and Practices from Tim Washburn, Director of Planning. Sacramento, CA: SAFCA.

Siegrist, M., and H. Gutscher. 2008. Natural hazards and motivation for mitigation behavior: People cannot predict the effect evoked by a severe flood. Risk Analysis 28(3): 771-778.

Soane, E., I. Schubert, P. Challenor, R. Lunn, S. Naredran, and S. Pollard. 2010. Flood perception and mitigation: The role of severity, agency, and experience in the purchase of flood protection, and the communication of flood information. Environment and Planning A 42(12): 3023-3038.

Terpstra, T., M. K. Lindell, and J. M. Gutteling. 2009. Does communicating (flood) risk affect (flood) risk perceptions? Results of a quasi-experimental study. Risk Analysis 29(8): 141-1155.

Tinker, T., and G. E. Galloway. 2008. How Do You Effectively Communicate Flood Risks? Looking to the Future. Available online at: http://www.emforum.org/vforum/BAH/How%20Do%20You%20Effectively%20Communicate%20Flood%20Risks.pdf. Accessed November 14, 2012.

Tobin, G. A. 1995. The levee love-affair—a stormy relationship. Water Resources Bulletin 31(3): 359-367.

Vrem, M. J. 2010. Communicating Risk: Ten key learnings from the FloodSmart campaign. Available online at: http://www.asfpmfoundation.org/pdf_ppt/2010_GFW_Forum_Background_Reading.pdf#page=66. Accessed November 20, 2012.

USGS (U.S. Geological Survey). 2010. Flood of April and Map 2008 in Northern Maine, Fact Sheet 2010-3003. Available online at: http://pubs.usgs.gov/fs/2010/3003/pdf/fs2010-3003.pdf. Accessed October 19, 2012.

Wood, M. M., D. S. Mileti, M. Kano, M. M. Kelley, R. Regan, and L. B. Bourque. 2012. Communicating actionable risk for terrorism and other hazards. Risk Analysis 32(4): 601-605.

8

Shared Responsibility in Moving to Flood Risk Management

The passage of the National Flood Insurance Act in 1968 and the establishment of the NFIP marked what can be considered the official shift in the United States from a principal focus on flood control to a focus on flood damage reduction with greater consideration of nonstructural measures and insurance, paired with structural measures, to manage flood risk. The movement toward flood risk management continued and was emphasized, for example, in the wake of Hurricane Katrina with the application of tools associated with risk management in other sectors to flood risk management. This movement is changing the manner in which the nation addresses its floodplain management issues and is requiring agencies to make, or begin to make, significant changes in the manner in which they carry out their programs. These changes are seen in FEMA's management of the NFIP and will be affecting their approach in the years immediately ahead. It is not a question of if, but when, the transition to risk-based approaches will be completed across all relevant parties.

The principal responsibility for NFIP-related actions is assigned by the National Flood Insurance Act to the Federal Emergency Management Agency (FEMA) (originally assigned to the Department of Housing and Urban Development or HUD). FEMA also has organic authorities to "lead the Nation's efforts to prepare for, protect against, respond to, recover from, and mitigate the risk of natural disasters, acts of terrorism, and other man-made disasters, including catastrophic incidents" (Post-Katrina Emergency Management Reform Act of 2006). Other federal agencies, states, and local communities are assigned responsibilities by the National Flood Insurance Act and Executive Order 11988 (1977) to support these goals in collaboration with FEMA. The National Flood Insurance Act specifically encourages states and local governments to make appropriate land-use adjustments to reduce the exposure of communities to flooding.

The National Flood Insurance Act also establishes the requirement for the development of a Unified National Program (UNP) for floodplain management. The initial UNP document was created by the Water Resources Council in 1976 and followed by a second in 1979. Updated plans were prepared by an interagency task forces operating under FEMA oversight in 1986 and 1994 and reflected agency consensus. The UNP urged cooperative efforts among the federal government, states, and the private sector as well as within the federal government itself.

Other federal agencies play important roles in managing the nation's flood risk. Since the early part of the 20th century, USACE has been assigned responsibility under various flood control and water resources development acts for the construction and, in some cases, operation of flood control works. In 2006, USACE, in coordination with FEMA, established the National Flood Risk Management Program in support of an effort to move the nation from a flood control approach to a flood risk management approach.[1] The National Oceanic and Atmospheric Administration (NOAA), Natural Resources Conservation Service (NRCS), U.S. Geological Survey (USGS), U.S. Forest Service, Bureau of Reclamation, U.S. Fish and Wildlife Service (FWS), the U.S. Environmental Protection Agency, the Tennessee Valley Authority, the Federal Energy Regulatory Commission (FERC), and U.S. Department of Housing and Urban Development (HUD) also have legislated responsibilities that directly or indirectly involve flood risk management.[2]

State, tribal, and local governments have responsibilities that are also critical to managing the nation's flood risk in general and behind levees in particular. For example, local and to a lesser degree, state governments control land use and exposure to risk. Local government plays an important role in levee-related risk communication (Chapter 7).

The various roles and responsibilities at all levels highlight the need for a coordinated, national approach that *shares responsibility* across federal, state, and local government and other entities in managing flood risk. In this chapter, challenges and opportunities related to a shared national strategy for levee-related flood risk and management are explored.

COORDINATION CHALLENGES AT THE FEDERAL LEVEL

Many federal agencies have responsibilities that involve management of flood risk and levee-related flood risk, in particular. Their roles are dictated by their legislated authorities and responsibilities as well as by presidential guidance. Because of the large number of agencies and complexity of legislation, coordination challenges exist that are manifested in many ways—from ensuring consistency in flood risk communication to compilation of a national levee inventory.

USACE and FEMA: Different Approaches to Levee-Related Risk Analysis

When HUD and FEMA initially began to develop procedural guidelines for levees in the NFIP, they relied heavily on the resources and advice of the USACE. When Title 44, Section 65.10, of the *Code of Federal Regulations* (44 CFR §65.10) was developed, it included references to USACE technical publications as measures of good practice in engineering. Although the USACE technical approaches have evolved over time, reference to these older documents remain in 44 CFR §65.10.

In 1996, when USACE began to move to a risk-based approach to analysis of flood hydrology and hydraulics and other aspects of flood system integrity determination, FEMA continued with its existing approach. In 2006, the Interagency Levee Policy Review Committee recommended that FEMA shift to a risk-based analysis procedure over the following 10-year period (ILPRC, 2006). In Chapter 3, this report recommends that FEMA move to use of a risk-based levee analysis approach. As noted in Chapter 2, the 2012 Biggert-Waters Act requires coordination between FEMA and USACE in the assessment of levees that exist in both programs: "information and data collected by or for the USACE under the Inspection of Completed Works Program is sufficient to satisfy the flood protection structure accreditation requirements." Since it appears that approximately 65 percent of NFIP-accredited and PAL levees are also in one of the USACE programs and subject to USACE risk-based inspections, it would not be efficient to establish two separate evaluation methodologies for ensuring the integrity of those levees. If FEMA chooses to continue its present approach for the analysis of other NFIP-accredited levees not in the USACE programs, levee owners and their engineers will be faced with two or more different approaches to analysis of levee integrity and risk.

The technical and programmatic differences between FEMA and USACE as they relate to the evaluation of flood hazards and levee systems have been and continue to be an issue that compromises interagency communica-

[1] See http://www.nfrmp.us/.

[2] A Federal Interagency Floodplain Management Task Force, established in 1975, promotes programs and policies that reduce flood losses and protect the environment. See http://www.fema.gov/national-flood-insurance-program-1/federal-interagency-floodplain-management-task-force.

tion and causes confusion and frustration in the local communities that work with the agencies and the public that is being served. Furthermore, the benefits of a more modern risk-based approach to assessing risk behind levees have been made clear. Both the complexities of coordinating the programs of multiple government entities and the benefits of conducting risk-based analyses and acquisition of the data needed to conduct these analyses are illustrated in recent events involving the City of Dallas, Texas (Box 8-1). **FEMA and USACE should jointly develop a common, risk-based approach to levee assessment in a timely manner and apply this approach to all levees assessed by the two agencies. This includes a joint methodology, procedure, and where feasible, the sharing of models and other risk analysis tools.** To achieve this goal, FEMA and USACE could craft a memorandum of understanding. Furthermore, it would be useful for other federal agencies with interest in risk-based assessment of levees and floodwalls to be invited to participate in the process.

Other Coordination Challenges Between FEMA and USACE

When accredited levees are found deficient by either FEMA and/or USACE, levee owners in affected communities attempt to take action to remediate the deficiency. Where the level of protection provided by the levee is acknowledged to be the same by both USACE and FEMA, coordination problems are minimized and the com-

BOX 8-1
Dallas Levees: Working to Determine the True Risk

The Dallas Levee System in Dallas, Texas, includes over 20 miles of levees that protect downtown Dallas and thousands of citizens against a 0.0125 percent (8,800-year) annual chance flood. In 2009, USACE conducted a periodic inspection of the levee system and informed the city and FEMA that the levees would not performed as designed and rated these levees unacceptable. This jeopardized the city's levee system accreditation status and FEMA began to remap the areas behind the levees, designating them as being in the Special Flood Hazard Area.

In response, the City of Dallas spent more than $25 million dollars studying the integrity of the levee system and devising a plan to remediate known deficiencies in the levee system. The city also completed an Emergency Action Plan and identified 150 million dollars of levee improvements and began to correct 198 maintenance deficiencies previously identified in the system and those identified by the city's engineering study.

Given the potential consequences of levee deaccreditation and in keeping with an agencywide effort to conduct risk-based analyses, USACE applied a new risk assessment process to the Dallas levee system. After an intensive study of the levees, made possible in part by the engineering investigations funded by the City of Dallas, USACE concluded that:

- The frequency of major floods was rarer than had been originally determined. A flood event that could overtop the levees was rated at 0.01 to 0.02 percent annual chance (1,000-year to 5,000-year).
- The duration of floods that might occur is much shorter than once thought. Peak flooding lasts hours and days, not months. As a result, the levees would be less likely to become saturated and fail.
- While levee slope slides have occurred over the years, the detailed analysis indicates that "these slides do not represent an unacceptable risk" (USACE, 2012).

The City of Dallas is continuing efforts to remediate levee deficiencies; however, as a result of USACE's modern risk analysis approach, the extent and cost to the city of remediation will be decreased. The City of Dallas expects to complete repairs in 2013 and intends to submit levee certification to FEMA so that the remapping of the area behind the levees reflects the new conditions and supports accreditation. The city plans to work with USACE to remove the unacceptable rating under the USACE periodic inspection program.

munity can attempt to remediate against a common standard. When, on the other hand, the levee was built by USACE to a higher standard than was necessary for entry into the NFIP, USACE may require the levee owner to remediate the deficiency not to the NFIP base flood elevation but to the original level of protection provided by USACE. When the community can only find resources to accomplish the mitigation to the lower level, USACE may determine that it will not permit work to move forward because that work would not bring the levee into USACE compliance and might make it more difficult to eventually achieve the higher USACE standard (Box 5-7).

Another potential issue arises when an accredited levee (granting the areas behind the levee exemption from the mandatory flood insurance purchase requirement), owned and maintained by USACE, is no longer certified by USACE. USACE may not permit local governments to accomplish the remediation, leaving the local governments with no way to bring the levees back to an accredited status.

In both of these cases, the agencies are operating within their regulatory authorities. However, the community is caught in the middle and faces the consequences. The utility of a joint FEMA-USACE procedure allowing quick attention to levee-related situations where regulatory authority creates intractable situations for the affected community is clear.

Levee Construction and Repair

The 1994 Interagency Floodplain Management Review Committee (IFMRC) report on the 1993 Mississippi River flood identified that federal agencies other than USACE were providing funds to communities to build or reconstruct damaged levees following flood events (e.g., the HUD and the Department of Commerce's Economic Development Administration). The IFMRC report recommended that having multiple agencies fund the construction or reconstruction of such levees created challenges later. If not built to certification standards, the levee would not be accredited or possibly ineligible for participation in the Flood Control and Coastal Emergency Act (P.L. 84-99) program despite the fact that it was built by a federal agency. As a solution, the IFMRC report recommended that USACE be designated as the federal agency responsible for the oversight and construction of levees by federal agencies. The consolidation of construction responsibility for levees that receive federal funding promotes consistent construction outcomes that align with USACE and FEMA standards.

A National Levee Inventory

Currently, the location and condition of levees in the United States is unclear—where they are, what they are protecting (assets and lives), and their condition (Chapter 6). This represents a significant gap in addressing the challenge of flood risk management behind levees: How can the problem be addressed if the extent and location of the problem is unclear?

Although the long-term plan of both USACE and FEMA is to develop a single database for covering the levees that operate under the oversight of the two agencies, this consolidation effort, which was initially recommended in the 2006 National Levee Policy Study, is proceeding slowly because of resource constraints (FEMA, 2006). The National Levee Database (NLD) operated by USACE, which will be the eventual single database, is limited in information concerning NFIP and other non-USACE levee systems. Eventually USACE may be able to bring into the database information about state, local, and individually operated levees that do not fall into either of the federal agency programs. However, at present, there is a significant need for completion of the consolidation of the federal levee inventories so that levees in the two federal programs can be identified and overlaps eliminated. The maintenance today of two databases is inefficient and creates a potential for disinformation at times when information about levees is badly needed.

The transfer of information about levees that are not part of the NFIP but are a part of FEMA's MLI database into USACE's National Levee Database is needed. (As indicated in Chapter 7, in areas covered by its Flood Insurance Rate Maps (FIRMs), FEMA has identified approximately 29,800 miles of levees, only 5,100 of which are part of the NFIP.) These non-NFIP levees, along with several thousand other levees may eventually be entered into the database. The addition of these nonfederal program levees will require the close cooperation of the states and local entities.

Risk Communication

In 2006 the Association of State Floodplain Managers (ASFPM) and the National Association of Flood and Stormwater Management Agencies (NAFSMA) jointly sponsored a National Flood Risk Summit, convening federal, state, and local flood experts and selected members of the public at Wye Island, Maryland. Flood risk communication was a key topic during discussions. The principal challenge identified regarding risk communication was the multiplicity of uncoordinated messages developed and communicated by federal agencies. The collective opinion of those in the Wye Island meeting was the need for increased coordination of this messaging. Discussion with relevant stakeholders during information-gathering efforts of this committee corroborate this opinion and indicate that the problem continues to exist.

Given the multiple entities that have either regulatory authority or considerable interest in levees, it is no surprise that inconsistent messaging from multiple sources exists and leads to confusion and misunderstanding. Inconsistent messaging is directly related to the lack of collaboration and coordination among entities in communicating risk to local stakeholders. Inconsistent messaging can involve definitions, for example, what defines a levee, what the appropriate definition of risk is, and concepts such as the reality of residual risk behind a levee. Simplicity of messaging is also important in communicating risk (Paveglio et al., 2009; Ballard et al., 2012) and becomes a challenge in the case of, for example, communicating an explanation of the one percent annual chance flood.

Flood inundation and other specialty maps for flood-prone areas are being prepared by multiple agencies to cover the same land area (FEMA, NRCS, FWS, USGS, USACE, and NOAA). Some of the overlap among these maps is a result of research and development activities being conducted by the agencies. In other cases, the overlap results from the necessity of supporting an agency mission. Furthermore, other messaging products prepared by federal agencies addressing the challenge of floodplain management or flood risk reduction are not always coordinated with a consistent message (Chapter 7). There is no one agency that has the responsibility to provide oversight and coordination of federal flood risk mapping products and other communication products. Similarly, there is little to no coordination of the collection of underlying data, such as with Lidar, by these agencies.

The utility of levee-related flood risk communication developed and presented at the federal level is challenged by uncoordinated messaging. **One federal message using consistent terminology, transparent data, and open discussion and decisions about the determination of flood risk is critical to inform the affected communities who, in turn, communicate and manage risk at the local level. FEMA should assume a leadership role in providing direction for research, development, and release of flood risk communication products and maps.** This role might include FEMA constituting a central leadership group or some type of coordinating body.

FEMA should communicate flood risk through a collaborative approach that works with and provides strong support to local communities. This additional emphasis entails local involvement (from individuals, communities, and states) in all phases of the risk communication process. This requires that FEMA interact with preestablished collaborative decision-making networks at the local level. Success lies in providing risk information to the local level where local governments and citizens have the responsibility to drive behavioral changes to reduce risk.

INCREASING AWARNESS OF SHARED FLOOD RISK RESPONSIBILITIES AT THE FEDERAL, STATE, AND LOCAL LEVELS

USACE and FEMA share responsibility for oversight of the conditions of levees that are part of both programs and ensuring that the integrity of the structures means necessary standards to ensure the safety of those behind levees. It has been difficult for FEMA and USACE to develop a comprehensive approach to reducing flood damages across the nation, and in the case of levees, to ensure that levees are given proper oversight by those responsible for them, the levee owners, and operators. However, promising coordination efforts between the two agencies exist, including three joint "task force" initiatives that promote better interagency cooperation and collaboration, including improving the results of levee-related activities in both agencies, and better aligning data collected in USACE levee inspection with requirements for accreditation purposes (FEMA, 2012a,b). This collaboration permits FEMA and USACE to work together to deal with joint challenges such as development of policies for vegetation on levees (USACE, 2009).

In most cases, states play a minimal role in dealing with the flood challenges facing communities in their states. Less than five states have inventories of the levees within their boundaries.[3] Few states have permitting or land-use regulations governing the construction of levees. Even though the National Flood Insurance Act strongly encouraged state participation in the program in support of its goals, most states defer action to FEMA and their communities. State-focused nongovernmental organizations such as the ASFPM and NAFSMA play a critical role in promoting the shared responsibility of the states.

Since flooding does not fall neatly into geographical borders, "communities" cannot be defined solely by jurisdictional boundaries (NRC, 2011). Regional bodies conducting risk communication to achieve common commitment to sustainable mitigation, can play an important role. For example, the cooperative decision-making frameworks built by metropolitan planning organizations[4] have been in place for several decades, and many have built networks of trust and credibility in communities.

The last Unified National Program for Floodplain Management was published by the Federal Interagency Floodplain Management Task Force in 1994. It identified, as did the 1994 Sharing the Challenge report on the 1993 Mississippi River floods (IFMRC, 1994), the need for action by the federal government to bring together all parties involved in development and implementation of national floodplain management programs to better define the responsibilities of each group and to develop a roadmap dealing with these challenges in the future. No action has been taken on recommendations of either effort.

The Federal Interagency Floodplain Management Task Force is a useful vehicle to promote the need for better definition of the shared responsibilities for floodplain management among all the parties.

Individuals and communities also share in the responsibility for taking those steps necessary to manage flood risks. The deaccreditation of thousands of miles of NFIP and USACE levees in the wake of map modernization indicates that community leaders were not monitoring the condition of the structures protecting their communities and not understanding their responsibility for ensuring that these levee systems were meeting accreditation requirements. However, it is reasonable to conclude that the deaccreditation of levees has countered this notion in NFIP communities around the nation. Use of modern risk methodology will provide more accurate information on risks, especially behind levees, information that will be useful at the community level. Communication of the flood risk behind levees can be most effectively accomplished by local governments (Chapter 7), but will require the active engagement of these officials, not only for the actual communication of the risk messages, but also for the implementation of land-use and building code regulations needed to deal with the residual risks that will be described by the use of new risk-based tools. Once properly informed by local governments of the risks they face behind levees, the property owners are responsible for the actions necessary to physically mitigate potential losses and transfer risk, for example, through the use of risk-based insurance.

IN CONCLUSION

This report provides advice on the current treatment of levees within the NFIP (Box 8-2). Key to this advice is the transition of FEMA to a modern, risk-based approach to flood risk analysis. This transition will, no doubt, be challenging. New or revised regulations, procedures, and training systems will need to be established, and clear, consistent messaging to those affected by the new procedures is critical. Legal challenges, as there are today, will most likely arise. The appropriate balance between public and private investment will continually be considered.

Despite these challenges, it is clear that a national move toward floodplain management anchored in a modern risk-based analysis is imperative. This is increasingly recognized in the floodplain management community and

---

[3] This statement is based on information in the Interagency Levee Policy Review Committee report (ILPRC, 2006) and inquiries to states during committee deliberations.

[4] Metro areas with a population of more than 50,000 are required by federal law (23 U.S.C. §134, Metropolitan transportation planning) to establish a metropolitan planning organization (MPO) to represent local, regional, and national interests in the transportation planning process (AMPO, 2009). The nation's 385 MPOs represent areas ranging from 50,000 to 19 million people totaling roughly 84 percent of the population, and with about half administered by a regional council (AMPO, 2009). MPOs are governed by democratically elected officials accountable to local and regional constituencies. These regional leaders and the staffs of the MPOs routinely work with representatives and officials responsible for other planning activities including land use, economic development, environmental protection, and more (AMPO, 2009).

BOX 8-2
Improving NFIP Policies and Practices: Conclusions and Recommendations

The committee identified multiple conclusions and recommendations to improve the policies and practices of the treatment of levees within FEMA's NFIP, spanning risk analysis, flood insurance, risk reduction, and risk communication. Advice falls within the central theme of the report—the need for a national move to flood risk management. Conclusions and recommendations are found throughout the report and consolidated here.

Recommendations to improve NFIP's flood risk analysis, the analytical piece of flood risk management, including advice on FEMA's Levee Analysis and Mapping Procedure, include:

- **Recommendation 1: The NFIP should move to a modern risk analysis that makes use of modern methods and computational mapping capacity to produce state-of-the-art risk estimates for all areas that are vulnerable to flooding.**
- **Recommendation 2: FEMA should move directly to a modern risk-based analysis for dealing with areas behind levees and not implement the Levee Analysis and Mapping Procedure (LAMP).**

Should the decision be made not to implement LAMP, Chapter 4 contains recommendations to pursue interim steps pending development and completion of a modern risk-based analysis.

Conclusions and recommendations related to flood insurance including flood insurance pricing behind levees and the mandatory flood insurance purchase requirement, both in the context of a modern-risk based analysis, include:

- **Conclusion 1: The method used to calculate the average historical loss year ensures that in the long run there will be inadequate premiums collected to cover costs in significant flood years.**
- **Conclusion 2: The NFIP is constructed using an actuarially sound formulaic approach for the full-risk classes of policies, but is financially unsound in the aggregate because of constraints (i.e., legislative mandates) that go beyond actuarial considerations.**
- **Recommendation 3: Rate setting for properties behind levees, accredited or not accredited, should be improved by using the modern risk analysis method employing advances in hydrology, meteorology, and geotechnical engineering to more precisely calculate the probability of water inundation levels and the associated damage estimates throughout the area behind the levee in a graduated fashion.**
- **Recommendation 4: To the extent possible to better achieve fiscal soundness, the properties that have been grandfathered into the NFIP or that receive discounts due to statutory considerations should have their pricing moved to the actuarially based prices using the detailed probability-of-inundation estimates and detailed economic damage models.**
- **Conclusion 3: The current policy of mandatory flood insurance purchase appears to be ineffective in achieving widespread purchase of NFIP flood insurance policies.**
- **Conclusion 4: The rate of compliance with the mandatory purchase requirement indicates challenges with lender enforcement and federal oversight of this lending.**
- **Conclusion 5: Mandatory purchase requirements have led many property owners to perceive that if they are not mandated to have insurance, then they are not susceptible to damage from floods.**
- **Conclusion 6: At this time, there is no sound reason to institute mandatory purchase of flood insurance in areas behind accredited levees.**
- **Conclusion 7: A modern risk-based analysis has the potential to impact the purchase of flood insurance, diversify the NFIP's exposure to flood risk, and generate a more fiscally sound program.**

continued

- **Recommendation 5: Once the risk-based approach has been put in place and matures, FEMA should review and study the necessity of the mandatory purchase requirement, behind levees and throughout the Special Flood Hazard Areas.**

Conclusions and recommendations related to implementing flood risk management strategies, exploring all available structural, nonstructural, and risk transfer mechanisms such as insurance, include:

- **Conclusion 8: There is a clear need for a comprehensive, tailored approach to risk management behind levees that (1) is designed and implemented at the local level; (2) involves federal and state agencies, communities, and households; (3) takes into account possible future conditions; and (4) relies on an effective portfolio of structural measures, nonstructural measures, and insurance to reduce the risk to those behind levees.**
- **Recommendation 6: To reduce the flood risk to those behind levees, FEMA should encourage communities to develop and implement multimeasure risk management strategies for areas behind accredited levees.**

Recommendations related to understanding and communicating flood risk behind levees, an integral part of flood risk management, include:

- **Recommendation 7: FEMA and others involved in risk communication at all levels should incorporate contemporary risk communication principles in the development of flood risk communication strategies and implementation efforts.**
- **Recommendation 8: FEMA should support evaluation of the success of risk communication efforts, including at the community level when appropriate, that are informed by appropriate assessment tools such as baseline information and predefined metrics.**

Moving to flood risk management is a shared responsibility at all levels: federal, state, and local government, private citizens, communities, and all other entities dealing with flood risk. Related conclusions and recommendations include:

- **Recommendation 9: FEMA and USACE should jointly develop a common, risk-based approach to levee assessment in a timely manner and apply this approach to all levees assessed by the two agencies. This includes a joint methodology, procedure, and where feasible, the sharing of models and other risk analysis tools.**
- **Conclusion 9: One federal message using consistent terminology, transparent data, and open discussion and decisions about the determination of flood risk is critical to inform the affected communities who, in turn, communicate and manage risk at the local level.**
- **Recommendation 10: FEMA should assume a leadership role in providing direction for research, development, and release of flood risk communication products and maps.**
- **Recommendation 11: FEMA should communicate flood risk through a collaborative approach that works with and provides strong support to local communities.**
- **Conclusion 10: The Federal Interagency Floodplain Management Task Force (FIFMTF) is a useful vehicle to promote the need for better definition of the shared responsibilities for floodplain management among all the parties.**

is well illustrated by USACE. Indeed, FEMA can learn from the experience of USACE in moving to a modern risk-based analytical approach. Specifically, USACE has converted its hydrologic and hydraulic analyses to a risk-based approach and instituted a nationwide risk management program to provide risk-based evaluations of levees as part of USACE's periodic inspection programs. In the NFIP, some analyses will be carried out by private engineering firms, many of which are already involved with USACE implementation of these new approaches.

The intersection between FEMA and USACE is only one piece of the necessary collaborative and coordinated approach to levee-related flood risk. Currently, these efforts at all levels are not as robust as they could be. Increased coordination and cooperation among and between governmental agencies, state and local entities as well as the public at large, would result in increased effectiveness of the NFIP, improved efficiency within the NFIP, and improve the nation's approach to floodplain management and treatment of levees. Promising developments do exist, as discussed above, but, more progress is necessary. A coordinated, national approach that *shares responsibility* across federal, state, and local entities is necessary to appropriately manage the nation's flood risk behind levees.

REFERENCES

AMPO (Association of Metropolitan Planning Organizations). 2009. Issue Brief. Washington, DC: AMPO.

Ballard, H. L., E. Evans, V. E. Sturtevant, and P. Jakes. 2012. The Evolution of Smokey Bear: Environmental Education About Wildfire for Youth. Journal of Environmental Education 43(4): 227-240.

FEMA (Federal Emergency Management Agency). 2006. The National Levee Challenge: Levees and the FEMA Flood Map Modernization Initiative. Washington, DC: FEMA.

FEMA. 2012a. Continued Coordination between FEMA and USACE, Fiscal Year 2012. Washington, DC: FEMA.

FEMA. 2012b. FEMA-USACE Levee Task Force. Draft Briefing Sheet, September 25, 2012. Washington, DC: FEMA.

IFMRC (Interagency Floodplain Management Review Committee). 1994. Sharing the Challenge: Floodplain Management into the 21st Century. Report of the Interagency Floodplain Administration Floodplain Management Task Force. Washington, DC: U.S. Government Printing Office.

ILPRC (Interagency Levee Policy Review Committee). 2006. Sharing the Challenge: Floodplain Management into the 21st Century. Washington, DC: ILPRC.

NRC (National Research Council). 2011. Building Community Disaster Resilience Through Private-Public Collaboration. Washington, DC: The National Academies Press.

Paveglio, T., M. S. Carroll, J. D. Absher, and T. Norton. 2009. Just Blowing Smoke? Residents' Social Construction of Communication About Wildfire. Environmental Communication 3(1): 76-94.

USACE (U.S. Army Corps of Engineers). 2009. Guidelines for Landscape Planting and Vegetation Management at Levees, Floodwalls, Embankment Dams, and Appurtenant Structures. ETL 1110-2-571. Washington, DC: USACE.

USACE. 2012. Talking Points on the Dallas Floodway System—DF & DFE, Southwest Division. Washington, DC: USACE.

Appendixes

Appendix A

Code of Federal Regulations Title 44: Emergency Management and Assistance, Section 65.10 Mapping of areas protected by levee systems or "44 CFR §65.10"[1]

(a) *General.* For purposes of the NFIP, FEMA will only recognize in its flood hazard and risk mapping effort those levee systems that meet, and continue to meet, minimum design, operation, and maintenance standards that are consistent with the level of protection sought through the comprehensive flood plain management criteria established by §60.3 of this subchapter. Accordingly, this section describes the types of information FEMA needs to recognize, on NFIP maps, that a levee system provides protection from the base flood. This information must be supplied to FEMA by the community or other party seeking recognition of such a levee system at the time a flood risk study or restudy is conducted, when a map revision under the provisions of part 65 of this subchapter is sought based on a levee system, and upon request by the Federal Insurance Administrator during the review of previously recognized structures. The FEMA review will be for the sole purpose of establishing appropriate risk zone determinations for NFIP maps and shall not constitute a determination by FEMA as to how a structure or system will perform in a flood event.

(b) *Design criteria.* For levees to be recognized by FEMA, evidence that adequate design and operation and maintenance systems are in place to provide reasonable assurance that protection from the base flood exists must be provided. The following requirements must be met:

(1) *Freeboard.* (i) Riverine levees must provide a minimum freeboard of three feet above the water-surface level of the base flood. An additional one foot above the minimum is required within 100 feet in either side of structures (such as bridges) riverward of the levee or wherever the flow is constricted. An additional one-half foot above the minimum at the upstream end of the levee, tapering to not less than the minimum at the downstream end of the levee, is also required.

(ii) Occasionally, exceptions to the minimum riverine freeboard requirement described in paragraph (b)(1)(i) of this section, may be approved. Appropriate engineering analyses demonstrating adequate protection with a lesser freeboard must be submitted to support a request for such an exception. The material presented must evaluate the uncertainty in the estimated base flood elevation profile and include, but not

[1] From the Electronic Code of Federal Regulations, current as of August 2, 2012, available http://ecfr.gpoaccess.gov/cgi/t/text/text-idx?c=ecfr&sid=60d5a4d4887bf9b97092a242096f95a2&rgn=div8&view=text&node=44:1.0.1.2.32.0.17.10&idno=44, accessed August 6, 2012.

necessarily be limited to an assessment of statistical confidence limits of the 100-year discharge; changes in stage-discharge relationships; and the sources, potential, and magnitude of debris, sediment, and ice accumulation. It must be also shown that the levee will remain structurally stable during the base flood when such additional loading considerations are imposed. Under no circumstances will freeboard of less than two feet be accepted.

(iii) For coastal levees, the freeboard must be established at one foot above the height of the one percent wave or the maximum wave runup (whichever is greater) associated with the 100-year stillwater surge elevation at the site.

(iv) Occasionally, exceptions to the minimum coastal levee freeboard requirement described in paragraph (b)(1)(iii) of this section, may be approved. Appropriate engineering analyses demonstrating adequate protection with a lesser freeboard must be submitted to support a request for such an exception. The material presented must evaluate the uncertainty in the estimated base flood loading conditions. Particular emphasis must be placed on the effects of wave attack and overtopping on the stability of the levee. Under no circumstances, however, will a freeboard of less than two feet above the 100-year stillwater surge elevation be accepted.

(2) *Closures.* All openings must be provided with closure devices that are structural parts of the system during operation and design according to sound engineering practice.

(3) *Embankment protection.* Engineering analyses must be submitted that demonstrate that no appreciable erosion of the levee embankment can be expected during the base flood, as a result of either currents or waves, and that anticipated erosion will not result in failure of the levee embankment or foundation directly or indirectly through reduction of the seepage path and subsequent instability. The factors to be addressed in such analyses include, but are not limited to: Expected flow velocities (especially in constricted areas); expected wind and wave action; ice loading; impact of debris; slope protection techniques; duration of flooding at various stages and velocities; embankment and foundation materials; levee alignment, bends, and transitions; and levee side slopes.

(4) *Embankment and foundation stability.* Engineering analyses that evaluate levee embankment stability must be submitted. The analyses provided shall evaluate expected seepage during loading conditions associated with the base flood and shall demonstrate that seepage into or through the levee foundation and embankment will not jeopardize embankment or foundation stability. An alternative analysis demonstrating that the levee is designed and constructed for stability against loading conditions for Case IV as defined in the U.S. Army Corps of Engineers (COE) manual, "Design and Construction of Levees" (EM 1110–2–1913, Chapter 6, Section II), may be used. The factors that shall be addressed in the analyses include: Depth of flooding, duration of flooding, embankment geometry and length of seepage path at critical locations, embankment and foundation materials, embankment compaction, penetrations, other design factors affecting seepage (such as drainage layers), and other design factors affecting embankment and foundation stability (such as berms).

(5) *Settlement.* Engineering analyses must be submitted that assess the potential and magnitude of future losses of freeboard as a result of levee settlement and demonstrate that freeboard will be maintained within the minimum standards set forth in paragraph (b)(1) of this section. This analysis must address embankment loads, compressibility of embankment soils, compressibility of foundation soils, age of the levee system, and construction compaction methods. In addition, detailed settlement analysis using procedures such as those described in the COE manual, "Soil Mechanics Design—Settlement Analysis" (EM 1100–2–1904) must be submitted.

(6) *Interior drainage.* An analysis must be submitted that identifies the source(s) of such flooding, the extent of the flooded area, and, if the average depth is greater than one foot, the water-surface elevation(s) of the base

flood. This analysis must be based on the joint probability of interior and exterior flooding and the capacity of facilities (such as drainage lines and pumps) for evacuating interior floodwaters.

(7) *Other design criteria.* In unique situations, such as those where the levee system has relatively high vulnerability, FEMA may require that other design criteria and analyses be submitted to show that the levees provide adequate protection. In such situations, sound engineering practice will be the standard on which FEMA will base its determinations. FEMA will also provide the rationale for requiring this additional information.

(c) *Operation plans and criteria.* For a levee system to be recognized, the operational criteria must be as described below. All closure devices or mechanical systems for internal drainage, whether manual or automatic, must be operated in accordance with an officially adopted operation manual, a copy of which must be provided to FEMA by the operator when levee or drainage system recognition is being sought or when the manual for a previously recognized system is revised in any manner. All operations must be under the jurisdiction of a Federal or State agency, an agency created by Federal or State law, or an agency of a community participating in the NFIP.

(1) *Closures.* Operation plans for closures must include the following:

(i) Documentation of the flood warning system, under the jurisdiction of Federal, State, or community officials, that will be used to trigger emergency operation activities and demonstration that sufficient flood warning time exists for the completed operation of all closure structures, including necessary sealing, before floodwaters reach the base of the closure.

(ii) A formal plan of operation including specific actions and assignments of responsibility by individual name or title.

(iii) Provisions for periodic operation, at not less than one-year intervals, of the closure structure for testing and training purposes.

(2) *Interior drainage systems.* Interior drainage systems associated with levee systems usually include storage areas, gravity outlets, pumping stations, or a combination thereof. These drainage systems will be recognized by FEMA on NFIP maps for flood protection purposes only if the following minimum criteria are included in the operation plan:

(i) Documentation of the flood warning system, under the jurisdiction of Federal, State, or community officials, that will be used to trigger emergency operation activities and demonstration that sufficient flood warning time exists to permit activatizon of mechanized portions of the drainage system.

(ii) A formal plan of operation including specific actions and assignments of responsibility by individual name or title.

(iii) Provision for manual backup for the activation of automatic systems.

(iv) Provisions for periodic inspection of interior drainage systems and periodic operation of any mechanized portions for testing and training purposes. No more than one year shall elapse between either the inspections or the operations.

(3) *Other operation plans and criteria.* Other operating plans and criteria may be required by FEMA to ensure that adequate protection is provided in specific situations. In such cases, sound emergency management practice will be the standard upon which FEMA determinations will be based.

(d) *Maintenance plans and criteria.* For levee systems to be recognized as providing protection from the base flood, the maintenance criteria must be as described herein. Levee systems must be maintained in accordance with an officially adopted maintenance plan, and a copy of this plan must be provided to FEMA by the owner of the levee system when recognition is being sought or when the plan for a previously recognized system is revised in any manner. All maintenance activities must be under the jurisdiction of a Federal or State agency, an agency created by Federal or State law, or an agency of a community participating in the NFIP that must assume ultimate responsibility for maintenance. This plan must document the formal procedure that ensures that the stability, height, and overall integrity of the levee and its associated structures and systems are maintained. At a minimum, maintenance plans shall specify the maintenance activities to be performed, the frequency of their performance, and the person by name or title responsible for their performance.

(e) *Certification requirements.* Data submitted to support that a given levee system complies with the structural requirements set forth in paragraphs (b)(1) through (7) of this section must be certified by a registered professional engineer. Also, certified as-built plans of the levee must be submitted. Certifications are subject to the definition given at §65.2 of this subchapter. In lieu of these structural requirements, a Federal agency with responsibility for levee design may certify that the levee has been adequately designed and constructed to provide protection against the base flood.

[51 FR 30316, Aug. 25, 1986]

Appendix B

Biographical Information, Committee on Levees and the National Flood Insurance Program: Improving Policies and Practices

Gerald E. Galloway, Jr. (NAE), *Chair*, is the Glenn L. Martin Institute Professor of Engineering in the Department of Civil and Environmental Engineering, and an affiliate professor of public policy, at the University of Maryland. Dr. Galloway is a civil engineer, geographer, and public administrator whose research and interests focus on U.S. national water policy, in general, and national floodplain management policy, in particular. Dr. Galloway had a 38-year military career that included positions of commander of the U.S. Army Corps of Engineers District in Vicksburg, Mississippi, and professor and founding head of the Department of Geography and Environmental Engineering at the U.S. Military Academy at West Point. Dr. Galloway was promoted to brigadier general in 1990 and retired from active duty in 1995. He has served on several national and state committees that have reviewed levee and dam safety issues and concerns, including some panels supported by the Federal Emergency Management Agency. He received his B.S. degree (civil engineering) from the U.S. Military Academy, his M.S.E. degree (civil engineering) from Princeton University, and his Ph.D. degree (geography) from the University of North Carolina.

Patrick L. Brockett is director of the Risk Management and Insurance Program and holds the Gus S. Wortham Memorial Chair in Risk Management and Insurance at the University of Texas. Dr. Brockett teaches courses and conducts research in a variety of fields including risk management and insurance; financial risk; actuarial science; decision analysis; management science/operations management and research; statistical analysis and business applications, and information theory. Dr. Brockett is the winner of the 2011 "Excellence in Teaching Award" given by the American Risk and Insurance Association. He was elected as a fellow of the Institute for Risk Management (2008), and is a fellow of the American Association for the Advancement of Science, the American Statistical Association, and the Institute of Mathematical Statistics. He also is an elected member of the International Statistical Institute. In 2006, he received the American Risk and Insurance Association Outstanding Achievement Award for furthering the science of risk management through the promotion of education, research, and communication. He served for 9 years as editor of *The Journal of Risk and Insurance*, the premier academic journal in risk management and insurance. Dr. Brockett was a member of the National Research Council Board on Mathematical Sciences and their Applications from 2005 to 2008. Dr. Brockett received his B.S. degree in mathematics from the California State University in Long Beach, and his M.S. and Ph.D. degrees in mathematics from the University of California, Irvine.

Susan L. Cutter is a Carolina Distinguished Professor of Geography at the University of South Carolina where she directs the Hazards and Vulnerability Research Institute. Her research interests are in disaster vulnerability/resilience science—what makes people and the places where they live vulnerable to extreme events and how vulnerability and resilience are measured, monitored, and assessed. She led a Hurricane Katrina post-event field team to examine the geographic extent of storm surge inundation along the Mississippi and Alabama coastline and its relationship to the social vulnerability of communities. She also provided testimony to Congress on hazards and vulnerability and was a member of the Corps of Engineers Interagency Performance Evaluation Task Force team evaluating the social impacts of the New Orleans and Southeast Louisiana Hurricane Protection System in response to Hurricane Katrina. Dr. Cutter serves on many national advisory boards and committees, including those of National Research Council, the American Association for the Advancement of Science, the National Science Foundation, and the Heinz III Center for Science, Economics, and the Environment. She was a founding member of the Executive Committee of the National Consortium for the Study of Terrorism and Responses to Terrorism (START, a Department of Homeland Security Center of Excellence focused on social and behavioral sciences). Dr. Cutter also holds the MunichRe Foundation Chair (2009-2012) on Social Vulnerability through the United Nations University–Institute for Environment and Human Security, in Bonn, Germany Dr. Cutter received her B.S. degree from the California State University and her M.S. and Ph.D. degrees from the University of Chicago.

David T. Ford is president of David Ford Consulting Engineers in Sacramento, California, a firm that specializes in flood and floodplain management. He is an internationally recognized expert in hydrologic engineering, flood risk analysis, and water resources planning and management. He has 40 years of project experience, including 22 as president of David Ford Consulting Engineers, Inc., and 12 as a senior hydraulic engineer at the U.S. Army Corps of Engineers' Hydrologic Engineering Center. He has served as a consultant to federal, state, and local water management agencies and to engineering firms worldwide. Dr. Ford has written and/or revised, in whole or in part, much Corps of Engineers guidance, including the Engineer Manual (EM) on risk-based analysis for flood risk management studies, the EM on hydrologic engineering requirements for flood risk studies, and the EM on hydrologic analysis of interior areas. He also has prior National Research Council committee experience, most recently with the Water Science and Technology Board Committee on Hydrologic Science. Ford has served on the faculty of the University of California and California State University. He received his B.S., M.S., and Ph.D. degrees in civil engineering from the University of Texas. Dr. Ford is a registered professional engineer in 15 states of the United States.

Clive Q. Goodwin is assistant vice president and manager, natural hazard peril underwriting, for FM Global Insurance Company, located in Johnston, Rhode Island. In this position, Goodwin manages worldwide underwriting of wind, flood, and collapse perils. This involves developing and maintaining strategies to capitalize on FM Global's engineering knowledge of these hazards to benefit clients in terms of risk improvement and insurance terms and conditions. Recently, Goodwin has been the leader of FM Global's efforts to collaborate with the U.S. Army Corps of Engineers, Federal Emergency Management Agency, and other agencies to highlight the concerns regarding the aging inventory of levees while supporting their efforts to change U.S. national policy concerning the levee risk. Goodwin holds a B.S. in mechanical engineering and metallurgy from the University of Manchester, Manchester, UK, and received a certified diploma in accounting and finance. Additionally, he is a chartered engineer and a member of the Institution of Mechanical Engineers and has served on the Industry Leaders Council of the American Society of Civil Engineers.

Karin M. Jacoby is president of Spica Consulting, LLC, a firm she founded in 2009 focused on reducing flood risk through strategic solutions that incorporate economic, human, and environmental considerations. She has more than 25 years of experience in water resources management, much of that in planning, design, and implementation of drainage, stormwater, and flood damage reduction projects and policies. Prior to that she established the Waterways Division at the City of Kansas City, Missouri; as an assistant city engineer/division manager, she provided oversight for a $600 million waterways program. Jacoby has served as the executive director of the Missouri and Associated Rivers Coalition (MOARC) since 2001, a long-standing regional nonprofit focused on beneficial management of water and related land resources. She is an active member of the National Waterways Conference, serving on its Legislative and Policy Committee. In 2008, the Assistant Secretary of the Army (Civil Works) appointed her to the congressionally authorized National Committee on Levee Safety. Jacoby received her B.S. degree in civil engineering from the University of Missouri at Rolla, her M.P.A. and J.D. degrees from the University of Missouri at Kansas City, and is a registered engineer and licensed attorney.

David I. Maurstad is a director and senior vice president with Optimal Solutions and Technologies, Inc., a provider of management consulting, integrated information technology, engineering services, and business process outsourcing in Washington, DC. Mr. Maurstad previously served as director of water policy and planning for a nationally recognized engineering firm specializing in flood mapping and floodplain management. He has more than 30 years of leadership experience with both the private insurance industry and federal, state, and local governments. Mr. Maurstad served as assistant administrator for the U.S. Department of Homeland Security Federal Emergency Management Agency (FEMA). In June 2004, he was appointed by President George W. Bush to provide leadership for some of the nation's leading multihazard risk reduction programs. In this role, he was the federal insurance administrator charged with the overall management of FEMA's National Flood Insurance Program. He previously served as the director of FEMA Region VIII from 2001 to 2004 coordinating federal, state, tribal, and local management of emergencies through planning, preparedness, mitigation, response, and recovery. A native Nebraskan, he served as Mayor of Beatrice, State Senator, and Lieutenant Governor. He received his B.S. degree in business administration and his M.B.A. degree from the University of Nebraska.

Martin W. McCann is president of Jack R. Benjamin and Associates, Inc., in Menlo Park, California. Dr. McCann also serves as a consulting professor of civil and environmental engineering at Stanford University. At Stanford, he is the former chairman of the National Performance of Dams Program, which he founded and which has created a national network to report dam safety incidents and archive this information for wide use by the geotechnical and seismic engineering communities. Dr. McCann's professional background and research have focused on probabilistic hazards analysis, including hydrologic events, risk assessment, reliability and uncertainty analysis, and systems analysis. Dr. McCann has served as a consultant to several government and private-sector groups in the United States and abroad. He currently also is serving as a member of the National Research Council Committee on Integrating Dam and Levee Safety and Community Resilience. Dr. McCann received his B.S. degree from Villanova University and his M.S. and Ph.D. degrees from Stanford University.

Andre D. McDonald is founder and president of Fort Bend Flood Management Association in Fort Bend County, Texas, a group of 17 flood management agencies and the consultants that service that industry. He has served as an appointed board member of Fort Bend County Levee Improvement District since 2000, and been president of that board since 2004. He has an extensive background in construction, primarily with management and operation of heavy civil construction engineering companies. His more than 30 years of experience include the day-to-day management with full profit and loss responsibilities of a heavy civil contracting company. The company was involved with direct construction and commissioning of underground utilities distribution and collection systems, earthwork, paving, and related infrastructure and other intraurban development. He has been part of the management and direct field operations of airport construction, mass earthworks, industrial plant construction, wastewater treatment plant construction, modernization of oil refineries and grassroots LNG facilities. Mr. McDonald studied engineering and business administration at Mississippi State University.

Earthea A. Nance is an assistant professor in the Department of Planning and Urban Studies at the University of New Orleans. Dr. Nance has over 18 years of experience in the areas of environmental planning and management, hazard mitigation, sustainable urban development, environmental remediation, water, wastewater, hazardous waste, and alternative energy. After Hurricane Katrina, Dr. Nance served as a Ford Foundation loaned-executive to the City of New Orleans, where she directed the city's hazard mitigation, environmental, and alternative energy divisions and authored the city's sustainability strategy. She is a consultant to the RAND Corporation on policy adaptation to climate change in New Orleans and advises the U.S. Environmental Protection Agency Science Advisory Board as a member of its environmental engineering committee. She also serves as a consultant to local environmental groups trying to understand the impacts of the Deepwater Horizon oil spill. Her research has examined the impacts of disasters on social and ecological diversity, the development of executive education in resilience and risk management, community-based environmental monitoring in Gulf Coast communities, and participatory water and sanitation systems in developing countries. Dr. Nance received her B.S. and M.S. degrees in civil and environmental engineering from the University of California at Davis and her Ph.D. degree in civil and environmental engineering from Stanford University.

Kenneth W. Potter is a professor in the Department of Civil and Environmental Engineering at the University of Wisconsin. Dr. Potter's areas of research include estimation of hydrologic risk, especially flood risk; adaptation of hydrologic design to climate change; assessment and mitigation of human impacts on aquatic systems; and restoration of aquatic systems. Dr. Potter is a fellow of the American Association for the Advancement of Science , a fellow of the American Geophysical Union, and a Woodrow Wilson fellow. He has extensive National Research Council (NRC) committee experience, as well. He is a former member of the Water Science and Technology Board, chaired the NRC Committee on American River Flood Frequencies and the NRC Committee on Integrated Observations for Hydrologic and Related Sciences, served as vice chair of the NRC Committee on Flood Control Alternatives in the American River Basin and, most recently, served on the NRC Committee on the New Orleans Regional Hurricane Protection Projects. Dr. Potter received his B.S. degree in geology from Louisiana State University and his Ph.D. degree in geography and environmental engineering from Johns Hopkins University.

J. David Rogers holds the Karl F. Hasselmann Chair in Geological Engineering at the Missouri University of Science & Technology in Rolla, Missouri. Dr. Rogers has extensive experience in evaluating the stability of natural slopes, embankments, stream channels, highways, and hydraulic structures. He has served as principal investigator for research funded by the National Science Foundation (NSF), U.S. Geological Survey (USGS), National Geospatial Intelligence Agency, Federal Highway Administration, Department of Defense, and several state departments of transportation. Dr. Rogers has served on numerous panels, including the Mississippi Delta Science & Engineering Special Team, the Coastal Louisiana Recovery Panel, the NSF Independent Levee Investigation Team and USGS Investigation Teams evaluating the impacts of Hurricanes Katrina and Rita, the NSF team evaluating the 2008 and 2011 Mississippi River floods, and the Resilient and Sustainable Infrastructure Networks team funded by NSF to make a 5-year examination of the California Bay Delta flood protection systems. Recently, he has also been working on development of computerized algorithms for regional landslide hazard mapping using geographic information systems in the United States, Africa, Asia, and the Hundu-Kush regions. Dr. Rogers received his B.S. degree in geology from Cal Poly Pomona, and his M.S. degree in civil engineering and his Ph.D. in geological and geotechnical engineering from the University of California, Berkeley. He served on the Berkeley faculty for 7 years prior to accepting his current position in 2001.

STAFF

Laura J. Helsabeck is a senior staff officer with the National Research Council's Water Science and Technology Board. Her interests include the use of scientific information to enhance water policy and management decisions pertaining to water quality and quantity. Since joining the National Research Council, she has served as the study director for a variety of studies including Sustainable Management in California's Bay Delta, Challenges and Opportunities in the Hydrologic Sciences, and Global Change and Extreme Hydrology: Testing Conventional Wisdom. Dr. Helsabeck received her B.A. from Clemson University, her M.S. from Vanderbilt University, and Ph.D. from The Ohio State University in environmental chemistry. Her dissertation work, Ibuprofen photolysis: Reaction kinetics, chemical mechanism, and byproduct analysis, was awarded the Ellen C. Gonter Environmental Chemistry Award by the American Chemical Society.

Michael J. Stoever is a research associate with the National Research Council's Water Science and Technology Board. He has worked on a number of studies including Desalination: A National Perspective, the Water Implications of Biofuels Production in the United States, and the Louisiana Coastal Protection and Restoration Program. He has also worked on NRC studies on the implementation progress of the Comprehensive Everglades Restoration Plan, the effect of water withdrawals on the St. Johns River, and achieving nutrient and sediment reduction goals in the Chesapeake Bay. Mr. Stoever received his B.A. in political science from the Richard Stockton College of New Jersey in Pomona.

Appendix C

Glossary of Terms

100-year flood: See one percent annual chance flood.

Accreditation and accredited levee system: A levee system for which evidence has been provided to the Federal Emergency Management Agency (FEMA) that demonstrates adequate design (i.e., certification), operation, and maintenance systems are in place to provide reasonable assurance that protection from the one percent annual chance flood exists. Design criteria, operation plans and criteria, and maintenance plans and criteria are discussed in the Title 44, Section 65.10, of the *Code of Federal Regulations* (44 CFR §65.10). See Appendix G.

Aleatory uncertainty: A type of irreducible uncertainty attributed to the inherent randomness of events in nature.

Base flood: See one percent annual chance flood.

Breach: Formation of a gap in the levee system through which water may flow uncontrolled onto the adjacent floodplain. A breach in the levee system may occur prior to or subsequent to overtopping.

Certification and certified levee system: A technical evaluation by a certificated professional engineer or federal agency provided to FEMA that demonstrates that the levee system meets the requirements of 44 CFR §65.10 to reduce risk from at least the one percent annual chance flood.

Digital Flood Insurance Rate Map (DFIRM): A Flood Insurance Rate Map that has been made available digitally.

Efficacy: The capacity or power for producing a desired outcome or effect.

Epistemic uncertainty: A type of uncertainty attributed to the lack of understanding about a physical process or system that must be modeled, referred to as knowledge-based uncertainty.

Executive Order 11988: Codified in 1977, the order requires federal agencies to "avoid to the extent possible the long and short term adverse impacts associated with the occupancy and modification of floodplains and to avoid direct or indirect support of floodplain developments wherever there is a practicable alternative." It required each agency to "provide leadership and take action to reduce the risk of flood loss, to minimize the impacts of floods on human safety, health, and welfare, and to restore and preserve the natural and beneficial values served by floodplains in carrying out its responsibilities."

FEMA Procedure Memorandum 34, Interim Guidance for Studies Including Levees: Provides FEMA staff, contactors, and mapping partners with guidance for the evaluation and mapping of levees and levee-affected areas as part of the FEMA Flood Map Modernization effort. See Appendix G.

FEMA Procedure Memorandum 43, Guidelines for Identifying Provisionally Accredited Levees (PALs): Provides a 2-year period for affected levee owners to complete those actions necessary to retain their accreditation. See Appendix G.

FEMA Procedure Memorandum 53, Guidelines for Notification and Mapping of Expiring Provisionally Accredited Levee (PAL) Designation: Provides guidance to FEMA field units regarding the steps that are to be taken if the required improvements to PALs have not been accomplished in the requisite 24-month period. See Appendix G.

Flood Hazard Boundary Map (FHBM): As defined by 44 CFR §59.1: "An official map of a community, issued by the Federal Insurance Administrator, where the boundaries of the flood, mudslide (i.e., mudflow) related erosion areas having special hazards have been designated as Zones A, M, and/or E."

Flood Insurance Rate Map (FIRM): As defined by 44 CFR §59.1: "An official map of a community, on which the Federal Insurance Administrator has delineated both the special hazard areas and the risk premium zones applicable to the community."

Flood insurance study (FIS): As defined by 44 CFR §59.1, (denoted as a flood elevation study therein): "an examination, evaluation, and determination of flood hazards and, if appropriate, corresponding water surface elevations, or an examination, evaluation and determination of mudslide (i.e., mudflow) and/or flood related erosion hazards."

Flood or flooding: As defined by 44 CFR §59.1, flooding is:

(a) A general and temporary condition of partial or complete inundation of normally dry land areas from:

(1) The overflow of inland or tidal waters.

(2) The unusual and rapid accumulation or runoff of surface waters from any source.

(3) Mudslides (i.e., mudflows) which are proximately caused by flooding as defined in paragraph (a)(2) of this definition and are akin to a river of liquid and flowing mud on the surfaces of normally dry land areas, as when earth is carried by a current of water and deposited along the path of the current.

(b) The collapse or subsidence of land along the shore of a lake or other body of water as a result of erosion or undermining caused by waves or currents of water exceeding anticipated cyclical levels or suddenly caused by an unusually high water level in a natural body of water, accompanied by a severe storm, or by an unanticipated force of nature, such as flash flood or an abnormal tidal surge, or by some similarly unusual and unforeseeable event which results in flooding as defined in paragraph (a)(1) of this definition."

Flood risk analysis: A subset of a risk analysis, a flood risk analysis is an analytical process that provides information about or quantifies probabilities and consequences of a flood event.[1]

Flood risk management: Comprehensive efforts to *continuously carry out analyses, assessments, and related activities to reduce flood risk.*

Flood risk reduction: As defined by FEMA,[2] "The goal of flood risk reduction is to reduce the risk to life and property, which includes existing structures and future construction, in the pre- and post-disaster environments. This is achieved through regulations, local ordinances, land use and building practices, and mitigation projects that reduce or eliminate long-term risk from flood hazards and their effects."

Flood zone: A geographic area defined by FEMA according to risk and designated by a community's FIRM.

Floodplain: As defined in 44 CFR §59.1 (also denoted as flood-prone area therein): "Any land area susceptible to being inundated by water from any source."

---

[1] Modified from NRC (2010)

[2] See http://www.fema.gov/what-mitigation#4.

Freeboard: As defined in 44 CFR §59.1, freeboard means "a factor of safety usually expressed in feet above a flood level for purposes of floodplain management. 'Freeboard' tends to compensate for the many unknown factors that could contribute to flood heights greater than the height calculated for a selected size flood and floodway conditions, such as wave action, bridge openings, and the hydrological effect of urbanization of the watershed."

Hazard mitigation: see Flood risk mitigation

Interior drainage: The drainage that is required to evacuate rainwater from areas behind levees. During periods when rivers are high on the levee, any accumulating rainfall must be pumped out of the leveed area or the area becomes subject to flooding.

Levee: As defined in 44 CFR §59.1: "A levee is a man-made structure, usually an earthen embankment, designed and constructed to contain, control, or divert the flow of water so as to provide protection from temporary flooding."

Levee Analysis and Mapping Procedure (LAMP): A new approach to provide a more precise assessment of flood risk in areas impacted by levees. Developed and circulated for public comment in early 2012, the LAMP was designed to be implementable without statutory or regulatory changes.

Levee failure: Poor performance of a levee through a variety of occurrences including breach before or after overtopping and malfunction of system components (e.g., gates, pumping plants, or floodwalls).

Levee fragility curves: As defined in Schultz et al. (2010), levee fragility curves are intended to be "functions that describe the conditional probability of system failure over the full range of loads to which that system might be exposed."

Levee reach: As defined in USACE (2010a): "A levee reach is a portion of a levee system (usually a length of a levee) that may be considered as a unit taken for analysis purposes to have approximately uniform representative properties. A levee reach will be the unique entity having properties different than other reaches of the levee system that may be used to evaluate the performance of a portion of the levee system. No maximum length is associated with a reach."

Levee segment: A levee segment is a discrete portion of a levee system that is operated and maintained by a single entity. A levee segment may comprise one or more levee features.

Levee system: As defined in 44 CFR §59.1, a levee system is: "a flood protection system which consists of a levee, or levees, and associated structures, such as closure and drainage devices, which are constructed and operated in accordance with sound engineering practices."

Leveed area: The lands from which floodwater is excluded by the levee system.

Level of protection: Normally seen as the flood recurrence interval against which a specific structure is designed to withstand.

Mitigation (structural and nonstructural): As defined by the U.S. Army Corps of Engineers (USACE, n.d.): "Structural measures such as dams, levees, and floodwalls alter the characteristics of the flood and reduce the probability of flooding in the location of interest. Nonstructural measures alter the impact or consequences of flooding and have little to no impact on the characteristics of the flood."

One percent annual chance flood: A term commonly used to describe a hydrologic event that has in any year a 1 in 100 chance of being equaled or exceeded.

Overtopping: The condition that occurs when floodwaters reach the crest of the levee and exceed its height. Overtopping might or might not be in conjunction with levee failure.

Practices: Methods or techniques used to implement policies.

Policies: Rule or principle used to guide decision making, typically defined in guidance documents.

Residual risk: The risk that remains after considering the mitigating effects of structural, nonstructural, and other risk reduction measures is known as residual risk. Residual risk is always present behind a levee, because no levee is failsafe.

Risk: The potential for adverse effects from the occurrence of a particular hazardous event, which is derived from the probable combination of physical hazards (physical characteristics), the exposure and vulnerabilities of people and property subject to danger or damage from the hazard, and the consequences (impact or damage) caused by the hazard. In insurance nomenclature, risk is the term used to designate the loss consequence of the realization of the uncertain peril, and it may be financial or nonfinancial in nature (e.g., reputational loss). The causes of loss are called a peril, and a condition or situation that increases either the likelihood or severity of the peril occurring is called a hazard.

Risk analysis: A detailed examination to understand the nature of adverse consequences from a particular event to human life, property, or the environmental; an analytical process that provides information about or quantifies probabilities and consequences of an unwanted event. Often, broad definitions of risk analysis include examination of risk communication, risk perception, and risk management alternatives.[3]

Severe repetitive loss properties: Section 1361A of the National Flood Insurance Act, as amended, 42 U.S.C. 4102a, defines it as a residential property that is covered under an NFIP insurance policy and has at least four claim payments and for which two of these payments in total have exceeded the market value of the property.

Special Flood Hazard Area (SFHA): As defined in 44 CFR §59.1 (denoted as an area of special flood hazard therein), the SFHA is: "the land in the flood plain within a community subject to a 1 percent or greater chance of flooding in any given year."

Standard project flood: The standard project flood is a derived discharge estimate that represents a flood that can be expected from the combination of the most severe meteorologic and hydrologic condition that are considered reasonable characteristic of a region (USACE, 1965).

Tolerable risk guidelines: As defined in USACE (2010b), tolerable risk guidelines "categorize the nature of risks in ways that can assist in assessing their acceptability or nonacceptability, and in prioritizing actions for reducing risks."

Uncertainty: As discussed in NRC (2010), uncertainty is "always present in our ability to predict what might occur in the future, and is present as well in our ability to reconstruct and understand what has happened in the past. This uncertainty arises from missing or incomplete observations and data; imperfect understanding of the physical and behavioral processes that determine the response of natural and built environments and the people within them; and our inability to synthesize data and knowledge into working models able to provide predictions where and when we need them."

Vulnerability: As defined in NRC (2012), vulnerability is "the potential for harm to the community and relates to physical assets (building design and strength), social capital (community structure, trust, and family networks), and political access (ability to get government help and affect policies and decisions). Vulnerability also refers to how sensitive a population may be to a hazard or to disruptions caused by the hazard."

REFERENCES

NRC (National Research Council). 2010. Review of the Department of Homeland Security's Approach to Risk Analysis Washington, DC: The National Academies Press.

NRC. 2012. Disaster Resilience: A National Imperative. Washington, DC: The National Academies Press.

Schultz, M. T., B. P. Gouldby, J. D. Simm, and J. L. Wibowo. 2010. Beyond the Factor of Safety: Developing Fragility Curves to Characterize System Reliability. Washington, DC: U.S. Army Corps of Engineers.

[3] Modified from NRC (2010).

USACE (U.S. Army Corps of Engineers). 1965. Standard Project Flood Determinations, EM 110-2-14-11. Washington, DC: USACE.

USACE. 2010a. USACE Process for the National Flood Insurance Program (NFIP) Levee System Evaluation. Available online at http://publications.usace.army.mil/publications/eng-circulars/EC_1110-2-6067.pdf. Accessed November 27, 2012.

USACE. 2010b. Proceedings of the Workshop Exploration of Tolerable Risk Guidelines for the USACE Levee Safety Program. Report 10-R-9. Washington, DC: USACE.

USACE. n.d. National Nonstructural Flood Proofing Committee. Available online at http://www.usace.army.mil/Missions/CivilWorks/ProjectPlanning/nfpc.aspx. Accessed November 15, 2012.

Appendix D

Contributors to the Report, Levees and the National Flood Insurance Program: Improving Policies and Practices

During the course of this study, numerous persons contributed to the development of this report. Some provided material and talks at the request of the committee, some provided unsolicited material, and others provided advice for the committee to consider. The committee would like to thank all of these persons for their interest and support for this effort.

Ray Alexander, Deputy Chief, Office of Homeland Security, U.S. Army Corps of Engineers (USACE)
Bryan Baker, Technical Project Manager, USACE
Mark Baker, Civil Engineer, U.S. National Parks Service
David N. Bascom, Program Specialist, Federal Emergency Management Agency (FEMA)
Perry Beider, Senior Advisor, Congressional Budget Office
Douglas A. Bellomo, Risk Analysis Division Director, FEMA
Michael Bishop, Contractor Lead, Midterm Levee inventory, Michael Baker Corporation William R. Blanton, Jr., Engineering Management Branch Chief, FEMA
Samuel D. Brody, Professor, Texas A&M University, College Station
Kelly Bronowicz, Program Manager, FEMA
Pernille Buch-Pedersen, Program Manager, Michael Baker Corporation
Michael K. Buckley, Vice President for Strategic Initiatives of Management and Consulting Services, Dewberry
Jae-won Chung, Postdoctoral Scholar, Missouri University of Science and Technology
Roger Churchwell, Director of Engineering, San Joaquin Area Flood Control Agency (SJAFCA)
Tammy L. Conforti, Levee Safety Program Manager, HQ, USACE
Dan Delich. National Levee Issues Alliance
Susan Dell'Osso, Project Director, Cambay Group
Siamak Esfandiary, Program Specialist, FEMA
G. Michael Fitzgerald, Galveston County Engineer, Galveston County, Texas
Steve Fitzgerald, Chief Engineer, Harris County Flood Control District
Greg Frank, Vice President, Costello, Inc.
Christopher Gallion, Superintendent, Velasco Drainage District
Sharmae Gambrel, Compliance Review Examiner, Federal Deposit Insurance Corporation (FDIC)
David Greenwood, Vice President, Michael Baker Corporation

Colonel Christopher Hall, St. Louis District Commander, USACE
Eric Halpin, Special Assistant for Dam and Levee Safety, USACE
Les Harder, Water Resources Technical Advisor, HDR, Inc.
Steve Hazelwood, Director, Dow Chemical Company
Noller Herbert, Conservation Engineering Division Director, Natural Resources Conservation Service (NCRS)
Claudia Hoeft, National Hydraulic Engineer, NCRS
Robert Howell, Engineering Practice Manager, USACE
Paul Huang, Mitigation Directorate Branch Chief, FEMA
Ryan Janda, Program Lead for Flood Mitigation Assistance (FMA), Severe Repetitive Loss (SRL), and Repetitive Flood Claims (RFC) programs, FEMA
Jeffrey Janecek, Assistant Engineer, Fort Bend County Drainage District
R. Donald Janssen, Senior Vice President, Planned Community Developers
Jill A. Jordan, Assistant City Manager, City of Dallas, Texas
Michael Jordan, Civil Engineer, USACE
Jim Keith, Program Analyst, U.S. Bureau of Reclamation
George Kidwell, Chairman, Velasco Drainage District
Sandra K. Knight, Deputy Federal Insurance and Mitigation Administrator for Mitigation, FEMA
Richard W. Krimm, former Associate Director, FEMA
Brian G. Koper, Civil Associate, Michael Baker Jr., Inc.
Christina Kuo, Senior Engineer, California Department of Water Resources
Kent Lang, Director, West Sacramento Area Flood Control Agency
Larry A. Larson, Director Emeritus, Association of State Floodplain Managers
Donna Lee, Technical Specialist, David Ford Consulting Engineers, Inc.
Scott Leimer, Galveston District Levee Safety Program Manager, USACE
Maria Lorenzo-Lee, Senior Engineer, California Department of Water Resources
Guy Lowes, Senior Civil Engineer, FEMA
Jay R. Lund, Director, Center for Watershed Sciences, University of California, Davis
Tim Mantz, Chief, Business Analysis Branch, FEMA
Rod Mayer, Assistant Deputy Director for FloodSAFE, California Department of Water Resources
Dave McMurray, Chairman, Upper Mississippi, Illinois, Missouri Rivers Association (UMIMRA)
Craig McRaney, Levee Safety Program Manager, Vicksburg District, USACE
Erwann Michel-Kerjan, Wharton Risk Center, Wharton School, University of Pennsylvania
Melissa Mullen, Acting Levee Safety Program Manager, Mississippi Valley Division, USACE
Corise Morrison, Executive Director, USAA
Jeffrey F. Mount, Professor, University of California, Davis
Don Natolli, Sacramento County Supervisor, Sacramento County, California
D. Andrew Neal, Actuary, FEMA
Leiann Nelson, Senior Underwriter, Risk Management Agency, U.S. Department of Agriculture (USDA)
Linda Pilgrim, Office of Counsel, FEMA
Ricardo Pineda, Floodplain Risk Management Branch Chief, California Department of Water Resources
Richard G. Reinhardt, Principal, MBK Engineering
Rhonda Robins, Senior Technical Specialist, David Ford Consulting Engineers
Nicholas Pinter, Professor, Southern Illinois University
Robert Quint, Chief of Staff, U.S. Bureau of Reclamation
Kathleen Schafer, Engineer, FEMA
Scott L. Shapiro, Partner, Central Valley Flood Control Association
Jeff Sparrow, Assistant Vice President, Michael Baker Corporation
Dave Stearrett, Chief, Floodplain Management Branch, FEMA
Rune Storesund, Consulting Engineer, Storesund Consulting
Ronald Stork, Policy Director, Friends of the River

Nancy L. Ward, Regional Director, Region IX, FEMA
Timothy Washburn, Director of Planning, Sacramento Area Flood Control Agency
Douglas P. Whipple, Director, Organizational Effectiveness, Dow Chemical Company
Jeffrey Woodward, Senior Lender Compliance Officer, FEMA
Roy E. Wright, Deputy Director, Risk Analysis Division, FEMA

Appendix E

Flood Zone Designations

Moderate to Low Risk Areas

In communities that participate in the NFIP, flood insurance is available to all property owners and renters in these zones:

ZONE	DESCRIPTION
B and X (shaded)	Area of moderate flood hazard, usually the area between the limits of the 100-year and 500-year floods. Are also used to designate base floodplains of lesser hazards, such as areas protected by levees from 100-year flood, or shallow flooding areas with average depths of less than one foot or drainage areas less than 1 square mile.
C and X (unshaded)	Area of minimal flood hazard, usually depicted on FIRMs as above the 500-year flood level.

High Risk Areas

In communities that participate in the NFIP, mandatory flood insurance purchase requirements apply to all of these zones:

ZONE	DESCRIPTION
A	Areas with a 1% annual chance of flooding and a 26% chance of flooding over the life of a 30-year mortgage. Because detailed analyses are not performed for such areas; no depths or base flood elevations are shown within these zones.
AE	The base floodplain where base flood elevations are provided. AE Zones are now used on new format FIRMs instead of A1-A30 Zones.
A1-30	hese are known as numbered A Zones (e.g., A7 or A14). This is the base floodplain where the FIRM shows a BFE (old format).
AH	reas with a 1% annual chance of shallow flooding, usually in the form of a pond, with an average depth ranging from 1 to 3 feet. These areas have a 26% chance of flooding over the life of a 30-year mortgage. Base flood elevations derived from detailed analyses are shown at selected intervals within these zones.
AO	River or stream flood hazard areas, and areas with a 1% or greater chance of shallow flooding each year, usually in the form of sheet flow, with an average depth ranging from 1 to 3 feet. These areas have a 26% chance of flooding over the life of a 30-year mortgage. Average flood depths derived from detailed analyses are shown within these zones.

SOURCE: FEMA's Map Service Center, available online at https://msc.fema.gov/webapp/wcs/stores/servlet/info?storeId=10001&catalogId=10001&langId=-1&content=floodZones&title=FEMA%2520Flood%2520Zone%2520Designations [accessed August 6, 2012].

AR	Areas with a temporarily increased flood risk due to the building or restoration of a flood control system (such as a levee or a dam). Mandatory flood insurance purchase requirements will apply, but rates will not exceed the rates for unnumbered A zones if the structure is built or restored in compliance with Zone AR floodplain management regulations.
A99	Areas with a 1% annual chance of flooding that will be protected by a Federal flood control system where construction has reached specified legal requirements. No depths or base flood elevations are shown within these zones.

High Risk - Coastal Areas

In communities that participate in the NFIP, mandatory flood insurance purchase requirements apply to all of these zones:

ZONE	DESCRIPTION
V	Coastal areas with a 1% or greater chance of flooding and an additional hazard associated with storm waves. These areas have a 26% chance of flooding over the life of a 30-year mortgage. No base flood elevations are shown within these zones.
VE, V1 - 30	Coastal areas with a 1% or greater chance of flooding and an additional hazard associated with storm waves. These areas have a 26% chance of flooding over the life of a 30-year mortgage. Base flood elevations derived from detailed analyses are shown at selected intervals within these zones.

Undetermined Risk Areas

ZONE	DESCRIPTION
D	Areas with possible but undetermined flood hazards. No flood hazard analysis has been conducted. Flood insurance rates are commensurate with the uncertainty of the flood risk.

Appendix F

Levee-Related Observations, Conclusions, and Recommendations from Previous Reports

Throughout the preparation of this report, the committee came across many previously published reports on the topic of levees and the National Flood Insurance Program that were of use. The observations, conclusions, recommendations put forth by selected works are listed below, in chronological order.

A Levee Policy for the National Flood Insurance Program (NRC, 1982)

Existing levees should be recognized for the purpose of reducing insurance rates where they provide protection against 25-year or larger floods and where they meet specified structural design criteria, including requisite freeboard.

New levees should be recognized for the purpose of reducing insurance rates where they provide protection against 100-year or larger floods and where they meet specified structural design criteria including, freeboard. A1l levees on which construction begins after a date to be determined by FEMA should be considered new levees.

All levees (existing and new) to be given credit for reducing flood risk in the NFIP must meet standard minimum engineering criteria with respect to geometric parameters, freeboard, soils and foundations, interior drainage, closure devices, and rights of way.

In its administration of design evaluations and construction conformance inspections, FEMA should first pursue the possibility of using the services of federal and state agencies having water resources experience. Where federal participation cannot be arranged, FEMA must use state and private sector capabilities.

Where responsible federal or state agencies have had continuous maintenance responsibilities on levees they designed and constructed, and will attest to their adequacy under FEMA standards, independent evaluations should not usually be required. Evaluations should be required, however, on levees that were designed and constructed by federal agencies but are currently being operated and maintained by others.

Where a federal or state agency does not evaluate a Levee, Levee evaluations should be done by "levee evaluation contractors" (LECs), private consulting firms designated by FEMA.

While FEMA should designate qualified private consulting firms to be LECs, the firms should work under contract to an applicant interested in having a levee evaluated for recognition by FEMA. All costs of the LECs work should be borne by the applicant.

Since recognition of 25-year levees for the purpose of reducing insurance rates would represent an important change in federal policy, FEMA should publicize the benefits, costs, and procedural details for levee recognition.

FEMA should inventory all levees previously credited as providing projection from the100-year flood, set priorities, and schedule communities for restudy to reevaluate the levees.

FEMA should develop a short and simple checklist that can be used to make a quick assessment of whether a levee meets recommended criteria.

FEMA should confirm its interim policy that does not recognize sand bags on top of levees and other types of human intervention, except for structural closures which are legitimate parts of planned levee systems, as augmenting levee systems design level of protection.

Federal Code 208.10, Title 33, "Local flood protection works; maintenance and operation of facilities should be modified where not well suited, supplemented for interior drainage, and adopted by FEMA as a guideline for operation and maintenance of levee systems.

A specific operation and maintenance plan, tailored to the local needs, must be formally adopted by the levee owner for a levee to be credited and continue to be credited in the NFIP.

The operation and maintenance plan must, provide for periodic inspections. The plan should be completed within a designated time period, meet the requirements of, and be acceptable to FEMA. Each inspection must be by a professional engineer retained by the levee owner and registered in the state in which the levee is located. A written report to the levee owner should be promptly prepared and certified by the professional engineer making the inspection. The report must specifically describe items found deficient and emerging potential future problem areas. Copies of the certified report should be sent, by the certifying professional engineer to the regional FEMA office. Also, a copy should be provided thelocal political entities which have responsibilities to FEMA for the levee-protected area. FEMA should follow up to assure corrections are made within a reasonable length of time. In instances where corrections are not made to critical deficiencies, FEI4A should withdraw recognition.

FEMA should require the elevation of new residential structures and the flood-proofing of other new buildings in all areas protected by levees unable to contain the 100-year flood.

FEMA should require purchase of flood insurance in all areas where the ground is lower than the unconfined 100-year flood level except where protected by a levee built to contain the 500-year flood.

Communities should regulate the placement of critical facilities (not regulated by some higher level of government) in all leveed areas in accordance with the procedures of Executive Order 11988.

Owners, tenants, and lenders occupying areas designated as protected by levees should be notified periodically by responsible local officials that their land in the levee protected area is still subject to flooding in the event of levee failure.

Local officials of any NFIP participating community protected by a levee, regardless of its size and reliability, should prepare and promulgate an action plan for warning and evacuation in the event of levee failure.

FEMA should help make local governments and special districts aware of the possibility of liability for actions or nonactions that aggravate flood hazards.

FEMA should, in appropriate cases, seek to recoup federal flood-related costs (including flood insurance payments, disaster assistance, etc.) from levee owners/operators when such costs arise from improper operation and maintenance of levee and associated interior drainage facilities.

Regardless of the level of protection provided, the levee-protected area should be disaggregated into flood risk zones and an actuarial rate be established for each zone that reflects the degree of protection actually provided by the levees.

Consistent with the design levels of protection recommended in Chapter 3, NFIP policyholders in areas behind existing 1evees that offer more than 25-year protection or new levees affording 100-year protection or greater should pay lower rates that reflect the reduced risk of property damage.

The levee flood risk zones should match the existing flood risk zones established for the regular Flood Insurance programs

FEMA should continue its efforts to establish an actuarial rate basis for the Flood insurance program and, as far as practical, convert its present rate schedule to actuarial rates. The actuarial rates as is done by the private insurance industry would be updated annually to reflect experience, claims paid, and the cost of doing business.

FEMA should contract for the development of a list of key categories concerning tire physical condition of a levee that would be used to evaluate the levee's ability to function effectively and concerning use of those factors to estimate geotechnical risk. An unsatisfactory rating would result in increased flood insurance premiums.

All levees, dikes, or floodwalls should be labeled as such on any new or revised FHBM. Also, the areas protected by levees providing100-year or greater protection should be delineated on the new or revised FHBMs as Zone LP. Areas protected by Levees not meeting this standard would continue to be mapped as Zone A.

The location of all levees, dikes, or floodwalls credited as providing 100-yearprotection or more should be clearly denoted on all future FIRMs.

Areas behind recognized 100-year levees that would be flooded (assuming no levee) by a 100-year flood should be designated as Zone ALP. Areas between the 100-year flood boundary and 500-year flood boundary should be designated as Zone BLP.

The locations of all credited levees, dikes, and floodwalls should be clearly denoted on all Floodway Maps. Areas behind credited levees providing 100-year protection that would be flooded during a 500 year flood should be shown on Floodway Maps.

FEMA should create an AL(No.) Zone that would designate an area protected by a levee with a frequency of protection defined by the (No.). For example, if the levee had an elevation equal to the 4O-year flood (plus required freeboard), then the Zone would be Al(40).

FEMA should not make it a mandatory requirement to include evacuation routes on maps If the information is available when the maps are being prepared and if it makes sense from a community-to-community mapping standpoint, then the evacuation routes could be included on the maps. In other cases, it may be expeditious for communities to develop their own special evacuation route maps.

Interior drainage situations in areas on the landside of levees should be analyzed in conjunction with the riverine flood analysis, and the areas flooded by interior drainage should be shown on FEMA maps in accordance with current practice. Appropriate consideration should be given to correlation of the event on the river and the event causing the interior drainage problem.

Assessing the Adequacy of the National Flood Insurance Program's One Percent Flood Standard 2006 (Galloway et al., 2006)

If implementation of the standard can be improved, FEMA should retain the one percent annual chance flood as the Federal standard for regulation of activity in the SFHA. The Nation needs to have a common standard for Federally imposed land use restrictions.

FEMA should take action to improve the implementation of the one percent standard for regulation of land use. Such actions as enhancement of public understanding of hazards, use of future-conditions hydrology to account for urbanization and climate change, reduction in floodway infringements, and greater attention to enforcement of existing NFIP provisions would greatly improve the effectiveness of NFIP related land use decisions.

States and their communities should exercise their responsibility to impose higher standards, where the health and safety of the population merits a higher standard for land use regulation. Concurrently, FEMA should examine the use of incentives, possibly through use of the Community Rating System (CRS), to reward States that exercise these responsibilities. Imposition of higher standards is well within the purview of the States and the communities that lie within the States and receive their land use authority from the States.

FEMA should seek legislative authority to require mandatory purchase of flood insurance by those living in the 0.2 percent floodplain if they hold a Federally insured mortgage or if they are to receive any disaster assistance from the Federal government in the case of a flood. The cost of this insurance should be determined actuarially, based on the reduced risk of living at a specific elevation within the 0.2 percent floodplain.

FEMA should not recognize levees under the NFIP unless they provide protection to the 0.2 percent (500-year flood) level. Levees in non urban areas should protect against the 1 percent or larger flood, depending on the economic costs and benefits of the levee.

FEMA should seek legislative authority to require mandatory purchase of flood insurance by those living behind accredited levees to address the residual risks they face and to ensure they are aware of this risk. Structures behind levees are subject to residual risks and should be insured against that risk.

FEMA should ensure that NFIP guidance and program activities clearly indicate that critical facilities should be located outside the 0.2 percent floodplain.

FEMA should improve the collection of policy and claims data, to assist in ongoing evaluation efforts, and should actively support Federal funding of efforts by NOAA to upgrade precipitation frequency estimates and flood data collection, and the U.S. Geological Survey (USGS) efforts to upgrade its stream gaging program. The accuracy of the Federal flood data is no better than the baseline information from which it is derived.

FEMA should ensure that consideration of natural and beneficial functions is fully integrated into all aspects of FEMA and NFIP actions influencing floodplain activity.

GAO Hurricane Katrina: Strategic Planning Needed to Guide Future Enhancements Beyond Interim Levee Repairs (GAO, 2006)

The Army Corps of Engineers should develop (1) a comprehensive strategy that includes an integrated approach for all projects and plans for rebuilding and strengthening the system and (2) an implementation plan that will achieve the specific level of protection in a cost-effective manner, within a reasonable time frame.

The Army Corps of Engineers should establish an evaluative organization like the Interagency Performance Evaluation Task Force, to assist in its efforts in developing a strategic plan, monitoring progress, and providing expert advice for constructing a stronger and well-integrated hurricane protection system.

The National Levee Challenge; Levees and the FEMA Map Modernization Initiative – 2006 Interagency Committee (FEMA, 2006)

FEMA should define a new flood insurance zone (Zone XL) for areas behind levees that provide 100-year protection and meet other requirements for inclusion in the NFIP and, in coordination with other agencies, identify the level of risk to those structures behind those levees. The XL zone would include those areas behind the levee that are subject to inundation by the 100-year flood in the without-levee condition. The area between the 100- and 500-year floodplain lines would be shown as a shaded X zone.

Levees would also be identified under a scientifically based levee risk classification system (e.g., high, medium, or low). This classification system would be based on several factors, including the following:

- Potential depth of flooding in the event of failure or overtopping,
- Type and density of development in protected areas behind the levee,
- Steps taken to ensure that levee failure does not occur during levee capacity exceedance (overtopping),
- Warning times,
- The number and types of egress so that people who may be inundated may move out of harm's way.

This levee risk classification would be designated on the Flood Insurance Rate Map (FIRM) and in the Flood Insurance Study report.

To ensure that levee systems maintain the conditions that qualified them for NFIP accreditation, FEMA should require as a condition of retaining FEMA recognition of levee status that sponsors:

- Conduct annual inspections of the levee system,
- Submit to FEMA biennially the results of the annual inspections, levee system operation and maintenance records, and an assessment of the levee system during any flood events that occurred within the reporting period,
- Submit to FEMA every 10 years a report, prepared by a registered Professional Engineer or Federal agency responsible for levee design, that recertifies the engineering and geotechnical conditions of the levee system,
- Certify levees as systems that include not only levees and floodwalls, but also the pumps, interior drainage systems, closures, penetrations, and transitions that provide systems integrity, and ensure that operation and maintenance plans cover all elements of the system.

To ensure that all components of a levee system meet required standards, FEMA should exclude embankments, such as roads and railroads, from inclusion in accredited levee systems unless those embankments meet the engineering criteria required of levees and are maintained and operated in accordance with the provisions of 44 CFR 65.10.

FEMA should revise 44 CFR 65.10 to incorporate the above recommendations and to deal with other needed changes in the administrative requirements of 44 CFR 65.10 as noted in the appendices to this report.

FEMA and USACE should continue their efforts to develop a joint database structure that will meet the needs of both agencies and other Federal and State organizations to maintain an inventory and assessment of flood damage reduction structures, including levees.

While agency-specific components of such a database will be needed, the management and fiscal benefits that accrue from a common core and adherence to the national spatial data infrastructure (NSDI) will be significant.

Working with its Federal, State, and local partners and with levee sponsors, FEMA should develop and implement a public awareness and outreach strategy that will improve public awareness and understanding of the hazards and risks associated with levees. This audience includes the Administration and Congress, other public officials, and private citizens.

As part of this public outreach effort, levee sponsors should be required to notify annually property owners in areas behind a levee about the residual risk that exists and the risk classification of the levee.

The relationships among institutions and organizations dealing with levees are not clearly defined and vary by agency and region.

The Federal Government's role in flood damage reduction through USACE, the Natural Resources Conservation Service (NRCS), the Bureau of Reclamation, the National Oceanic and Atmospheric Administration (NOAA), and the U.S. Geological Survey (USGS) is generally understood. FEMA's role in floodplain management is evolving and is less understood by the public. Similarly, the role of State and local organizations in carrying out floodplain-related activities is evolving.

Success in Map Mod as well as flood damage reduction will depend on the development of close working relationships among representatives of the agencies involved.

Efforts should be made to closely coordinate the activities of relevant Federal agencies in floodplain management and related mapping programs.

FEMA and USACE should review existing authorities and funding available to assist communities in performing certification analysis and make recommendations to the Office of Management and Budget (OMB) if the funding and authorities are not available.

FEMA and USACE also should develop expedited processes to assist communities in remediation of publicly-owned levees and seek OMB and Congressional approval of these processes and the need for resource commitments to support them.

FEMA should consider seeking legislative action to permit use of A99 designation and procedures by communities that are able to self-fund levee upgrades

To ensure that the base data used to support the hydraulics and hydrology (H&H) analysis are up to date and reflect 21st-century computation technologies, FEMA, in coordination with other Federal agencies that use these data and techniques, should support funding of NOAA efforts to upgrade precipitation frequency estimates, upgrades to the USGS gaging program, and the Advisory Committee on Water Information Subcommittee on Hydrology revision of Bulletin 17B, which guides flood recurrence interval determinations.

Recognizing the national and international movement toward use of risk analysis in dealing with floods and their consequences, the significant strides that have been made in developing risk assignment techniques, and the current use of these techniques by USACE in levee design, FEMA should modify 44 CFR 65.10 to phase out, over the next 10 years, use of the freeboard-based approach and should substitute the risk analysis methodology for levee-height determination. During the transition period, FEMA should permit either approach to be used in levee design and recertification analysis

Strong consideration should be given to not recognizing, for NFIP purposes, levees that provide protection to highly urbanized areas unless they provide protection against floods greater than 100 years (e.g., the 500-year flood). There is a long history of recommendations and analyses to support protection to 500-year or greater levels. Given the large number of existing levees, the transition would not be quick or easy. Nevertheless, FEMA should take steps to immediately examine this issue in detail and determine if a change in the standard for recognition should take place.

Non-Federal levees exist at the sufferance of the States and counties in which they are located. Federal levees, for the most part, are passed to State and local sponsors following their completion for operation and maintenance.

State and local agencies must play a significant role in dealing with levee issues as they arise. Federal agencies should recognize the State and local roles and develop incentives and support mechanisms to ensure that State and local agencies can effectively carry out their responsibilities.

Property owners behind levees continue to face the residual risk of flooding, yet they are not required to share in the mitigation of this risk through insuring their property against the dangers of the residual risk. FEMA should address the challenge of the residual risk in the following ways:

- Seeking legislative change to require property owners to purchase some level of flood insurance for structures behind levees (new Zone XL).
- Requiring communities to establish special early warning systems and to develop flood warning – preparedness plans for those areas that are protected by levees.

The current H&H approach for calculation of the BFE ensures that structures being designed to comply with the BFE are out of date by the time they are constructed. Future H&H conditions are not directly addressed. FEMA, working closely with other Federal agencies, States and communities, should examine how best to deal with climate change, sea-level rise, and future development. Such efforts will require the commitment of significant resources. It also represents a conceptual shift from specifying the mapping criteria to reflect present conditions to establishing criteria based on possible, but not fully quantifiable, future conditions.

When communities are not able to document immediately the recertifiability of currently certified levees and the levees are not known to be at risk, FEMA should notify the communities of the need for a more detailed examination of the levees and provide a specific schedule for that activity. Communities should be required to notify all residents who are living in areas behind the levees of the re-examination and identify for them the residual risks and the need for information about the levees. This action would closely parallel ongoing FEMA efforts to permit completion and adoption of new countywide maps that contain levees needing further study.

A major challenge facing those responsible for levee certification is conduct of appropriate and rapid geotechnical assessments of levee integrity. These assessments are critical to providing assurances of levee safety. However, such assessments, depending on the nature of the material and the cross section of the levee, are very costly. The bulk of the costs are related to the number and depth of soil borings. No methods or technologies are currently available to replace soil borings, and little effort is underway in the Research and Development (R&D) community to deal with the challenge. FEMA, USACE and USGS should support R&D efforts focused on improvement of rapid assessment of levee geotechnical integrity and should jointly recommend to the National Science Foundation that attention be given to this area of research.

ASFPM: National Flood Policy Challenges – Levees: The Double-Edged Sword (ASFPM, 2007)

The ASFPM urges FEMA and the U.S. Army Corps of Engineers, along with other federal water resources agencies, to revisit and revise the definition of levee so that it includes elements of function, risk, and vulnerability. This effort should include defining a levee, dam, or incidental work that modifies flood flows and the interrelationships among these definitions. The Federal Interagency Task Force on Floodplain Management is one potential vehicle to undertake this task.

The ASFPM believes that the Corps of Engineers should be tasked as the lead agency to develop and maintain a comprehensive inventory of current and future levees. This would start with federal levees and ultimately include non-federal and private levees.

Levees should be used as a structure of last resort and only after other measures, especially nonstructural ones, have been fully considered. Levees should not be used as a means to facilitate the development of currently undeveloped floodprone lands.

Federal investments in levees should not be made for a structure that provides less than 500-year protection, and the Corps of Engineers planning process of maximizing the NED should explicitly incorporate this public safety standard as a lower boundary for federal investment.

Levees should not be constructed in floodways and, to the maximum extent possible, when constructed or reconstructed levees should be set back from rivers to allow the river to function more naturally and to provide for the protection or restoration of riparian and wetland resources between the river bank and the levee.

The ASFPM urges Congress and the Administration to adopt a policy that the 500-year level of protection for levee design is the minimal standard for purposes of flood insurance and other federal investment."

Current levees that provide less than 500-year protection but meet all the requirements for design, maintenance, and operation, and are recognized by federal programs as meeting the standards for 100-year protection, could be provided grandfathered status. Criteria should be developed to determine when and if protection provided by a specific levee would need to be upgraded and how that would be achieved.

Benefit/cost analysis is an appropriate tool with which to evaluate and contrast federal projects, but it should be bounded by a strong public safety design standard, which for federally supported levees should be the 500-year level of protection.

The design of levees should include improved methods of providing resiliency, most notably the inclusion of designed fail-resistant spillways built into many levees so that when the levee design is exceeded, excess flow spills through that area, preventing catastrophic overtopping or failure of the structure.

The impacts of any new, rehabilitated, or reconstructed levee that would result in the transfer of damage or in adverse economic, social, or environmental consequences must be mitigated.

The local sponsor must demonstrate the financial and staffing capability to provide operation and maintenance for the life of the structure—before the project is approved, constructed, re-constructed, or recognized as providing a certain level of flood protection.

Congress should fund the National Research Council to engage experts to evaluate and propose modifications to levee design, operation, and maintenance standards. These efforts should include review of previous National Academies reports, and the extent to which previous recommendations have been addressed.

Written guidance is needed on what constitutes a "proper" inspection, what is needed for certification to enable the NFIP to recognize the levee, and what the actual consequences are to the levee owner if the levee is not properly maintained to meet these requirements. Both the Corps of Engineers and FEMA have guidance for requirements of programs that come into play with these issues, and the guidance from each agency must be consistent and correlated with the other agency's guidance.

A federal policy should be clearly articulated that the periodic certification and inspection of levees, including related operation and maintenance, is the responsibility of the levee owner and that transferring this responsibility to the federal government is inappropriate. Participation in federal programs of repair, insurance, and disaster relief must be contingent on levee owner compliance with these elements.

A federal policy should be clearly articulated that the periodic certification and inspection of levees, including related operation and maintenance, is the responsibility of the levee owner and that transferring this responsibility to the federal government is inappropriate. Participation in federal programs of repair, insurance, and disaster relief must be contingent on levee owner compliance with these elements.

A state-administered national levee safety program is needed to protect the federal interest in public health, safety, and fiscal responsibility, as well as to protect public safety and costs related to all levees not in the federal system. Such a program must be fully integrated with state and local programs of flood risk management, especially floodplain management and dam safety, and should use a state delegation model similar to that used to implement the Clean Water Act, rather than function as an independent program like the existing National Dam Safety Program. State capability in this area is critical and can be developed most effectively through federal legislation that provides incentives and disincentives for states to accept delegation for the development and implementation of effective state levee safety programs.

FEMA should require that all communities with an NFIP-recognized levee have a multihazard mitigation plan that considers how other hazards affect the safety of their levee (e.g., earthquake, subsidence, river sedimentation, erosion, etc.) and appropriate emergency action plans (EAPs) with action steps to account for any of these factors that affect the safety of the levee. FEMA should require that this plan be updated at least every five years, including accounting for any changes in flood flows caused by increased watershed development. The potential for catastrophic consequences of levee failure or overtopping should be included in levee planning, design, regulations, and insurance considerations.

The area that would be inundated when a levee fails or is overtopped, or when internal drainage systems are overwhelmed or incapacitated should be mapped as a residual risk flood hazard area and depicted on Flood Insurance Rate Maps.

Emergency action plans (EAPs) that address flood warning and evacuation should be required for all residual risk areas behind levees in order to protect lives and minimize property damage. These plans, and the periodic exercise of them, should be a requirement of any federal or state program that recognizes the levee as providing protection.

The purchase of flood insurance and implementation of appropriate development standards should be mandatory for all property protected by levees, to reflect the potential for the catastrophic consequences of levee failure.

Communication of the residual risk behind levees on a regular basis should be an explicit component of all aspects of proposed and current levee activities. It should include notification to all property owners of the risk (e.g., a notice in an annual water bill or tax bill) along with other measures such as posting signs in all land areas at risk behind the levees. All communication should state clearly that the area behind the levee is provided with some level of protection by levees, that the levees may fail or be overtopped, and that the area is a floodplain, with indications of the depth of flooding when the levee fails or is overtopped. Communication to the property owners should provide clear information on their role if an evacuation is ordered.

The liability of owners of structural flood control projects, such as levees and dams should be communicated to the owners of those structures on a periodic basis.

FEMA and the Corps of Engineers should evaluate and eliminate practices that cause increased flood damage or that lead to induced flooding (the transfer of flooding to other property that is primarily open space) unless property owners agree to a permanent flooding easement in return for this intrusion of flooding on their property.

The cumulative impacts of levees within a system or watershed should be evaluated before any levees are permitted, so those impacts are considered and mitigated, including increasing the design height to account for increased flood levels.

Levee construction, repair, and reconstruction should account for the protection of existing natural functions to avoid adverse impacts to the natural system. In addition, during repair or reconstruction of the levee, these natural functions should be restored to the maximum extent that is practical to account for past adverse impacts.

A California Challenge—Flooding in the Central Valley (State of California DWR, 2007)

Provide the highest level of risk reduction feasible to existing urban areas where thousands of people are at unacceptably high risk. The Panel believes that this level of protection should be equivalent to protection against the Standard Project Flood, which represents a flood that can be expected from the most severe combination of meteorologic and hydrologic conditions that are considered reasonably characteristic of the region. Providing this level of protection does not, by itself, prevent the failure of the system or of individual levees; nor does it guarantee that the Standard Project Flood cannot be exceeded in rare circumstances. One hundred year protection is not an acceptable level of protection for urban areas.

Develop an implementation plan for providing this reasonably high level of protection for all urban areas. The needed level of flood protection should be phased in with at least a 200-year level of flood protection to be achieved by 2020, and Standard Project Flood protection by 2030. Priority should be given to urban areas in deep floodplains.

In less populated areas, provide for protection against less severe floods (e.g. less than 200-year protection) as economically and environmentally justified, and maintain that lower level of protection into the future.

Ensure that any flood protection provided is sustainable fiscally and physically over time.

Manage the floodplain by focusing new development outside of the floodplain or in low-risk locations within protected areas of the floodplain, supporting the use of undeveloped and unprotected land for agriculture and other low-intensity land uses.8 Floodplain management should be accompanied by requirements for local governments to adopt and enforce needed land-use controls, financial and technical support to enable them to do so, and appropriate penalties if local governments fail to manage development to reduce flood risk. The state should continue to support the Federal Emergency Management Agency's levee policy and assist them in accelerating completion and adoption of updated flood maps. This would ensure that any new development in areas behind inadequate levees takes place under the land-use provisions mandated by the National Flood Insurance Program, as a minimum.

Site, where feasible, new levees or major rehabilitation of levees at a distance from the river and from existing levees. This would provide a degree of redundancy in the system, increase the land available for habitat and flood storage, reduce operation and maintenance costs, and help to ensure the integrity of the structures. Levees built this decade will be in place for decades to come, and now is the time to begin building structures that will last. Where re-siting is not feasible, the existing flood system should be modified to mitigate the impacts of floods that exceed the design level of the system.

Mitigate potential financial losses to those behind levees and to those in the non-leveed 500-year floodplain shown on Federal Emergency Management Agency flood maps through institution of mandatory purchase of flood insurance, or through inclusion of flood insurance in homeowners' policies of those within these areas.

Share the liability for flood damages among state and local governments. This would ensure that any local governments making land-use decisions that could increase potential flood damages share not only the benefits of that development, but also any liability incurred from potential flood consequences should those decisions prove to have been unwise.

Communicate to the public and each property owner in the floodplain the specific risks of occupying areas at risk of flooding, and provide steps property owners can take to reduce their exposure to flood damages.

Communicate to the public and each property owner in the floodplain the specific risks of occupying areas at risk of flooding, and provide steps property owners can take to reduce their exposure to flood damages.

Supplement the structural protection provided with floodproofing, elevation of homes and businesses, land-use regulations, and other non-structural approaches to reduce the residual risk that will continue to exist. Support this with emergency response systems including the development of post-disaster sheltering and redevelopment plans and the exercising of floodplain evacuation plans on a regular basis. Dealing with flooding in the Central Valley will require a close examination of existing governmental institutions and how they work together. The lessons learned from the New Orleans disaster point out the disconnects that develop when too many agencies are involved in the decision-making process and no one agency has overall direction. Large flood events exploit those disconnects. California must address this difficult issue, especially in terms of the large number of overlapping roles, responsibilities and accountabilities of reclamation districts, and state and local governments. Without reforming the institutions that manage flood protection, large investments in infrastructure are likely to be wasted.

Hurricane Protection System Peer Review (Battelle, 2007)

USACE should seek Administration support for and Congressional approval to:

- Develop a single authorization for the HPS similar to that provided for the Mississippi River and Tributaries (MR&T) project.
- Authorize flexibility for reprogramming.
- Ensure multi-year or timely appropriations; recognize the consequences of delays.
- Provide expedited review of issues.
- Ensure close cooperation with local governments and understanding of the problems associated with lack of cooperation

Nonstructural elements should become an equal partner in completion of the HPS. USACE should receive increased funding to carry out programs of nonstructural technical assistance for communities and proactively provide technical assistance to FEMA and local authorities with respect to elevating structures, floodproofing, planning for interior drainage, and other flood management considerations. This is especially the case for interior polderization of protected areas which could reduce the spatial extent of flood risk.

TFH should work closely with local officials on full implementation of effective floodplain management and enforce that provision of Section 402 of the Water Resources Development Act (WRDA) of 1986 requiring development of local floodplain management plans. There is a clear opportunity for USACE to provide needed technical assistance, so that rebuilding occurs in a manner that minimizes the potential for future flood dam-age to public buildings and infrastructure and private homes and businesses.

USACE should provide maps of residual risk for early distribution to residents and officials. Risk communication maps should be accompanied with information on personal mitigation measures to include flood avoidance, financial protection, and life-safety concerns. USACE must ensure that residents and officials understand the residual risk and the level of protection that is actually being provided.

Given the experience and the strategies the Japanese and Dutch use for protection, TFH should consider establishment of a centralized planning and inspection capability for O&M. Fragmented planning efforts provide significant opportunities for system disconnects and potential failures.

USACE should include high-tech monitoring systems for key elements of the completed HPS. Technology now makes it possible to remotely sense levees, floodwalls and other protective system problems as they occur. Some systems are currently under consideration by TFH.

Given the tight schedule for completion of the 100-year system, TFH should not divert human and physical resources from the principal task of finishing the HPS to the 100-year level to provide interim improvements. Since the HPS will not be at a 100-year level until the 2011 hurricane season at the earliest, some have expressed concern over the need for interim protection in components of IHNC and West Bank areas. While these concerns are important, completion of the system should have priority.

USACE should seek an accelerated decision on MRGO to allow early movement forward on dealing with hurricane protection in the area of the MRGO-GIWW and improving protection to the IHNC. Such a decision will save time, eliminate the need to plan against alternatives, and speed completion of the HPS.

TFH, in current construction activities, should take into account the potential footprints of subsequent construction under LaCPR so as to minimize repeat disruptions, multiple right-of-way acquisition, etc. Major long-term cost savings can be generated and significant public push back can be avoided by dealing with real estate issues only once.

Observations:

Provision of risk reduction (protection) against the 100-year annual chance (100-year) event is not sufficient to prevent possible loss of life and social disruption in a major urban area. For example, there is a 26% chance of the occurrence of such an event with the life of a 30-year mortgage. Urban areas need a higher level of protection. Providing 100-year protection may make it difficult to subsequently provide this higher level of protection.

It is not clear that there is agreement that the HPS being built reflects the understanding of all stakeholders and Congress as to what level of protection is supposed to be provided. TFH indicated that its first priority was to provide 100-year certified levees for all areas rather than 100-year or 'as originally authorized,' whichever is higher. This is a sound decision, but USACE should take action to obtain written agreement from stakeholders and the Congress that the 100-year first path it is following is in fact the path expected.

Many USACE technical manuals are badly out of date. Manuals being used in levee design are Dutch, because appropriate USACE manuals did not exist or had not been revised in many years. USACE O&M programs must include adequate funding to support continuous maintenance and upgrade of the documents and the necessary research and development (R&D) that maintains currency.

The New Orleans Hurricane Protection: What Went Wrong and Why (ASCE, 2007)

Keep safety at the forefront of public priorities by having all responsible agencies re-evaluate their policies and practices to ensure that protection of public safety, health, and welfare is the top priority for infrequent but potentially devastating impacts from hurricanes and flooding. Also, encourage Congress to establish and fund a mechanism for a nationwide levee safety program, similar to that which is in place for dams.

Quantify and periodically update the assessment of risk. This approach should be extended to all areas in the United States that are vulnerable to major losses from hurricanes and flooding.

Determine the level of acceptable risk in the community through quality interactive public risk communication programs in New Orleans and other areas threatened by hurricanes and flooding. Once determined, manage the risks accordingly.

Correct the system's deficiencies by establishing mechanisms to incorporate changing information, making the levees survivable if overtopped, strengthening the I-walls and levees, and upgrading the pumping stations.

Assign to a single entity or individual (a licensed engineer) the responsibility of managing critical hurricane and flood protection systems such as the one in New Orleans.

Implement more effective mechanisms for coordination and cooperation. (For example, those responsible for maintenance of the system must collaborate with system designers and must upgrade their inspection, repair, and operations processes to ensure that the system is hurricane- and flood-ready.

Upgrade engineering design procedures and practice to place greater emphasis on safety.

Engage independent experts in high-level reviews of all critical life safety structures, including hurricane and flood-protection systems

Hurricane Protection Decision Chronology – 2008 (USACE, 2008)

Reflections on Project Decision-Making:

1. Concerns about project cost growth, constrained federal and local budgets, delays in project completion, and the possible need for reauthorization if major changes were proposed, help to explain District decisions to construct the project according to original designs and datum benchmarks.
2. There was no Corps organizational process that required and provided funding for a continuing assessment of project performance capability during the postauthorization implementation period.
3. There is no evidence in the project record indicating that project engineers believed that the decisions made would threaten engineering reliability.
4. The only recurring organizational provision for systematically reporting the expected performance capability of the project was the annual Budget Justification Sheet.

Reflections on the Future:

1. The importance of sharing knowledge
2. Need for flexibility and adaptation in planning, design, and implementation

National Committee on Levee Safety (NCLS, 2009)

Establish a National Levee Safety Commission to provide national leadership and comprehensive and consistent approaches to levee safety including standards, research and development, technical materials and assistance, training, public involvement and education, facilitation of the alignment of federal programs and design, delegation and oversight of a delegated program to states.

Expand and Maintain the National Levee Database to include a one-time US Army Corps of Engineers inventory and inspection of all non-federal levees. Baseline information will be included and maintained in an expanded National Levee Database (NLD) in order that critical safety issues, true costs of good levee stewardship, and the state of individual levees can inform priorities and provide data for needed risk-informed assessments and decision-making.

Adopt a Hazard Potential Classification System as a first step in identifying and prioritizing hazard in leveed areas. Due to a lack of data regarding probability of failure, initial classifications should be based solely on consequences in order to assist in setting priorities, criteria, and requirements as the NSLP is being established.

Develop and Adopt National Levee Safety Standards that will assist in ensuring that the best engineering practices are available and implemented throughout the nation at all levels of government.

Develop Tolerable Risk Guidelines in order to facilitate an understanding of the options to reduce identified risks, how uncertainty affects this understanding, and to better inform levee construction/enhancement decisions and weigh nonstructural alternatives to flood risk management in a risk-informed context. Change "Levee Certification" to "Compliance Determination" to better articulate the intent that "certification" under the National Flood Insurance Program (NFIP) requirements does not constitute a safety guarantee or warranty. The purpose of this change is to more clearly communicate residual risks of living and working in leveed areas.

Change "Levee Certification" to "Compliance Determination" to better articulate the intent that "certification" under the National Flood Insurance Program (NFIP) requirements does not constitute a safety guarantee or warranty. The purpose of this change is to more clearly communicate residual risks of living and working in leveed areas.

Subject Levee Certifications (Compliance Determinations) under FEMA's National Flood Insurance Program to Peer Review in order to increase confidence in technical determinations of compliance.

Swiftly Address Growing Concerns Regarding Liability for Damages Resulting from Levee Failures through exploration of a range of measures aimed at reducing the potential liability of engineering firms and/or government agencies that perform engineering services for levee systems (e.g. inspections, evaluations, design, construction administration, certification, or flood fighting). Congress should address this liability concern as a first priority in order to help ensure state and local interest in developing levee safety programs, and to prevent much needed levee repairs, rehabilitation and certification from coming to a halt.

Develop a Comprehensive National Public Involvement and Education/Awareness Campaign to Communicate Risk and Change Behavior in Leveed Areas as an essential element of levee safety by improving public understanding of the role of levees, associated risks, and individual responsibilities to empower people to make risk-informed choices.

Provide Comprehensive Technical Materials and Direct Technical Assistance crucial to the successful implementation of consistent national standards to states, local communities and owner/operators.

Develop a National Levee Safety Training Program including a combination of courses, materials, curricula, conferences, and direct assistance resulting in an increase in the level of expertise and knowledge in all aspects of levee safety. This would include the development of curricula and certification requirements for a Certified Levee Professional program.

Develop and Implement Measures to More Closely Harmonize Levee Safety Activities with Environmental Protection Requirements to ensure that critical levee operations and maintenance is not delayed and that, where possible without compromising human safety, environmentally-friendly practices and techniques are developed and used.

Conduct a Research and Development Program that will continually advance state-of-the-art technologies and practices for levee safety and conduct critical operations and maintenance activities in as cost-effective and environmentally-friendly manner as possible.

Design and Delegate Program Responsibilities to States to assist states and local governments develop effective levee safety programs focused on continual and periodic inspections, emergency evacuation, mitigation, public involvement and risk communication/awareness, etc.

Establish a Levee Safety Grant Program to assist states and local communities develop and maintain the institutional capacity, necessary expertise, and program framework to quickly initiate and maintain levee safety program activities and requirements.

Establish the National Levee Rehabilitation, Improvement, and Flood Mitigation Fund to aid in the rehabilitation, improvement or removal of aging or deficient national levee infrastructure. Investment (cost-shared) is recommended to be applied to the combination of activities, both structural and non-structural, that combined, would maximize overall risk reduction and initially be focused in areas with the greatest risk to human safety.

Explore Potential Incentives and Disincentives for good levee behavior through alignment of existing federal programs.

Mandate Purchase of Risk-Based Flood Insurance in Leveed Areas to reduce financial flood damages and increase understanding of communities and individuals that levees do not eliminate risk from flooding.

Augment FEMA's Mapping Program to improve risk identification and communication in leveed areas and consolidate critical information about flood risk.

Align FEMA's Community Rating System (CRS) to Reward Development of State Levee Safety Programs by providing further incentives to communities to exceed minimum program requirements and benefit from lower risk-based flood insurance rates to individuals who live in leveed areas.

The New Orleans Hurricane Protection System: Assessing Pre-Katrina Vulnerability and Improving Mitigation and Preparedness (NRC, 2009)

Hydrologic Realities and the Limits of Protective Structures:

- There are many inherent hydrologic vulnerabilities of living in the greater New Orleans metropolitan region, especially in areas below sea level. Post-Katrina repairs and strengthening have reduced some of these vulnerabilities. Nevertheless, because of the possibility of levee/floodwall overtopping—or more importantly, levee/floodwall failure—the risks of inundation and flooding never can be fully eliminated by protective structures no matter how large or sturdy those structures may be.

Future Footprint of the Hurricane Protection System:

- The pre-Katrina footprint of the New Orleans HPS consisted of roughly 350 miles of protective structures including levees, I-walls, and T-walls. There was undue optimism about the ability of this extensive network of protective structures to provide reliable flood protection. Future construction of protective structures for the region should proceed with this point firmly in mind and in the context of a more comprehensive and resilient hurricane protection plan.

Nonstructural Aspects and Options:

- Comprehensive flood planning and risk management for the New Orleans metro region will be based on a combination of structural and nonstructural measures, the latter including relocation options, floodproofing and elevation of structures, and evacuation studies and plans. Better risk communication also must be part of more effective mitigation and an improved state of preparedness. Structural measures such as levees and floodwalls should not be viewed as substitutes or replacements for nonstructural measures, but rather as complementary parts to a multi-tiered hurricane protection solution.
- Relocation: the planning and design for upgrading the current hurricane protection system should discourage settlement in areas that are most vulnerable to flooding due to hurricane storm surge. The voluntary relocation of people and neighborhoods out of particularly vulnerable areas—with adequate resources designed to improve their safety in less vulnerable areas—should be considered as a viable public policy option.

Floodproofing Measures:

- When voluntary relocations are not viable, floodproofing measures will be an essential complement to protective structures—such as levees and floodwalls—in improving public safety in the New Orleans region from hurricanes and induced storm surge. This committee especially endorses the practice of elevating the first floor of buildings to at least the 100-year flood level, and preferably to a more conservative elevation. The more conservative elevation reflects a subsequent finding in this report regarding the inadequacy of the 100-year flood as a flood protection standard for a large urban center such as New Orleans.
- Critical public and private infrastructure—electric power, water, gas, telecommunications, and flood water

collection and pumping facilities—should be strengthened through reliable construction, and ensuring reliable interdependencies among critical infrastructure systems.

Evacuation:

- The disaster response plan for New Orleans, although extensive and instrumental in successfully evacuating a very large portion of the New Orleans metropolitan area population, was inadequate for the Katrina event. Thus, there is a need for more extensive and systematic evacuation studies, plans, and communication of evacuation plans. A comprehensive evacuation program should include not only well designed and tested evacuation plans, protocols, and criteria for evacuation warnings, but also alternatives such as improved local and regional shelters that could make evacuations less imposing. It also should consider longer-term strategies that can enhance the efficiency of evacuations, such as locating facilities for the ill and elderly away from more vulnerable areas that may be subject to frequent evacuations.

Risk Communication:

- Before Katrina, there was a limited understanding and appreciation of the residual risks of living behind levees. Improvements in future hurricane preparedness and response will depend partly upon improved public understanding of these risks. In order to enhance the communication and appreciation of these risks, it will be important to extend the efforts of the IPET and to refine, simplify, and communicate consistently the risks of hurricanes and storm surge to the region's residents, including how those risks vary across the region. Effective communication of the risk-based findings from the entire IPET report will be enhanced by creating a professional summary and compilation of the entire IPET draft report with layman's terminology (see earlier recommendation in this report).

The 100-Year Level of Flood Protection:

- The 100-year level of flood protection is a crucial flood insurance standard. It has been applied widely across the nation and it is being used in some circumstances in reconstruction and planning activities in the New Orleans region. For areas in which catastrophic levee failure is not a major public safety concern, and where large floods would not imperil evacuation routes, the 100-year standard may be appropriate. For heavily-populated urban areas, where the failure of protective structures would be catastrophic—such as New Orleans—this standard is inadequate.

Independent Review for Engineering and Design:

- It is important to enlist periodic external review in the design, construction and maintenance of large, complex civil engineering projects such as the New Orleans hurricane protection system. A "second opinion" allows an opportunity to ensure that calculations are reliable, methods employed are credible and appropriate, designs are adequate and safe, potential blind spots are minimized, and so on. An outside external review group also may be able to state politically sensitive findings or facts that a government agency may be reluctant to. Such a review team should be adequately independent of the authority that identified.

Periodic Assessments and Updates of Concepts, Methods, and Data:

- Changing environmental conditions can affect the performance and operation of large hurricane and flood protection projects. Advances in scientific and engineering theories and methods may render assumptions on which these projects were based partly or fully obsolete. Because of these changes and the important implications they may have for expected performance of protective structures, a process should be implemented to ensure periodic review of underlying environmental, scientific, and engineering factors that affect New Orleans hurricane protection system performance. The process for incorporating new scientific information into large flood protection projects, like the New Orleans hurricane protection system, can be affected by congressional reauthorization requirements. Changes or clarifications to congressional policies and reauthorizations as they relate to large construction projects may be necessary to effectively implement findings of periodic scientific reviews.

The Future of Hurricane Risk Analysis for New Orleans and the Gulf Coast Region:

- The IPET has conducted a landmark assessment of the New Orleans HPS that could serve as a platform for future and ongoing assessments of vulnerability, levels of protection, subsidence rates, geological studies, risk assessments, and so on. As the IPET investigations come to an end, many of the external experts that participated in the studies will return to their respective careers outside the Corps of Engineers. Much of the IPET "institutional memory" therefore may not be infused into Corps of Engineers New Orleans District office, the State of Louisiana, or the City of New Orleans. It is essential that these analyses be extended and subsequently built upon by the Corps of Engineers and others, including the FEMA, NOAA, the State of Louisiana, New Orleans regional municipalities, and the region's universities, engineers, and scientists. To facilitate future work that builds on the IPET studies, a publicly accessible archive of all data, models, model results, and model products from the IPET project should be created.

REFERENCES

ASCE (American Society of Civil Engineers). 2007. The New Orleans Hurricane Protection System: What Went Wrong and Why. Reston, VA: ASCE.

ASFPM (Association of State Floodplain Managers). 2007. National Flood Policy Challenges Levees: The Double-Edged Sword. Madison, WI: ASFPM.

FEMA (Federal Emergency Management Agency). 2006. The National Levee Challenge: Levees and the FEMA Flood Map Modernization Initiative. Washington, DC: FEMA.

Galloway, G. E., G. B. Baecher, D. Plasencia, K. G. Coulton, J. Louthain, M. Bagha, and A. R. Levy. 2006. Assessing the Adequacy of the National Flood Insurance Program's One Percent Flood Standard. Washington, DC: American Institutes for Research. Available online at http://www.fema.gov/library/viewRecord.do?id=2595.

GAO (U.S. Government Accountability Office). 2006. Hurricane Katrina: Strategic Planning Needed to Guide Future Enhancements Beyond Interim Levee Repairs. Washington, DC: GAO.

NCLS (National Committee on Levee Safety). 2009. Recommendations for a National Levee Safety Program. Washington, DC: NCLS.

NRC (National Research Council). 1982. A Levee Policy for the National Flood Insurance Program. Washington, DC: National Academy Press.

NRC. 2009. The New Orleans Hurricane Protection System: Assessing Pre-Katrina Vulnerability and Improving Mitigation and Preparedness. Washington, DC: The National Academies Press.

State of California, Department of Water Resources. 2007. A California Challenge: Flooding in the Central Valley. Sacramento, CA: Department of Water Resources.

USACE (U.S. Army Corps of Engineers). 2008. Decision-Making Chronology for the Lake Pontchartrain and Vicinity Hurricane Protection Project. Washington, DC: USACE.

Appendix G

FEMA Procedure Memorandums 34, 43, and 53

Federal Emergency Management Agency

Washington, D.C. 20472

August 22, 2005

MEMORANDUM FOR: Regional Directors
Regions I - X

FROM: David I. Maurstad, Acting Director
Mitigation Division

SUBJECT: Procedure Memorandum 34 — Interim Guidance for Studies Including Levees

Background: Throughout the United States, levees protect numerous communities and large expanses of agricultural land from floods. Their importance in mitigating flood hazards and their relevance to the National Flood Insurance Program (NFIP) are indisputable. However, riverine and coastal levees, in the aggregate, stretch for tens of thousands of miles, and information on their location, structural integrity, and certification often is outdated or missing altogether.

Issue: To address this challenge, a Levee Coordination Committee—including representatives from FEMA, other Federal agencies, and States—is examining current levee regulations and assisting in the development of a long-term policy that protects citizens and property, while accommodating the needs of the NFIP. This memorandum helps to clarify the entities responsible for providing information on levees identified during a mapping project.

Action Taken: Until the new policy is developed, this memo provides interim guidance to minimize delays in near-term mapping studies. The attached flow chart supplements FEMA's procedure memorandums 30 and 32. This information is in conformance with Section 65.10 of the NFIP regulations.

Supplement to Procedure Memo 30—FEMA Levee Inventory System.

Mapping partners – CTPs, IDIQs, OFAs, etc. -- should continue providing information about levees located in or adjacent to study areas. Information should be provided via the FEMA Levee Inventory System (FLIS) according to Procedure Memorandum 30 and the instructions available on the FLIS Web site located at http://flis.pbsjdfirm.com. The FLIS will be accessed via the MIP after release 3.0.

Levee coordinates should be gathered at a level of detail consistent with GIS accuracy and digital Flood Insurance Rate Map (FIRM) standards. Mapping partners who do not already have access to the FLIS can contact the National Service Provider at (703) 960-8800.

August 22, 2005

Page 2 of 2 – Procedure Memorandum 34

Supplement to Procedure Memo 32—Levee Review Protocol.
The protocol for levee reviews, particularly the details provided in Table 1 of Procedure Memorandum 32, is revised according to the attached flow chart.

Identification of Levees
It is critical that all levees within the scope of the mapping project be identified early in the mapping project, ideally no later than the scoping meeting. The role of all mapping partners, including coordination with the State and other Federal partners (e.g., U.S. Army Corps of Engineers), related to review of levee certification should be clearly identified as part of the scoping process. When levees are identified at the scoping meeting the community must be informed of the data requirements for FEMA to recognize a levee as providing protection from the 1-percent-annual-chance flood (base flood) on the FIRM. In accordance with 44 CFR Section 65.10(a), it is the responsibility of the community or other party seeking recognition of a levee system at the time of a flood risk study or restudy to provide the data outlined in 44 CFR Section 65.10. FEMA will not be conducting detailed examinations of levees to determine how a structure or system will perform in a flood event. In addition, the community or party seeking recognition should be provided with a deadline for submitting the data and informed that if the data are not submitted by the deadline, the levee cannot be recognized as providing protection from the base flood as part of the current mapping effort. However, a revision could be initiated once data are available.

Early identification of levees allows the mapping partner to outline to the community, or party seeking recognition, their responsibilities and FEMA's expectations to minimize study delays. In order to aid our mapping partners in properly assessing how to handle levee mapping issues, we have generated the below flowchart.

cc: See Distribution List

Distribution List (electronic distribution only):

Office of the Mitigation Division Director
Risk Assessment Branch
Risk Identification Branch
Flood Insurance and Mitigation Divisions in FEMA Regional Offices
Office of Legislative Affairs
Office of General Counsel
National Service Provider
Systems Engineering and Technical Assistance Contractor
Map Service Center

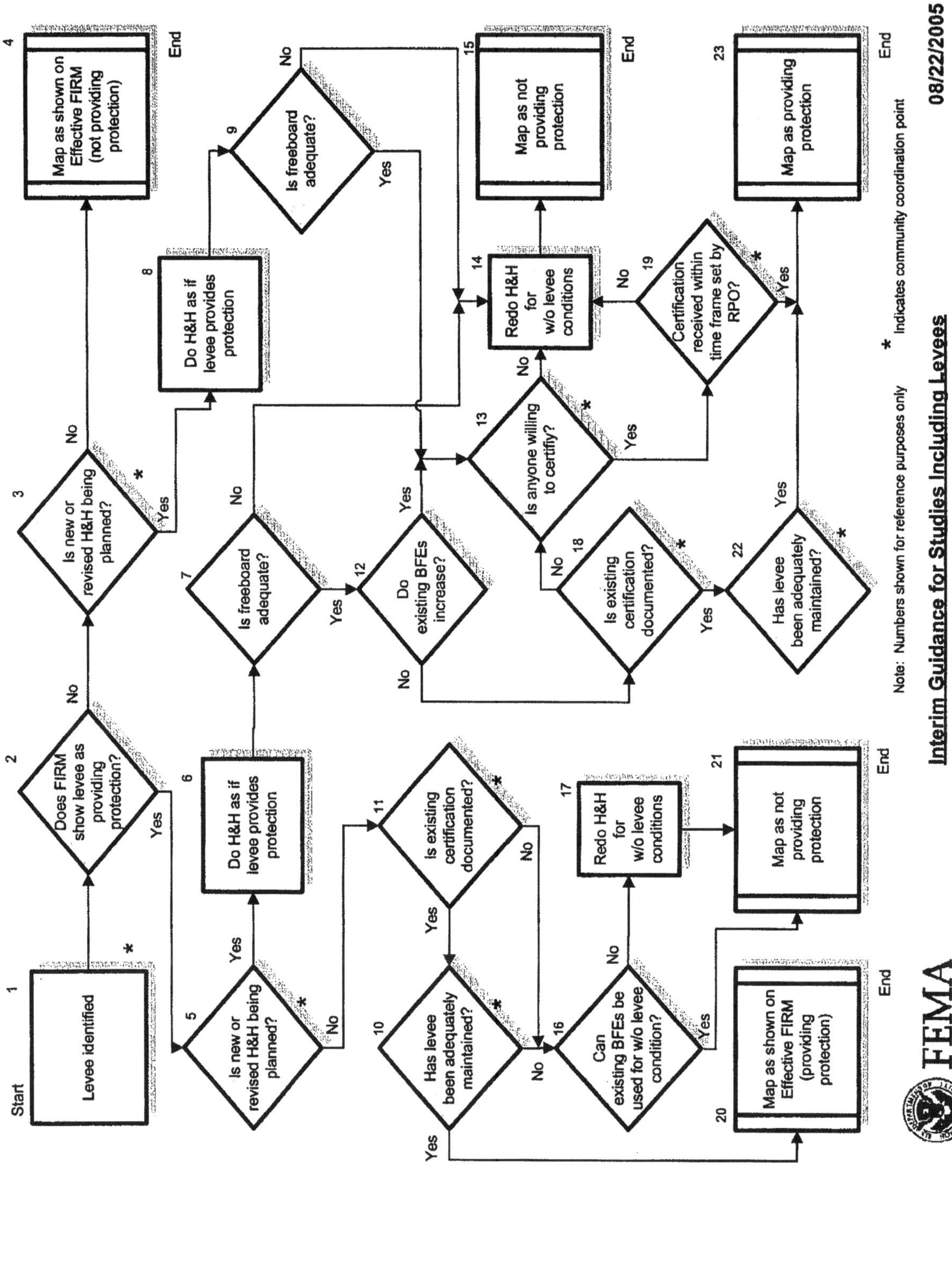
Start
1 Levee identified
2 Does FIRM show levee as providing protection?
3 Is new or revised H&H being planned?
4 Map as shown on Effective FIRM (not providing protection)
End
5 Is new or revised H&H being planned?
6 Do H&H as if levee provides protection
7 Is freeboard adequate?
8 Do H&H as if levee provides protection
9 Is freeboard adequate?
10 Has levee been adequately maintained?
11 Is existing certification documented?
12 Do existing BFEs increase?
13 Is anyone willing to certify?
14 Redo H&H for w/o levee conditions
15 Map as not providing protection
End
16 Can existing BFEs be used for w/o levee condition?
17 Redo H&H for w/o levee conditions
18 Is existing certification documented?
19 Certification received within time frame set by RPO?
20 Map as shown on Effective FIRM (providing protection)
End
21 Map as not providing protection
End
22 Has levee been adequately maintained?
23 Map as providing protection
End
Yes
No
Note: Numbers shown for reference purposes only
* Indicates community coordination point
FEMA
08/22/2005
Interim Guidance for Studies Including Levees

U.S. Department of Homeland Security
500 C Street, SW
Washington, DC 20472

FEMA

March 16, 2007
(Originally Issued on September 25, 2006)

MEMORANDUM FOR: Mitigation Division Directors
Regions I - X

FROM: David I. Maurstad
Director
Mitigation Division

SUBJECT: Revised Procedure Memorandum No. 43 - Guidelines for Identifying Provisionally Accredited Levees

Background: Early in the implementation of Flood Map Modernization (Map Mod), the Department of Homeland Security's Federal Emergency Management Agency (FEMA) recognized that the role of levees in flood risk reduction would be an important part of the efforts of Map Mod. Further, it was acknowledged that the condition of levees had not been assessed since they were originally mapped as providing base (1-percent-annual-chance) flood protection. Because of this, FEMA initiated a revised process to gain a better understanding of the actual flood risks for those citizens living and working behind levees nationwide.

On August 22, 2005, FEMA issued Procedure Memorandum No. 34 - Interim Guidance for Studies Including Levees. The purpose of the memorandum was to help clarify the responsibility of community officials or other parties seeking recognition of a levee by providing information identified during a study/mapping project. Often, documentation regarding levee design, accreditation, and the impacts on flood hazard mapping is outdated or missing altogether. To remedy this, Procedure Memorandum No. 34 provides interim guidance on procedures to minimize delays in near-term studies/mapping projects, to help our mapping partners properly assess how to handle levee mapping issues.

Issue: Levee owners or communities have the responsibility to provide documentation that a levee meets the requirements of Title 44 of the Code of Federal Regulations, Section 65.10 of the National Flood Insurance Program regulations (44 CFR Section 65.10), as part of a study/mapping project. Without the required documentation necessary to comply with 44 CFR Section 65.10, the area behind the levee will be re-delineated and mapped as Special Flood Hazard Area on the Digital Flood Insurance Rate Map (DFIRM). Procedure Memorandum No. 34 allows for the issuance of a deadline to the community for submitting the required documentation.

Page 2 of 4 Revised Procedure Memorandum No. 43

While 44 CFR Section 65.10 documentation is being compiled, the release of more up-to-date DFIRM panels for other parts of a community or county may be delayed. To minimize the impact on the Map Mod goals of mapping areas landward of levees, mapping partners should be provided with guidance that will allow preliminary and effective DFIRMs to be issued while the levee owner or community is given a reasonable amount of time to compile and submit data and documentation to show compliance with the requirements of 44 CFR Section 65.10. Guidance should also be provided to the mapping partners that allows, in specific situations, the preliminary DFIRM to be issued while providing the communities and levee owners with a specified timeframe to show compliance with 44 CFR Section 65.10 by correcting any maintenance deficiencies associated with the levee.

Action Taken: To minimize the impact of the levee recognition and certification process on Map Mod goals, guidelines have been developed that will allow mapping partners to issue preliminary and effective versions of DFIRMs while the levee owners or communities are compiling the full documentation required to show compliance with 44 CFR Section 65.10. The guidelines also explain that mapping partners can issue preliminary DFIRMs while providing the communities and levee owners with a specified timeframe to correct any maintenance deficiencies associated with a levee to and show compliance with 44 CFR Section 65.10. These guidelines are summarized in the attached document entitled "Guidelines for Identifying Provisionally Accredited Levees (PALs)."

The attached document describes the criteria for five scenarios intended to determine when a levee does or does not qualify for the PAL designation. FEMA has established a specified timeframe in which the community or levee owner may use to fulfill the remaining requirements for 44 CFR Section 65.10 before the levee is shown on the DFIRM as not providing base flood protection. The attached guidance also describes an additional process for maintenance deficient levees that do not currently qualify for the PAL designation. FEMA has established a separate specified timeframe for these levees, which allows the community or levee owner time to correct any maintenance deficiencies associated with a levee. If the levee qualifies for the PAL designation, FEMA will provide the community 90 days to sign and return an agreement indicating that the full documentation for 44 CFR Section 65.10 will be provided within 24 months of the signed agreement. If the signed agreement is not returned to FEMA within 90 days, or if the levee does not meet the PAL requirements (except for the maintenance deficient levees), the community is no longer eligible for the PAL designation, and the area landward of the levee will be remapped as Zone AE or Zone A, depending on the type of study performed for the area.

For levees that are included in the U.S. Army Corps of Engineers (USACE) Federal Program, FEMA will actively coordinate with the appropriate USACE district to determine which projects do not provide protection from the base flood. In a collaborative effort, existing data or project-specific information will be evaluated to identify and validate non-accredited levees in the USACE's inventory. As part of the USACE's recent survey of their levee inventory, levee projects have been identified to be no longer eligible for Public Law (PL) 84-99 rehabilitation assistance, based on the project's last inspection. However, many of these levee projects have been identified to be eligible for a one-time-only "maintenance deficiency correction period," established to allow public sponsors/levee owners to correct levee maintenance deficiencies before the levee is placed in an inactive status in the USACE Rehabilitation & Inspection Program and becomes ineligible for PL 84-99 rehabilitation assistance. The USACE has developed a written notification process to inform communities or levee owners of this status after it has coordinated with FEMA. Copies of the USACE notification letter will be provided to FEMA. If a community or levee owner receives this notification letter, the area landward of the identified levee will be mapped as Zone AE or Zone A, as appropriate.

Effective on the date of this Procedure Memorandum, levees that meet the PAL requirement (levees presently shown as providing base flood protection on the effective FIRM), for which the community or levee owner cannot readily provide full documentation of 44 CFR Section 65.10, will be identified on the FIRM with a map note. This note, placed landward of the levee, will indicate that the levee is provisionally accredited and any existing Zone X (shaded) area is provisional. If there is no existing Zone X (shaded) area on the effective FIRM, then the mapping partner should define the provisional Zone X (shaded) area using the best available data. Specific procedures and guidance for evaluating and mapping levees is provided in Appendix H of *Guidelines and Specifications for Flood Hazard Mapping Partners.*

The following note must be applied at several locations, point to the levee, and be placed landward of the levee in or near the Zone X (shaded) area:

> WARNING: Provisionally Accredited Levee. For explanation, see the Notes to Users.

The applicable Note to Users would read as follows:

> WARNING: This levee, dike, or other structure has been provisionally accredited and mapped as providing protection from the 1-percent-annual-chance flood. To maintain accreditation, the levee owner or community is required to submit documentation necessary to comply with 44 CFR Section 65.10 by (__________,____). Because of the risk of overtopping or failure of the structure, communities should take proper precautions to protect lives and minimize damages in these areas, such as issuing an evacuation plan and encouraging property owners to purchase flood insurance.

The five scenarios for determining whether the levee qualifies for the PAL designation are described in the attachment. The document also summarizes the process for coordinating with community officials and others to acquire the appropriate levee documentation, while moving forward with the production of countywide mapping for communities with levees.

Attachment

Guidelines for Identifying Provisionally Accredited Levees (PALs)

Page 4 of 4 Revised Procedure Memorandum No. 43

cc: See Distribution List

Distribution List (electronic distribution only):

Directors, Regions I - X

Office of the Mitigation Division Director

Risk Analysis Branch

Risk Reduction Branch

Risk Insurance Branch

Federal Insurance and Mitigation Divisions in FEMA Regional Offices

Office of Legislative Affairs

Office of General Counsel

National Service Provider

Independent Verification and Validation Contractor

Map Service Center

Indefinite Delivery Indefinite Quantity Contractors

Cooperating Technical Partners

Guidelines for Identifying Provisionally Accredited Levees (PALs)

Introduction

The Federal Emergency Management Agency (FEMA) issued Procedure Memorandum No. 34 (PM 34) on August 22, 2005, to provide interim guidance for processing studies/mapping projects for communities with levees and to define the roles of all FEMA contractors and mapping partners in meeting the regulatory requirements of the National Flood Insurance Program (NFIP) as cited at Title 44, Chapter 1, Section 65.10, of the Code of Federal Regulations, Section 65.10 (44 CFR Section 65.10). PM 34 reiterates that the community or other parties seeking recognition of a levee or levee system are responsible for providing information to demonstrate that the levee provides protection from the base (1-percent-annual-chance) flood. Therefore, when a study/mapping project is initiated under the Flood Map Modernization (Map Mod) program, FEMA will request that the community, levee owner, and/or local project sponsor to provide the data described in 44 CFR Section 65.10.

The requirement for complying with 44 CFR Section 65.10 is the responsibility of the community, levee owner, and/or local project sponsor, and they often find it difficult to provide the full documentation promptly. The lack of readily available data to comply with 44 CFR Section 65.10 has, in some cases, caused studies/ mapping projects to be delayed or placed on hold until the required information can be compiled and provided to FEMA.

Providing communities with up-to-date, accurate, and reliable flood hazard information on Digital Flood Insurance Rate Maps (DFIRMs) is one of the primary goals of Map Mod. Because levees are shown on the currently effective Flood Insurance Rate Maps (FIRMs) for over one-quarter of the counties being mapped under Map Mod, the issue of whether the levee provides 1-percent-annual-chance flood protection must be addressed. While 44 CFR Section 65.10 documentation is being compiled, the existing FIRMs remain in effect, showing the area behind the levee as protected from the 1-percent-annual-chance flood and potentially delaying the release of more up-to-date information for other parts of the community. As a result, communities would potentially be using outdated flood hazard information to regulate floodplain development. In addition, because the existing FIRMs are in effect, there may be no requirements for the purchase of flood insurance in areas that actually are floodprone.

This guidelines document outlines five scenarios that will allow the mapping for selected studies/mapping projects for communities with levees to move forward before the full documentation required in 44 CFR Section 65.10 is available. With this process, the FEMA Regional Offices, FEMA contractors, and mapping partners can issue preliminary and effective DFIRMs while providing the communities and levee owners a specified timeframe for the submittal of the documentation necessary to show compliance with 44 CFR Section 65.10. In addition, for specific situations, the Regional Offices, contractors, and mapping partners can move forward with the study/mapping project until the point where the Letters of Final Determination would be issued, while communities and/or levee owners are given a specified timeframe to address maintenance deficiencies identified by the U.S. Army Corps of Engineers (USACE).

The USACE has initiated a national levee inventory and assessment program to identify the condition, location, level of protection, and maintenance activities for all levees within its jurisdiction. This inventory will assist in the assessment of the risk to public safety associated with levees and levee systems across the Nation. The USACE and FEMA are working together throughout the inventory and assessment phase to coordinate this effort with Map Mod activities. The inventory data collected to date will be used by FEMA and the USACE to categorize levees for which the full documentation required by 44 CFR Section 65.10 is NOT readily available into the five scenarios described below.

Definitions

A levee is defined as a manmade structure, usually an earthen embankment, designed and constructed in accordance with sound engineering practices to contain, control, or divert the flow of water so as to provide protection from temporary flooding. The term does not include structures that are otherwise defined as dams in the *Federal Guidelines for Dam Safety*.

For the purpose of this guidelines document, levees are identified as being USACE or non-USACE levees. Levees within the USACE program include:

- Levees built by the USACE that were authorized for construction by the U.S. Congress or by USACE continuing authorities (e.g., Section 205);
- Levee projects constructed by non-Federal interests or other (non-USACE) Federal agencies and incorporated into the USACE Federal system by specific congressional action;
- Federal projects that are either operated and maintained by the USACE or turned over to a local sponsor for operation and maintenance; and
- Non-Federal projects within the Rehabilitation and Inspection Program (RIP), Public Law 84-99.

Non-USACE levees are defined to include:

- Levees not authorized by the U.S. Congress or other Federal agency authority;
- Levees built by other Federal agencies and not incorporated into the USACE Federal system;
- Locally built and maintained levees built by a local community; and
- Privately built by a nonpublic organization or individuals and maintained by a local community.

A "levee owner" can be a Federal or State agency, a water management or flood control district, a local community, a levee district, a nonpublic organization, or an individual. The "party responsible for operating and maintaining the levee" must be a Federal or

State agency, an agency created by Federal or State law, or an agency of a community participating in the NFIP.

This document summarizes an approach for identifying a Provisionally Accredited Levee (PAL) and for provisionally mapping the area behind (landward of) such a levee as Zone X (shaded), pending FEMA's receipt of all data and documentation required to show compliance with 44 CFR Section 65.10. FEMA established this approach to allow the mapping process to move forward when levees meet the criteria for an applicable scenario identified below, and to give communities and levee owners a specified timeframe to submit all the documentation necessary to show compliance with 44 CFR Section 65.10. The PAL requirements are provided below.

Scenario A – Levees not in the USACE program that are shown as providing base flood protection

Communication with levee owner and/or community:

The FEMA Regional Office will send a letter (template letter for Scenario A) to the appropriate levee owner or community to identify the levees for which 44 CFR Section 65.10 documentation is needed and will provide a copy of this letter to the appropriate USACE district office. The FEMA letter will describe the PAL option and an opportunity for a one-time-only, 1-year "maintenance deficiency correction period" associated with maintenance-deficient levees. This letter will also request that the community/levee owner submit, within 90 days, one of the following:

- A signed agreement stating that, to the best of the community's/levee owner's knowledge, the levee in question meets 44 CFR Section 65.10 requirements and all requirements for a PAL application package. See criteria for PAL Scenario A1 below.
- A signed letter stating that the community/levee owner has been notified of the one-time-only, 1-year "maintenance deficiency correction period" and agrees to proceed according to the associated process and requirements. See criteria for PAL Scenario A2 below. This one-time-only "maintenance deficiency correction period" will expire 1 year from the 91st day following the date of the initial notification letter. For the purposes this document, this signed letter will be identified as the "maintenance deficiency letter."

Procedures for processing PAL Scenario A1 and A2 levees are described in detail below. To help FEMA contractors and mapping partners properly assess how to handle levee mapping issues, a flowchart has been generated that depicts the possible steps for Scenario A levees. This flowchart is presented as Figure 2 at the end of this guidelines document.

Scenario A1: Once the FEMA Regional Office sends the initial notification letter for Scenario A, the community/levee owner will have 90 days to return either the PAL application package or the maintenance deficiency letter. If the community/levee owner believes that the levee meets 44 CFR Section 65.10 requirements at that time, then they may qualify for Scenario A1.

If the full documentation required to show compliance with 44 CFR Section 65.10 are readily available when the initial notification letter is sent, FEMA will request that the community/levee owner provide these documents within 30 days. If additional time is required to gather the proper documentation, the community/levee owner will choose to submit the PAL application package. For any community/levee owner that chooses the PAL option, the requirements for 44 CFR Section 65.10 must be submitted within 24 months of the 91st day following the date of the initial notification letter. Certification by a Registered Professional Engineer must accompany the submitted 44 CFR Section 65.10 data in compliance with Paragraph 65.10(e). In addition, the community/levee owner must submit a progress report to FEMA after 12 months to document progress toward obtaining 44 CFR Section 65.10 data and documentation.

Several conditions exist that may require FEMA to take immediate action to rescind the PAL designation and revise the DFIRM for the area landward of the levee. If any of the following conditions apply, FEMA will direct the contractor or mapping partner to remap the area landward of the levee as Zone AE or Zone A, depending on the type of study performed for the area:

- Neither the signed PAL agreement nor a request for a maintenance deficiency correction period is returned to FEMA before the 91st day following the date of the notification letter;
- The full documentation required for compliance with 44 CFR Section 65.10 is not provided within 24 months of the 91st day following the date of the initial notification letter; or
- The 12-month progress report is not provided to FEMA, and the FEMA Regional Office believes the PAL agreement should be rescinded.

When the FEMA Regional Office sends the initial notification letter, the following attachments must be included:

- A description of the requirements to meet 44 CFR Section 65.10, entitled "Mapping of Areas Protected by Levee Systems"; and
- An agreement to accept the PAL option (agreement for Scenario A1), for the community/levee owner to sign and return to FEMA before the 91st day following the date of the initial notification letter.

The PAL application package requirements for Scenario A1, to be submitted by levee owner or community, are:

- An agreement signed by the community/levee owner stating that the PAL designation is warranted because the levee now meets the requirements of 44 CFR Section 65.10;

- A copy of the adopted operation and maintenance plan for the levee; and
- Records of levee maintenance and operation, as well as tests of the mechanized interior drainage systems, if applicable.

Scenario A2: Once the FEMA Regional Office sends the initial notification letter for Scenario A, the community and/or levee owner will be given 90 days to return either the PAL application package or the maintenance deficiency letter. If the community/levee owner believes that the levee meets 44 CFR Section 65.10 requirements with the exception of maintenance deficiencies, then they may qualify for Scenario A2.

Once the community/levee owner determines that maintenance deficiencies exist, the community/levee owner will have 90 days from the date of the initial notification letter to submit a signed letter requesting the maintenance deficiency correction period. At a minimum, this letter must clearly state:

- The only grounds for the levee in question not currently meeting the 44 CFR Section 65.10 requirements or PAL requirements are maintenance issues; and
- Within the 1-year "maintenance deficiency correction period," the community/ levee owner can remedy the maintenance deficiencies and submit one of the following:
    - All documentation necessary to comply with the requirements listed in 44 CFR Section 65.10; or
    - A request for a PAL designation and the entire PAL application package (PAL application requirements listed below).

If the community/levee owner submits a response before the 91st day following the date of the initial notification letter, the FEMA Regional Office will notify the community/ levee owner that the current study/mapping project will move forward and that the area landward of the levee will be remapped and shown as Zone AE or Zone A, depending on the type of study performed for the area. The notification will state that the Letter of Final Determination (LFD) and effective DFIRM will be delayed until the 1-year correction period has elapsed. For FEMA to remove the Special Flood Hazard Area (SFHA) designation landward of the levee, the community and/or levee owner must submit the following within the 1-year correction period:

- All the requirements listed in 44 CFR Section 65.10; or
- A request for a PAL designation and the entire PAL application package (PAL application package requirements listed below).

If all the data and documents required to comply with 44 CFR Section 65.10 are submitted before the 1-year correction period has elapsed, FEMA will issue the LFD and show the levee on the effective DFIRM as accredited. However, if a request for a PAL designation and a PAL application package are submitted and approved before the 1-year correction period has elapsed, then FEMA will issue the LFD and show the levee on the effective DFIRM as provisionally accredited. In addition, for the PAL option, the

community/levee owner must provide a progress report to the FEMA Regional Office after 12 months to document progress toward obtaining 44 CFR Section 65.10 data
If any of the following alternatives occur, FEMA will issue the LFD and an effective DFIRM that shows the areas landward of the levee will be remapped and shown as Zone AE or Zone A, depending on the type of study performed for the area:

- The community/levee owner does not submit a signed response letter before the 91st day following the date of the initial notification letter.
- The community/levee owner is granted the 1-year correction period, but does not submit the required data within the 1-year correction period.
- The submitted deficiency correction data are determined to be inadequate.
- A request for a PAL designation and the entire PAL application package is not submitted and approved before the 1-year correction period has elapsed.
- The 12-month PAL progress report is not provided to FEMA, and the FEMA Regional Office believes the PAL designation should be rescinded.
- The full documentation necessary to comply with 44 CFR Section 65.10 is not provided within 24 months of the final day of the correction period.
- The data and documentation submitted to meet the requirements of 44 CFR Section 65.10 or the PAL application is determined to be inadequate.

When the FEMA Regional Office sends the initial notification letter, the following attachments must be included:

- A description of the requirements to meet 44 CFR Section 65.10, entitled "Mapping of Areas Protected by Levee Systems;" and
- An agreement to accept the PAL option (agreement for Scenario A2), for the community/levee owner to sign and return to FEMA.

In response to the initial notification letter, the community/levee owner is to submit the following to FEMA to meet the PAL application package requirements for Scenario A2:

- An agreement signed by the community/levee owner stating that the PAL designation is warranted because the levee now meets the requirements of 44 CFR Section 65.10;
- A copy of the adopted operation and maintenance plan for the levee; and
- Records of levee maintenance and operation, as well as tests of the mechanized interior drainage systems, if applicable.

Scenario B – Levees in the USACE program that are eligible for PAL

Levees in the USACE program that meet the following criteria are eligible for the PAL designation.

Criteria to meet Scenario B:

- The effective FIRM shows the levee as providing protection from the base flood;
- No available information indicates the levee does not provide base flood protection; and
- The project inspection rating is within an acceptable range (as defined by the USACE).

Communication with levee owner, community, and/or local project sponsor:

The FEMA Regional Office will send a letter (template letter for Scenario B) to the appropriate community/levee owner/local project sponsor identifying the levees that meet the above criteria and qualify for the PAL option. This letter will describe the PAL option and request that the community/levee owner/local project sponsor sign an agreement stating that, to the best of their knowledge, the levees in question meet 44 CFR Section 65.10 requirements.

If full documentation to comply with the requirements of 44 CFR Section 65.10 is readily available when the initial notification letter is sent, the FEMA Regional Office will request that the community/levee owner/local project sponsor provide these documents within 30 days. If the community/levee owner/local project sponsor requires time to gather the proper documentation, they will choose to submit the PAL application package. For any community/levee owner/local project sponsor that chooses the PAL, the documentation required to comply with 44 CFR Section 65.10 requirements must be submitted within 24 months of the 91st day following the date of the initial notification letter. Certification by a Registered Professional Engineer must accompany the submitted 44 CFR Section 65.10 data in compliance with Paragraph 65.10(e). As an alternative, USACE may also certify that the levee has been adequately designed and constructed to provide protection against the base flood. In addition, the community/levee owner/local project sponsor must submit a progress report to FEMA after 12 months to document progress toward obtaining data and documentation to comply with 44 CFR Section 65.10

Several conditions could occur that may result in the PAL designation being rescinded and FEMA taking immediate action to revise the DFIRM in the area landward of the levee. If any of the following conditions apply, FEMA will remap the area landward of the levee will be remapped and shown as Zone AE or Zone A, depending on the type of study performed for the area:

- The signed PAL agreement is not returned to FEMA within 90 days of the initial notification letter.
- The full documentation for 44 CFR Section 65.10 is not provided within 24 months of the final day of the 90-day agreement period.
- The 12-month PAL progress report is not provided to FEMA, and the FEMA Regional Office believes rescission is necessary.
- The data submitted to meet the requirements of 44 CFR Section 65.10 or the PAL application are determined to be inadequate.

When the FEMA Regional Office sends the initial notification letter, the following attachments must be included:

- A description of the requirements to meet 44 CFR Section 65.10, entitled "Mapping of Areas Protected by Levee Systems"; and
- An agreement to accept the PAL option (agreement for Scenario B), for the community/levee owner/local project sponsor to sign and send back to FEMA before the 91st day following the date of the initial notification letter

In response to the initial notification letter, the community/levee owner/local project sponsor is to submit the following to FEMA to meet the PAL application package requirements for Scenario B:

- An agreement signed by the community/levee owner/local project sponsor stating that the PAL designation is warranted because the levee now meets the requirements of 44 CFR Section 65.10;

Scenario C – Levees in the USACE program with known deficiencies that are shown as providing base flood protection

For levees in the USACE program that are shown on the effective FIRM as providing 1-percent-annual-chance flood protection but have known deficiencies, the criteria below are to be followed. Two scenarios are possible. To help FEMA contractors and mapping partners properly assess how to handle levee mapping issues, a flowchart has been generated that depicts the possible steps for Scenario C levees. This flowchart is presented as Figure 3 at the end of this guidelines document.

Scenario C1:

Criteria to meet Scenario C1:

- The USACE has determined that the levee's recent inspection ratings are listed as fair, poor, or unacceptable;
- The USACE has determined that the project status in the RIP has been switched from active to inactive; and
- The USACE has not provided a 1-year maintenance deficiency correction period for the levee.

Communication with levee owner, community, and/or local project sponsor:

The FEMA Regional Office will coordinate with the appropriate USACE District office regarding levee projects in the USACE inventory that have received an inspection rating of fair, poor, or unacceptable and have been placed in inactive status in the USACE Rehabilitation and Inspection Program (RIP). Once these projects have been identified, the USACE will send a notification letter to the community/levee owner/local project sponsor to inform them that the levee status has been switched from active to inactive in the RIP and is no longer eligible for PL 84-99 rehabilitation assistance because of maintenance deficiencies. These deficiencies will not allow the levee to meet the minimum requirements of the 44 CFR Section 65.10; thus, the levee does not provide 1-percent-annual-chance flood protection. The deficiencies will be identified in the USACE letter. The USACE District office will provide a copy of this letter to the FEMA Regional Office. These levee systems will not be eligible for the PAL option.

The FEMA Regional Office then will send a letter (template letter for Scenario C) to the community/levee owner/local project sponsor stating that this is a follow-up to the notification they received from the USACE. The FEMA letter will clearly state that the deficiencies in the levee have been determined in coordination with the USACE, and FEMA will remap the area landward of the levee to show it as Zone AE or Zone A, depending on the type of study performed for the area, if the levee does not provide 1-percent-annual-chance flood protection. The USACE letter will be attached to the FEMA letter as background information.

Scenario C2:

<u>Criteria to meet Scenario C2:</u>

- The levee has received an unacceptable, fair, or poor inspection rating;
- The levee was in an active status in the USACE RIP prior to September 30, 2005 (FY06); and
- The USACE has offered a one-time-only, 1-year "maintenance deficiency correction period" to remedy the maintenance deficiencies of the levee.

<u>Communication with levee owner, community, and/or local project sponsor:</u>

Once these projects have been identified, the USACE will send a notification letter to the community/levee owner/local project sponsor to inform them of the levee's specific maintenance deficiencies. This letter will also inform the community/levee owner/local project sponsor that they are eligible for the one-time-only, 1-year "maintenance deficiency correction period," which provides them 1 year to resolve any levee maintenance deficiencies. The USACE District office will provide a copy of this letter to the FEMA Regional Office.

The FEMA Regional Office then will send a letter (template letter for Scenario C2) to the community/levee owner/local project sponsor stating that this is a follow-up to the notification they received from the USACE. The letter will explain the PAL option (Scenario C2) and that FEMA will move forward with the current study/mapping project and will remap the area landward of the levee that will be inundated by the 1-percent-annual-chance flood as Zone AE or Zone A, depending on the type of study performed for the area. The letter also will state that even though FEMA is moving forward with the mapping, the LFD and effective DFIRM will be delayed until the 1-year correction period has elapsed. For FEMA to remove the SFHA designation landward of the levee, the following requirements must be met within the 1-year correction period:

- Evidence has been provided to show that the maintenance deficiencies have been remedied. This evidence will be provided to the FEMA Regional Office by the appropriate USACE District office.
- All of the requirements listed in 44 CFR Section 65.10 have been addressed, or a request for a PAL designation and the entire PAL application package has been submitted.

The FEMA Regional Office will coordinate with the appropriate USACE District regarding levee projects to evaluate and determine the adequacy of any data submitted within the 1-year period.

If all the data and documentation required to comply with 44 CFR Section 65.10 are submitted before the 1-year correction period has elapsed, FEMA will issue the LFD and show the levee on the effective DFIRM as accredited. Alternatively, if a request for a PAL designation and a PAL application package are submitted and approved before the 1-year correction period has elapsed, then FEMA will issue the LFD and show the levee on the effective DFIRM as provisionally accredited. In addition, to the community/levee owner/local project sponsor must submit a progress report to FEMA after 12 months to document progress toward obtaining documentation and data to comply with 44 CFR Section 65.10.

If any of the following alternatives occur, FEMA will issue the LFD and an effective DFIRM that shows the areas landward of the levee will be remapped and shown as Zone AE or Zone A, depending on the type of study performed for the area:

- The community/levee owner/local project sponsor is granted the 1-year correction period, but does not submit the required data within the 1-year correction period.
- The submitted deficiency correction data is determined to be inadequate.
- The 12-month PAL progress report is not provided to FEMA, and the FEMA Regional Office believes the PAL designation should be rescinded.
- A request for a PAL designation and the entire PAL application package is not submitted and approved before the 1-year correction period has elapsed.
- The full documentation necessary to comply with 44 CFR Section 65.10 is not provided within 24 months of the final day of the correction period.
- The data submitted to meet the requirements of 44 CFR Section 65.10 or the PAL application is determined to be inadequate.

When the FEMA Regional Office sends the initial notification letter, the following attachments must be included:

- A description of the requirements to meet 44 CFR Section 65.10, entitled "Mapping of Areas Protected by Levee Systems;" and

- An agreement to accept the PAL option (agreement for Scenario C2), for the community/levee owner/local project sponsor to sign and return to FEMA.

The PAL application package requirements for Scenario C2, to be submitted by levee owner, community, and/or local project sponsor, are:

- An agreement signed by the community/levee owner/local project sponsor stating that the PAL designation is warranted because the levee now meets the requirements of 44 CFR Section 65.10;

Scenario D – Levees in the USACE program that are shown as not providing base flood protection

Communication with levee owner, community, and/or local project sponsor:

For levees in the USACE program that are not currently shown as providing 1-percent-annual flood protection, no letter will be sent. In this case, there is no issue with how to map the area behind the levee, because it already has been determined that the levee does not provide 1-percent-annual flood protection or the levee has not gone through the certification process.. The DFIRM will continue to show the levee as not providing 1-percent-annual flood protection unless it is determined that the levee actually does provide this level of protection.

Scenario E – Levees in the USACE program that do not meet an adequate level of protection, as determined by the USACE in coordination with FEMA, but are shown as providing base flood protection

For levees in the USACE program that are shown as providing base flood protection but do not meet an adequate level of protection as determined by the USACE in coordination with FEMA, the following criteria will be followed.

Criteria to meet Scenario E:

- The levee is shown as providing protection but does not provide an adequate level (1-percent-annual-chance or greater) of flood protection, as indicated by the USACE levee inventory data and validated through coordination between the USACE district office and the FEMA Regional Office; and
- The levee inspection rating is NOT listed as fair, poor, or unacceptable, but the levee may have failed or experienced overtopping by a flood event less than the 1-percent-annual-chance flood.

The FEMA Regional Office will verify the engineering and mapping data used to produce the effective FIRM and determine whether it is the most up-to-date information, based on the best available data. However, the FEMA Regional Office will also determine if better data are available than the data used to produce the effective FIRM. The FEMA Regional Office will coordinate with the USACE district office to either verify the current flood data are the best available or provide the more recent and accurate data. The USACE district office will use the best available data, as identified by the FEMA Regional Office, to determine whether the levee provides an adequate level of protection.

Communication with levee owner, community, and/or local project sponsor:

When the USACE district office in coordination with the FEMA Regional Office, determines that a levee in the USACE program does not provide an adequate level of protection, the community/levee owner/local project sponsor will be notified by letter (template letter for Scenario E) from the FEMA Regional Office that "in coordination with the USACE, it has been determined that your levee no longer provides protection from the base flood." The reasons the levee no longer provides 1-percent-annual-chance flood protection will be identified in this letter. The community/levee owner/local project sponsor will be instructed to contact the FEMA Regional Office if they have any questions or if they can provide the documentation and data necessary to show compliance with 44 CFR Section 65.10. If the community/levee owner/local project sponsor does not provide the required documentation and data, the area landward of the levee will be mapped as Zone AE or Zone A, depending on the type of study performed for the area.

Mapping of the areas with and without the PAL designation

Levees and levee systems that meet the 44 CFR Section 65.10 criteria will continue to be mapped as providing protection from the 1-percent-annual-chance flood, and the PAL designation is not applicable. The area landward of the levee will be mapped as a Zone X (shaded) with the following note:

> WARNING! This area is shown as being protected from the 1-percent-annual-chance flood hazard by levee, dike, or other structure. Overtopping or failure of this structure is possible, which could result in destructive flood elevations and high-velocity floodwaters. There is a chance that large floods will occur that are greater than the level of protection provided by the levee. Communities should issue evacuation plans and encourage property owners behind these structures to purchase flood insurance, even if the structure is currently shown as providing protection for the 1-percent-annual-chance flood.

For levees and levee systems that are eligible for the PAL designation, the area landward of levees can still be mapped as Zone X (shaded), with the following note applied at several locations in or near the zone:

> WARNING: Provisionally Accredited Levee. For explanation, see the Notes to Users.

The following accompanying note in the Notes to Users should only be applied to DFIRM panels that depict a levee or levee system with a PAL designation:

> WARNING: This levee, dike, or other structure has been provisionally accredited and mapped as providing protection from the 1-percent–annual-chance flood. To maintain accreditation, the levee owner or community is required to submit documentation necessary to comply with 44 CFR Section 65.10 by (__________, ____). Because of the risk of overtopping or failure of the structure, communities should take proper precautions to protect lives and minimize damages in these areas, such as issuing an evacuation plan and encouraging property owners to purchase flood insurance.

The DFIRM in Figure 1 shows a levee on the east side of the river and Zone X (shaded) landward of the levee, with the warning note pointing to the levee. If a Zone X (shaded) area is already depicted on the effective FIRM/DFIRM, then the revised levee note can be added to the existing Zone X (shaded) area, as shown in Figure 1. If no Zone X (shaded) area exists on the effective FIRM/DFIRM, then the mapping partner should define the provisional Zone X (shaded) area using the best available data.

Flood Insurance Study Report Requirements

The Flood Insurance Study report should not be revised to identify those levees and levee systems that are eligible for the PAL designation.

Figure 1. Example of Zone X (Shaded) for the PAL Option

LEVEES
IDENTIFYING THE RISK

Provisionally Accredited Levees

Answers to Questions About Procedure Memorandum No. 43

Providing communities with up-to-date, accurate, and reliable flood hazard and risk information on Digital Flood Insurance Rate Maps (DFIRMs) is one of the primary goals of the Flood Map Modernization (Map Mod) effort undertaken by the Department of Homeland Security, Federal Emergency Management Agency (FEMA). Levee systems have been identified in over one-fourth of the counties that will receive modernized maps—Digital Flood Insurance Rate Maps (DFIRMs)—as part of Map Mod. Therefore, FEMA has been working, and continues to work with Federal, State, and local professionals and technical partners to determine the flood protection and risk-reduction capabilities of the Nation's levee systems and to accurately reflect the flood hazard and risk in "levee-impacted" areas on the DFIRMs. As part of the Map Mod effort, FEMA reviewed its existing guidance regarding the submittal of data and documentation to meet National Flood Insurance Program (NFIP) requirements for the evaluation and mapping of levee-impacted areas. As a result of this review, FEMA issued three Procedure Memorandums to clarify these requirements. The questions and answers below are provided to further explain these requirements.

Q: What is a levee system?

A: A levee system is a flood protection system that consists of a levee, or levees, and associated structures, such as closure and drainage devices, which are constructed and operated in accordance with sound engineering practices. A levee is a manmade structure, usually an earthen embankment, designed and constructed in accordance with sound engineering practices to contain, control, or divert the flow of water so as to provide protection from temporary flooding

Levee systems are designed to provide a *specific level of protection*. They can be overtopped or fail in larger flood events. They require regular maintenance and periodic upgrades to retain their level of protection. When levee systems do fail, they often fail catastrophically, and the resulting damage, including loss of life, may be more significant than if the levee system had not been built. Everyone should understand the risk to life and property that exists behind levee systems—risk that even the best flood protection system cannot eliminate completely.

Q: What regulations apply to the evaluation and mapping of levee systems and levee-impacted areas?

A: The regulatory requirements of the NFIP are cited at Title 44, Chapter 1, Section 65.10 of the Code of Federal Regulations (44 CFR Section 65.10). According to 44 CFR Section 65.10, it is the community, levee owner, and/or local project sponsor's responsibility to submit the data and documentation showing that a levee system complies with these requirements, including the development and maintenance of an operation and maintenance plan. You may access 44 CFR Section 65.10 through the FEMA Web site at www.fema.gov/plan/prevent/fhm/lv_fpm.shtm.

Q: What are the procedural requirements for evaluating and mapping levee-impacted areas?

A: FEMA has issued three Procedure Memorandums that provide guidance for mapping levee-impacted areas—Procedure Memorandum No. 34 (PM 34) —*Interim Guidance for Studies Including Levees,* Procedure Memorandum No. 43 (PM 43)—*Guidelines for Identifying Provisionally Accredited Levees*, and Procedure Memorandum No.45 (PM 45)—*Revisions to Accredited Levee and Provisionally Accredited Levee Notations* —as clarification to Appendix H of FEMA's *Guidelines and Specifications for Flood Hazard Mapping Partners.* PM 43 describes the various PAL scenarios and how each may be mapped. Information on PM 45, which provides updated information on the map notes that are to appear on DFIRM panels showing accredited and provisionally accredited levee systems, is provided on a separate Fact Sheet. You also may access these three Procedure Memorandums and other related guidance through the FEMA Web site at www.fema.gov/plan/prevent/fhm/lv_fpm.shtm.

Q: Who is responsible for complying with the requirements of 44 CFR Section 65.10?
A: Compliance with 44 CFR Section 65.10 requirements rests with communities, levee owners, and/or local project sponsors—not FEMA. A levee system owner can be a Federal or State agency, a water management or flood control district, a local community, a levee district, a non-public organization, or an individual. The party responsible for operating and maintaining the levee system must be a Federal or State agency, an agency created by Federal or State law, or an agency of a community participating in the NFIP. FEMA's responsibility is solely to review the data and documentation provided and either accredit the levee system with providing 1-percent-annual-chance flood protection on the DFIRM or, when the levee system is shown to be inadequate, to reflect the increased risk of flooding to people and structures in the levee-impacted areas on the DFIRM.

Q: Why did FEMA issue PM 34?
A: Documentation regarding levee design, structural integrity, and other requirements for accrediting a levee system with providing 1-percent-annual-chance flood protection often is outdated or missing altogether. To help clarify the entities responsible for providing data and documentation on levees systems identified during a study/mapping project, FEMA issued PM 34 on August 22, 2005. PM 34 clarifies that it is the levee owner or community's responsibility to provide data and documentation to show that a levee system meets the requirements of 44 CFR Section 65.10 as part of a study/mapping project. In addition, PM 34 provides clarification on procedures to minimize delays in near-term studies/mapping projects and to aid project teams in properly assessing how to handle levee system-related mapping issues.

Q: When and why did FEMA issue PM 43?
A: FEMA originally issued PM 43 on September 25, 2006. PM 43 provides guidance to FEMA contractors and mapping partners on issuing preliminary and, in some cases, effective DFIRMs, while providing communities and levee owners with additional time to compile and submit the data and documentation necessary to demonstrate compliance with 44 CFR Section 65.10. On March 16, 2007, FEMA issued a revised version of PM 43 to include guidance on evaluating levees systems that the U.S. Army Corps of Engineers has determined to be maintenance deficient and to offer a one-time-only 1-year "maintenance deficiency correction period."

Q: When is a levee system designated as a Provisionally Accredited Levee, or PAL, system?
A: The PAL designation is used for a levee system when FEMA has previously accredited the system with providing 1-percent-annual-chance flood protection on an effective Flood Insurance Rate Map (FIRM) and FEMA is awaiting data and/or documentation that will demonstrate the levee system's compliance with 44 CFR Section 65.10 of the NFIP regulations. A PAL is shown on a DFIRM as providing 1-percent-annual-chance flood protection, and the area impacted by the PAL system is shown as Zone X (shaded) except for areas of residual flooding, such as ponding areas, which will be shown as a Special Flood Hazard Areas (SFHAs).

Q: What happens when a levee system meets the PAL requirements of PM 43?
A: For levee systems that meet the PAL requirement (levee systems presently shown as providing 1-percent-annual-chance flood protection on the effective FIRM and for which the community or levee owner cannot readily provide all data and documentation required by 44 CFR Section 65.10), FEMA will place a note on the DFIRM panel landward of the levee system to indicate FEMA has provisionally accredited the levee system and the designation of any existing Zone X (shaded) area is provisional. FEMA will also add an explanatory note to the Notes to Users section of the map frame.

Before FEMA will designate a levee system as a PAL system, the community or levee owner will need to sign and return an agreement. By signing the agreement, the levee owner/community indicates the levee system currently complies with the requirements of 44 CFR Section 65.10 and that the data and documentation required for compliance with 44 CFR Section 65.10 will be provided within a specified timeframe. The timeframe will depending on the levee system's status (i.e., within 24 months of the 91st day following the date of the initial notification letter or within 24 months of the final day of the correction period for levees that have been offered the 1-year maintenance deficiency correction period).

Q: How will a PAL be identified on a DFIRM?

A: To identify the PAL, FEMA has been applying and may continue to apply the note below at several locations on DFIRM panels that will become effective before December 1, 2008. The note points to the levee system and is placed on the landward side of the levee system on the affected DFIRM panel(s) in or near the Zone X (shaded) area:

> WARNING: Provisionally Accredited Levee. For explanation, see the Notes to Users.

The following note has been or will be added to the Notes to Users on DFIRM panels that will become effective before December 1, 2008:

> WARNING: This levee, dike, or other structure has been provisionally accredited and mapped as providing protection from the 1-percent-annual-chance flood. To maintain accreditation, the levee owner or community is required to submit documentation necessary to comply with 44 CFR Section 65.10 by (__________,____). Because of the risk of overtopping or failure of the structure, communities should take proper precautions to protect lives and minimize damages in these areas, such as issuing an evacuation plan and encouraging property owners to purchase flood insurance.

The notes that will appear on the DFIRM panels that will become effective after December 1, 2008, are documented in PM No. 45 and in an accompanying Fact Sheet titled "Levee Notes on FEMA Maps: Answers to Questions About Procedure Memorandum No. 45."

Q: How does FEMA determine if a levee system meets the PAL requirements of PM 43?

A: For a levee system to be eligible for PAL designation, the levee system must be shown as providing protection from the 1-percent-annual-chance flood on the effective FIRM. Additional PAL requirements include the submittal of a PAL application package and a 12-month progress report. If applicable, there are other potential requirements, including a letter requesting a maintenance deficiency correction period and submittal of data demonstrating that maintenance deficiencies have been corrected (as appropriate). Specific timeframes for these requirements vary depending on the levee's status; however, more detailed information can be found in the guidance document, titled "Guidelines for Identifying Provisionally Accredited Levees." that accompanied PM 43. This document contains descriptions of different mapping scenarios and is accessible through the FEMA Web site at www.fema.gov/plan/prevent/fhm/lv_fpm.shtm.

For levee systems in the U.S. Army Corps of Engineers (USACE) Program that are shown on the effective FIRM as providing 1-percent-annual-chance flood protection but have known deficiencies, FEMA will coordinate with the appropriate USACE district to determine if the USACE will offer the one-time-only, 1-year maintenance deficiency period.

Q: What if a levee system qualifies for the maintenance deficiency correction period as specified in PM 43?

A: For levee systems not in the USACE Program, if the community/levee owner believes that the levee system meets 44 CFR Section 65.10 requirements with the exception of maintenance deficiencies, then the community/levee owner may qualify for a one-time-only 1-year maintenance deficiency correction period. The community/levee owner will have 90 days from the date of the initial notification letter from FEMA to submit a signed letter requesting the maintenance deficiency correction period.

The community/levee owner then has 12 months to submit 44 CFR Section 65.10-compliant data and documentation or a completed PAL application package (if additional time is needed to compile 44 CFR Section 65.10-compliant data and documentation) to show the levee system as accredited. If the community/levee owner does not provide 44 CFR Section 65.10-compliant data and documentation or a completed PAL application within the 12-month period, then FEMA will issue an effective DFIRM showing the area landward of the levee as a Special Flood Hazard Area (SFHA), labeled Zone AE or Zone A, depending on the type of engineering study performed for the area.

The USACE determines whether the one-time-only, 1-year maintenance deficiency correction period will be offered for a levee system in the USACE Program. If the USACE does offer the correction period for the levee system, FEMA will de-accredit the levee system and remap the levee-impacted area to show it as a high-risk SFHA, labeled Zone AE or Zone A, depending on the type of engineering study performed for the area. If the USACE does offer the correction period for the levee system, the community/levee owner has 1 year to either submit data and documentation for 44 CFR Section 65.10 compliance or request and be approved for PAL designation (if additional time is needed). If neither is received, then FEMA will de-accredit the levee system and will issue an effective DFIRM showing the levee-impacted area as an SFHA, labeled Zone AE or Zone A, depending on the type of engineering study performed for the area.

Q: What qualifies as a USACE Program levee system?
A: Levee systems within the USACE Program are defined to include the following:

- Levee systems built by the USACE that were authorized for construction by the U.S. Congress or by USACE continuing authorities (e.g., Section 205);
- Levee system projects constructed by non-Federal interests or other (non-USACE) Federal agencies and incorporated into the USACE Federal system by specific congressional action; and
- Federal projects that are either operated and maintained by the USACE or turned over to a local sponsor for operation and maintenance; and Non-Federal projects within the Rehabilitation and Inspection Program (RIP), Public Law 84-99.

Q: What qualifies as a Non-USACE Program Levee?
A: Non-Federal levees are defined to include the following:

- Levee systems not authorized by the U.S. Congress or other Federal agency authority;
- Levee systems built by other Federal agencies and not incorporated into the USACE Federal system;
- Locally built and maintained levee systems built by a local community; and
- Levee systems that are privately built by a nonpublic organization or individuals and maintained by a local community.

Q: Does adhering to PM 43 delay the release of new DFIRMs?
A: The PM 43 process allows FEMA to issue the preliminary and effective DFIRMs while providing communities and levee owners a specified timeframe to submit the data and documentation necessary to show compliance with 44 CFR Section 65.10. For levee systems with maintenance deficiencies (that are otherwise believed to comply with the requirements of 44 CFR Section 65.10), the release of new DFIRMs may be delayed up to 1 year to provide the community/levee owner with additional time to correct these deficiencies.

It is important that community officials and citizens have the most accurate and up-to-date information to make decisions based on the flood risk that exists in areas behind levee systems. PM 43 allows for community officials and the public to have the most current flood hazard and risk information while the community or levee owner is given a reasonable amount of time to compile and submit data and documentation to show compliance with the requirements of 44 CFR Section 65.10.

FEMA

APR 24 2009

MEMORANDUM FOR: Mitigation Division Directors
Regions I - X

FROM: Doug Bellomo, Director
Risk Analysis Division

SUBJECT: Procedure Memorandum No. 53
Guidance for Notification and Mapping of Existing Provisionally Accredited Levee Designations

EFFECTIVE DATE: Effective April 24, 2009, for all current, expiring, and Future Provisionally Accredited Levee Designations

Background: To clarify the guidance for the evaluation and mapping of levee systems provided in Appendix H of *Guidelines and Specifications for Flood Hazard Mapping Partners*, the Department of Homeland Security's (DHS), Federal Emergency Management Agency (FEMA), issued Procedure Memorandum No. 43 (PM 43) on September 25, 2006. PM 43 introduced the concept of the Provisionally Accredited Levee (PAL) system designation and established a reasonable timeframe (24 months) for a community or levee system owner to supply data and documentation demonstrating that a levee system owner to supply data and documentation demonstrating that a levee system meets the requirements of the National Flood Insurance Program (NFIP) regulations cited in the Code of Federal Regulations (CFR) at Title 44, Chapter 1, Section 65.10 (44 CFR Section 65.10). FEMA also implemented certain timeframes (12 months) where communities and/or levee system owners are to submit progress reports during the 24-month PAL data and documentation submittal period. To address subsequent issues and provide further clarification of requirements, FEMA issued a revised version of PM 43 on March 16, 2007. On May 12, 2008, FEMA issued PM 45 to provide updated guidance on the notes that will appear on DFIRM panels on which levee systems and levee-impacted areas are shown.

Issue: As the 24-month deadline approaches for communities and/or levee system owners to submit 44 CFR Section 65.10-compliant data and documentation for the PAL systems, FEMA recognized that a plan for mapping levee-impacted areas after the PAL period expiration date

needed to be developed in association with a public outreach and awareness strategy for these areas. The plan needs to cover both accredited levee systems and de-accredited levee systems. In developing the plan, FEMA will continue to consider stakeholders' concerns that communities and levee system owners are confronted with serious challenges in providing 44 CFR Section 65.10-compliant data and documentation.

Action Taken:
Guidelines have been produced to address these needs. The attached document, titled "Guidelines for Notification and Mapping of Expiring Provisionally Accredited Levee Designations," summarizes planned activities related to the mapping of levee-impacted areas after the PAL period expires.

The attached guidance document is to be implemented by all Regions, contractors, and mapping partners as PAL expiration dates are reached.

Attachment:

"Guidelines for Notification and Mapping of Expiring Provisionally Accredited Levee Designations"

cc: See Distribution List

Distribution List (electronic distribution only):

Office of the Acting Assistant Administrator for Mitigation

Risk Analysis Division

Risk Reduction Division

Risk Insurance Division

Regional Mitigation Division

Legislative Affairs Division

Office of Chief Counsel

Mapping Partners

Program Management Contractor

National Service Provider

Independent Verification and Validation Contractor

Map Service Center Contractor

Indefinite Delivery Indefinite Quantity Contractors

Customer and Data Services Contractor

Production and Technical Services Contractors

Guidelines for Notification and Mapping of Expiring Provisionally Accredited Levee Designations

Notification of Provisionally Accredited Levee (PAL) Period Expiration

In advance of PAL designations expiring, each FEMA Regional Office will remind the levee system owners, communities, and other stakeholders of the impending expiration date. This notification will be accomplished by sending two letters during the 24-month expiration period to all appropriate parties. The Regional Offices will send the first letter 90 days before the PAL period expiration date, and Regional Offices will send the second letter 30 days before the PAL period expiration date. The attached letter templates are provided to support this Regional communication effort.

The Regional Offices will address the letters to each levee system owner signatory of the PAL Agreement, and will send copies of the letters to the following (at a minimum):

- Chief Executive Officers of all affected communities;
- Floodplain Administrators of all affected communities;
- State National Flood Insurance Program (NFIP) Coordinators;
- State levee safety officials, where appropriate;
- District offices of U.S. Senators and U.S. Representatives; and
- U.S. Army Corps of Engineers (USACE) District offices, if the PAL system is in USACE program. A USACE program levee system is defined in the Guidelines for Identifying Provisionally Accredited Levees (PALs), dated March 16, 2007.

It is important that appropriate Congressional offices are aware of this communication effort in advance of the PAL period expiration date. Therefore, in addition to sending copies of the letters, the Regional offices shall notify their Regional Legislative Affairs Division who may coordinate additional outreach efforts, as appropriate.

The Regional Offices may utilize certified mail with return receipts or fax the letters to the levee system owners and affected communities, following up with a telephone call, to confirm receipt of the original or faxes letters. FEMA Regions, or their assigned contractor will file return receipts, fax reports, or phone logs documenting the follow-up telephone calls in the case file for each affected community. FEMA Regions, or their assigned contractor also will assure that the documents are included with other technical and administrative support data in the Technical Support Data Notebook for the mapping project that commences upon the expiration of the PAL period.

Mapping Process for Expired PAL Designations

Upon expiration of a PAL period, the FEMA Regional Office will determine if the PAL system will be accredited or de-accredited. This determination will be based on several conditions:

- Receipt of a 44 CFR Section 65.10-compliant data and documentation submittal for the levee system;
- A Regional Office-funded review and approval or denial of the submittal based on completeness when compared with the 44 CFR Section 65.10-compliant levee system checklist; and
- Resolution of conflicting data and documentation that may be submitted by other stakeholders.

FEMA will initiate a mapping project to revise the levee-impacted areas immediately upon determining that the PAL system is to be accredited or de-accredited. Depending on the scope and cost of the mapping project necessary to either accredit or de-accredit the system, the process can take up to 12 months before a Preliminary Digital Flood Insurance Rate Map (DFIRM) can be issued that reflects either the newly designated accredited or de-accredited levee system.

Physical Map Revisions (PMRs) will be used to revise DFIRMs that are impacted by FEMA revising the map notations (Map notes and Notes to Users) and floodplain boundary delineations for levee-impacted areas. There are some possible exceptions to this mapping project format. For example, a countywide study/mapping project may be warranted when the majority of the DFIRM panels are being impacted and another study is being processed concurrently within the County. However, it is not acceptable to delay the issuance of a PMR for the levee-impacted areas, to accommodate an ongoing study/mapping project for the purpose of issuing the revised map and report in Countywide Format. In addition, for a single panel mapping project to accredit a levee system, the Letter of Map Revision (LOMR) process may be appropriate, and the Regional Office should consult with FEMA Headquarters to make this determination.

For a de-accredited levee system, FEMA will not revise the floodplain and map notations associated with a PAL system using the LOMR process. There are too many risks associated with outreach and communication to be handled in a simplified process such as a LOMR. Further, from the start of the mapping project for a de-accredited levee system until the final effective date of the revised flood map, no less than 18 months shall pass to allow for proper outreach and due process. Any exceptions to either of these criteria must be coordinated with FEMA Headquarters.

Remapping of Area Impacted by an Accredited Levee System

For communities submitting 44 CFR Section 65.10-compliant data and documentation, the PAL designation will be revised on the DFIRM to identify the levee system as accredited. the notations in the Notes to Users regarding the PAL system must be removed and replaced with the current Notes to Users provided in Procedure Memorandum No. 45, making anything less than a full panel revision potentially confusing to users and, therefore, impractical. Procedures for evaluation and mapping of levee-impacted areas are found in Appendix H of FEMA's *Guidelines and Specifications for Flood Hazard Mapping Partners* (the *Guidelines*) and in Procedure Memorandum No. 52 – *Guidance for Mapping Processes Associated with Levee Systems*. The standard processing and outreach practices that are associated with mapping projects shall be followed.

Remapping of Area Impacted by a De-Accredited Levee System

At the end of 24-month PAL period, if the 44 CFR Section 65.10-compliant data and documentation submittal is not provided to FEMA, or if evidence is received by FEMA that the levee does not meet 65.10, or if the submittal is inadequate as determined by the Regional Office-funded review, FEMA will initiate a mapping project to de-accredit the levee system. Procedures for evaluation and mapping of levee-impacted areas are found in Appendix H of the *Guidelines* and in Procedure Memorandum No. 52. Procedure Memorandum No. 52 provides additional procedures for notification and outreach activities that shall be taken for de-accredited levee systems. This includes establishing a 90-day comment period (if an approximate Zone A area is mapped and no new Base Flood Elevations (BFEs) are proposed) or 90-day appeal period (if new or modified BFEs are proposed) following the issuance of the Preliminary DFIRM.

Levee Inventory

During the remapping of the area impacted by an accredited or de-accredited levee system procedures must be followed to ensure proper inventory and documentation occurs. FEMA's Mid-term Levee Inventory (MLI) and DFIRM database should be used as the appropriate systems to capture any updates for these mapping projects.

PAL Period Extensions Will Not Be Granted

FEMA will not grant extensions to the 24-month PAL period. Per Procedure Memorandum No. 43 – *Guidelines for Identifying Provisionally Accredited Levees*, in order for a levee to be provisionally accredited, the community or levee owner must commit to FEMA in writing to submit the full 44 CFR Section 65.10-compliant data and documentation for the levee within 24 months of this signed agreement. If the submittal is not received or is determined to be inadequate upon the expiration of the 24-month PAL period, FEMA will conclude that the levee is not 65.10-compliant and initiate a mapping project to publish a DFIRM that fully characterizes the appropriate flood hazard and risk associated with the de-accredited levee.

Accreditation Opportunities for Communities and Levee System Owners

FEMA will initiate a map revision immediately after the expiration of the PAL period. However, if a community or levee system owner can provide 44 CFR Section 65.10-compliant data and documentation for a de-accredited levee system, as determined by a Regional Office-funded review and approval based on completeness when compared with the 44 CFR Section 65.10-compliant levee system checklist, the levee system must be accredited and mapped accordingly, either within the current DFIRM update, or as soon as possible after the DFIRM becomes final. Thus, the timing of the submittal will drive the process.

If a community or levee system owner can provide 44 CFR Section 65.10-compliant data and documentation prior to a community adopting the DFIRM or prior to the four month period that would precede the effective date of the mapping project, whichever comes first, FEMA must revise the DFIRM and process it following the accreditation guidance above. While this process does not provide an extension to the PAL period, it allows communities and levee system owners additional time to provide the 44 CFR Section 65.10-compliant data and documentation submittal to FEMA before the DFIRM is finalized and becomes effective. This may cause FEMA to spend funds towards de-accreditation, only to receive the submittal and change direction toward accreditation during the mapping project. However, the need to be consistent nationwide in initiating map revisions in a timely manner following the expiration of a PAL period offsets concerns related to costs.

If a community or levee system owner provides the FEMA Regional Office with the 44 CFR Section 65.10-compliant data and documentation submittal after the end of the appeal or comment period and after the community adopts the new DFIRM, FEMA will accredit the levee and map it accordingly as soon as possible after the DFIRM becomes final. There are several mapping options available to FEMA one the DFIRM becomes effective. Consequently, the FEMA Regional Office will consult with FEMA Headquarters staff to choose the mapping process that will accurately reflect the accredited levee on the effective DFIRM in an expeditious manner.

(Date)

[Mr./Ms.] [Name of Community CEO]
[Title]
[City/County]
[Address]
[City, State Zip]

[90-Day / 30-Day] Notification of Provisionally Accredited Levee Period Expiration

Dear **[Mr./Ms.] [Name of Community CEO]**:

This letter is to remind you of the upcoming **(Date Data and Documentation Showing 65.10 Full Compliance is Due)** deadline for **(City/County Name and Levee System owner/operator)** to submit all data and documentation required to demonstrate that **[levee system known as (levee system name)] / [levee systems identified on the enclosed table]** are in full compliance with the National Flood Insurance Program (NFIP) regulations cited in the Code of Federal Regulations (CFR) at Title 44, Chapter 1, Section 65.10 (44 CFR 65.10). On **(Date of FEMA Acknowledgement for Signed PAL Agreement)**, the Federal Emergency Management Agency (FEMA) of the Department of Homeland Security sent your community a letter acknowledging and accepting your signed Provisionally Accredited Levee (PAL) agreement (copy enclosed) for the **[levee system known as (Levee System Name)] / [levee systems identified on the enclosed table]**. That letter informed you that **[this levee system / these levee systems]** would be designated as **[a PAL system / PAL systems]** on the new Digital Flood Insurance Rate Map (DFIRM) for **<<County>>, <<State>>** during a 24-month PAL period that started on **(Date 2-Year PAL Period Started)**. The PAL designation is shown for [this levee / these levees/ on the [Preliminary / Effective] DIRM for **(County Name, State Name)** and Incorporated Areas.

The **[levee system was / the levee systems were]** shown with the PAL designation during the 24-month PAL period to convey to map users that verification of the **[levee system's / levee**

systems'] accreditation status was underway. FEMA previously recommended that the levee system owner/operator(s) and the impacted communities implement outreach efforts to inform affected property owners that an assessment of the levee system[s] was underway. FEMA also encouraged the purchase of flood insurance in the areas impacted by the PAL system[s], although property owners are not federally mandated to purchase flood insurance policies in these areas.

In accordance with 44 CFR 65.10, it is the responsibility of the community or other party seeking recognition of a levee system, to provide the data and documentation definite and outlined in 44 CFR 65.10. Specifically, the design and construction data provided must be certified by a Registered Professional Engineer or by a Federal agency with responsibility for levee design.

If you are able to submit the required data and documentation and it is found to be acceptable by FEMA, FEMA will initiate a map revision to accredit the levee system[s]. FEMA will map the areas impacted by the levee system[s] as a moderate-risk area, designated Zone X (shaded). The mandatory flood insurance purchase requirements of the NFIP do not apply to structures in areas designated as Zone X (shaded), nor do the NFIP minimum floodplain management requirements.

If you are unable to submit the required data and documentation by **(Date Data and Documentation Showing 65.10 Full Compliance is Due)** or if the submitted data and documentation are determined to be inadequate, FEMA will initiate a map revision to de-accredit the levee system[s] and map the impacted areas on the landward side of the levee system[s]. These areas will be remapped as Special Flood Hazard Areas (SFHAs), designated **[Zone A / Zone AE]**. The SFHA is the area that is subject to flooding during the 1-percent-annual-chance flood. The mandatory flood insurance purchase and minimum floodplain management requirements of the NFIP apply to structures within the SFHA.

The mapping process from start to finish in situations where a levee system is proposed to be de-accredited will take no less than 18 months. Data showing that the criteria of 44 CFR 65.10 have been met can be submitted at any time during that period or after. Data submitted prior to a community adopting the DFIRM or prior to the four month period that would precede the effective date of the mapping project, whichever comes first, will be incorporated into the maps prior to them becoming final.

Communities are encouraged to increase floodplain management efforts and promote the purchase of flood insurance for all homes and businesses in areas impacted by levee systems, including areas that are impacted by levee systems that have been certified as providing 1-percent-annual-chance flood protection. Risk is dynamic and may not be completely eliminated by any levee system. Levee systems are designed and built to provide a certain level of flood

protection and to reduce the risks associated with flooding events in general. Levee systems can be overtopped or fail during flood events that exceed the design-level storm. Additionally, it is highly recommend that you consider this risk in your local emergency management plans, including creating evacuation plans for this area.

Please send all complete 44 CFR 65.10 data and documentation submissions to my attention on or before **(Date Data and Documentation Due)**. If you have questions or need additional information regarding the flood mapping for your community, please contact me by telephone at **(Regional Contact Telephone Number)**, or by e-mail at **(Regional Contact E-Mail Address)**.

Sincerely,

(Regional Contact Name), (Title)
Mitigation Division

Enclosures: **[PAL Status Table]**
Signed PAL Agreement(s)

cc: **(Name and Title of Floodplain Administrator)**
(Name and Title of State NFIP Coordinator)
(Name and Title of Impacted Community/Communities CEO)
(Name and Title of Senators/Congressional)
(Name and Title of USACE District Office Contact) (for B levees only)

Appendix H

USACE System-Wide Improvement Frameworks Program

DEPARTMENT OF THE ARMY
U.S. ARMY CORPS OF ENGINEERS
441 G STREET, NW
WASHINGTON, DC 20314-1000

REPLY TO
ATTENTION OF

CECW-HS NOV 29 2011

MEMORANDUM FOR COMMANDERS, MAJOR SUBORDINATE COMMANDS AND DISTRICTS

SUBJECT: Policy for Development and Implementation of System-Wide Improvement Frameworks (SWIFs)

1. References.

a. Engineer Regulation (ER) 500-1-1, Emergency Employment of Army and Other Resources - Civil Emergency Management Program, 30 Sep 2001.

b. Memorandum, HQ USACE (CECW-HS), 16 Nov 2007, subject: Levee Safety Program Implementation.

c. Memorandum, HQ USACE (CECW-HS), 9 Jan 2009, subject: Temporary extension of P.L. 84-99 Rehabilitation Eligibility for Non-Federal Sponsors Implementing System-wide Improvements.

d. Memorandum, HQ USACE (CECW-CE), not yet released, subject: Policy Guidance Letter (PGL) - Process for Requesting a Variance from Vegetation Standards for Levees and Floodwalls.

2. Definitions.

a. A "levee system" consists of one or more segments of earthen embankment or floodwall, and all appurtenant structures (such as closures, berms, pumping stations, culverts, and interior drainage), which are interconnected and necessary to reasonably reduce the potential of floodwater entering a defined area.

b. An "unacceptable inspection item" is an inspected item on the U.S. Army Corps of Engineers (USACE) levee inspection checklist. An unacceptable item or a combination of unacceptable items may lead to an overall levee-system rating of unacceptable.

c. A SWIF is a plan developed by the levee sponsor(s) and accepted by the USACE to implement system-wide improvements to a levee system (or multiple levee systems within a watershed) to address system-wide issues, including correction of unacceptable inspection items, in a prioritized way to optimize flood risk reduction.

CECW-HS
SUBJECT: Policy for Development and Implementation of System-Wide Improvement Frameworks (SWIFs)

d. An Interim Risk Reduction Measures (IRRM) plan contains actions to reduce life safety risks posed by a levee system while long-term solutions are being pursued. Example actions may include Emergency Action Plans, special modifications to evacuation plans that account for additional risk and potentially reduced response times, heightened public communication campaigns to inform citizens of the increased flood risk, and/or increased levee monitoring during flood events.

3. Background.

a. The USACE Public Law (P.L.) 84-99 program, pursuant to 33 U.S.C. 701n, is a voluntary program that includes the repair and restoration of participating flood risk reduction projects, such as levee systems. All levee systems that participate in P.L. 84-99 are inspected by USACE and rated against nationally consistent standards that USACE determined to be essential for the reliable performance of the levee system. Levee systems that have received an "Acceptable" or "Minimally Acceptable" overall system rating on the last periodic or routine/continuing eligibility inspection are "Active" in P.L. 84-99 and, consequently, are eligible to receive rehabilitation assistance from USACE to repair or restore levee systems to pre-disaster condition if they are damaged by a flood event. Levee systems that receive an "Unacceptable" overall system rating or that choose to no longer participate in the program are placed in "Inactive" status and are not eligible for rehabilitation assistance under P.L. 84-99.

b. In some cases, the items on a levee system found to be "Unacceptable" or "Minimally Acceptable" might be complex to correct. Developing and implementing solutions to address such deficiencies might require a multi-year effort and coordination between multiple entities. This may be especially true when resources protected under the Endangered Species Act or Tribal treaty rights could be impacted by any changes to the levee system. USACE is making the SWIF process available to levee sponsors facing such challenges as a way to facilitate the development of solutions to satisfy the multiple requirements that apply to their levee systems while allowing levee sponsors participating in the SWIF process to remain eligible for P.L. 84-99 rehabilitation assistance funding while addressing deficiencies.

c. Levee sponsors are responsible for operation, maintenance, repair, replacement, and rehabilitation of the levee system. The SWIF does not alter those responsibilities. The SWIF establishes a process using an interagency approach within which levee sponsors engage with federal, state, local and Tribal agencies and organizations in longer-term system-wide improvement efforts to optimize flood risk reduction by identifying solutions that efficiently use resources, prioritize improvements and corrective actions based on risk, and establish frameworks for coordinating overlapping or complementary programs and requirements.

CECW-HS
SUBJECT: Policy for Development and Implementation of System-Wide Improvement Frameworks (SWIFs)

d. Environmental compliance and consideration of other requirements, such as those imposed by treaties with Tribes, should be integrated into the SWIF process. USACE is responsible for assuring compliance with all applicable environmental requirements before it makes any decisions that would affect the environment or other resources; however, the levee sponsor involved in the SWIF process may be required to provide background information/documentation, mitigation, or other measures necessary to fulfill environmental compliance responsibilities as a condition of their participation in the P.L. 84-99 rehabilitation assistance program.

e. The SWIF can be used to address deficiencies or issues that cannot be accomplished through routine corrective actions, including:

(1) A levee system that has engineering deficiencies (e.g., change in hydraulic conditions that reduces level of authorized protection) in addition to "Unacceptable" inspection items;

(2) Improvements that involve multiple levee systems;

(3) Complex natural resource considerations that require additional time and coordination to ensure that the imperatives of both levee safety and environmental requirements are adequately served;

(4) Additional time and coordination to protect the rights of Tribes pursuant to treaty and statute; and,

f. This memorandum revises and supersedes the policy in Reference 1.c. above and will be incorporated into the revised Engineer Regulation (ER) 500-1-1, in Reference 1.a.

4. Purpose. The purpose of this policy is threefold and includes:

a. To facilitate interagency collaboration to address complex levee system deficiencies and encourage the establishment of interagency teams to jointly identify and implement regionally appropriate, science based solutions and tools to help reduce risk associated with levees or levee systems while ensuring compliance with other Federal laws, such as the Endangered Species Act, as appropriate;

b. To provide requirements and outline the process for the submittal and acceptance of a SWIF that will assist levee sponsors in attaining compliance with USACE standards; and,

CECW-HS
SUBJECT: Policy for Development and Implementation of System-Wide Improvement Frameworks (SWIFs)

c. To provide a mechanism for levee sponsors to maintain or regain eligibility for federal rehabilitation assistance under P.L. 84-99 while they are developing and implementing a system-wide improvement framework.

5. Applicability. This policy applies to all Headquarters USACE (HQ USACE) elements, Major Subordinate Commands (MSCs), USACE districts, and field operating activities that are responsible for Civil Works projects.

6. Eligibility for Rehabilitation Assistance Under P.L. 84-99.

a. Initial Eligibility for Rehabilitation of Flood Risk Reduction Projects. Pursuant to 33 U.S.C. 701n (P.L. 84-99) and ER 500-1-1, USACE administers a program for rehabilitation of flood risk reduction projects, such as levee systems, and for federally-authorized hurricane or shore protection projects. Federally-authorized, locally-operated and maintained levee systems are automatically placed in an "Active" status and are eligible for rehabilitation assistance upon construction completion. Non-federal levee systems constructed, operated and maintained by a local public entity may be placed in an "Active" status following an initial eligibility inspection by USACE that determines the levee system meets the minimum eligibility requirements and technical criteria.

b. Maintaining Eligibility.

(1) Eligible levee systems inspected by USACE are rated against nationally consistent standards that USACE determined are essential for the reliable performance of the levee system. Levee systems that receive an "Acceptable" or "Minimally Acceptable" overall inspection rating maintain an "Active" status and are eligible to receive rehabilitation assistance from USACE to repair flood damages. Levee systems that receive an "Unacceptable" overall inspection rating are immediately placed in "Inactive" status and are not eligible for rehabilitation assistance until the "Unacceptable" inspection items have been corrected and the system receives a rating of "Acceptable" or "Minimally Acceptable" through a re-inspection by the responsible USACE district office. Levee systems engaged in the SWIF process or a variance request will continue to be inspected against the standard applied at the time the request was submitted, either the national standard or a variance or other deviation from that standard.

(2) The concepts for developing a SWIF may be useful in situations where a levee sponsor wants to prevent a future "Unacceptable" rating. For these situations, the approval process outlined in paragraph 8 is not required; however, this process cannot be used to extend "Minimally Acceptable" ratings for levees with "Unacceptable" inspection items beyond the current two-year maximum period as specified in the USACE levee inspection checklist.

CECW-HS
SUBJECT: Policy for Development and Implementation of System-Wide Improvement Frameworks (SWIFs)

c. Transitioning "Acceptable" or "Minimally Acceptable" Levees. Levees sponsors with levees that are "Active" in the rehabilitation assistance program under an existing vegetation variance or deviation from the standard that want to use the SWIF process to transition to a new vegetation inspection standard through the vegetation variance request process, or that would like to systematically improve the condition of participating levees, may maintain their P.L. 84-99 rehabilitation assistance eligibility as long as they continue to meet the milestones set forth in their applicable SWIF.

d. Reinstating Eligibility While Developing and Implementing a SWIF. Levee sponsors that receive an overall levee system inspection rating of "Unacceptable" or have been "Inactive" in the rehabilitation program may regain eligibility for P.L. 84-99 rehabilitation assistance through the SWIF process. Upon approval by USACE of the letter of intent, requirements described below, the levee sponsor will receive an initial of up to two-year reinstatement of eligibility for P.L. 84-99 rehabilitation assistance. Continued eligibility will be determined annually based on milestones described in the subsequent SWIF. Levee sponsors who have never been eligible for rehabilitation assistance under P.L. 84-99 cannot gain P.L. 84-99 rehabilitation assistance eligibility through the SWIF process.

7. Requirements for Development and Submittal of a SWIF. The development of a SWIF is a two-step process consisting of (1) a Letter of Intent from the sponsor briefly describing levee system deficiencies and justification for how a system-wide approach will optimize flood risk reduction, and (2) development of a SWIF for addressing deficiencies and reducing flood risk. Once a Letter of Intent has been approved by USACE, a levee sponsor has up to two years to develop a SWIF plan. Eligibility after this two-year period will be dependent on the levee sponsor's progress in achieving the milestones defined in the SWIF. The SWIF plan is intended to be a specific document that guides sponsor activities, including anticipated milestones, but may also be adaptable and should be revised if conditions or needs change during implementation. The requirements for the Letter of Intent and SWIF are described as follows:

a. Requirements for Submitting a Letter of Intent for a SWIF. A Letter of Intent must be signed by all associated levee sponsors for each levee system involved in developing the SWIF and must include the following:

(1) Identification of levee system or systems to be covered by the SWIF, including system name and system identification number as listed in the National Levee Database;

(2) Brief description of deficiencies or issues that will be included in the SWIF and discussion of how a system-wide approach will improve and optimize overall flood risk reduction. This includes identifying any conditions not within the control of the levee sponsor(s) that prevents them from correcting "Unacceptable" inspection items in a timely manner;

CECW-HS
SUBJECT: Policy for Development and Implementation of System-Wide Improvement Frameworks (SWIFs)

(3) Demonstration that significant non-federal resources have been, or will be, committed for developing and/or implementing the SWIF (e.g., state legislative action, bond financing);

(4) Anticipated interim risk reduction measures that will be implemented throughout the SWIF process, including overall risk communication approach that addresses the risk to life increased by system-wide deficiencies;

(5) Brief description of existing or planned interagency collaborative efforts that will contribute positively to SWIF development, implementation and oversight; and

(6) List of anticipated state and federal permits and consultation requirements, needed to implement the SWIF.

b. Requirements for Submittal of a SWIF. SWIFs are developed and implemented by levee sponsor(s), reviewed and accepted by USACE, and monitored by a USACE district to address system-wide issues in a prioritized way to optimize system-wide risk reduction. As a minimum for acceptance by USACE, the levee sponsor's SWIF must include the following:

(1) Identification of levee system or systems covered by the system-wide improvement framework, including system name and identification number as listed in the National Levee Database;

(2) Description of proposed levee improvement and justification on how the SWIF optimizes flood risk reduction;

(3) A plan and schedule for interagency collaboration, including environmental and/or Tribal consultation if applicable, in the implementation of the SWIF;

(4) Documentation of specific agreements, such as project specific agreements, between levee sponsors and USACE or other agencies/organizations related to implementation of levee modifications, under Section 408 or other overlapping USACE policies and studies, applicable to the levee systems identified in the system-wide improvement framework;

(5) Documentation of any regional considerations, approaches, and tools to be used during implementation of the system-wide improvement framework;

(6) Description of interim maintenance standards that will be implemented during the SWIF to mitigate conditions of uncorrected "Unacceptable" inspection items;

CECW-HS
SUBJECT: Policy for Development and Implementation of System-Wide Improvement Frameworks (SWIFs)

(7) IRRM plan, including a risk communication plan that addresses the risk to life increased by system-wide deficiencies;

(8) Schedules and milestones that will be used to monitor progress and to determine continued eligibility for P.L. 84-99 rehabilitation assistance while the SWIF is being implemented; and

(9) For those levee systems shown as accredited on the Federal Emergency Management Agency (FEMA) Flood Insurance Rate Map that are part of the SWIF, demonstration that FEMA has been informed that these levee systems with "Unacceptable" inspection items are being addressed in a system-wide improvement framework. Please note that an extension of eligibility for rehabilitation assistance through the SWIF process by USACE does not constitute an extension of accreditation for FEMA purposes. FEMA determines how a SWIF may or may not impact accreditation.

8. Approval Process. The approval authority for reinstating eligibility for rehabilitation assistance under P.L. 84-99 via a Letter of Intent, and for acceptance of a SWIF is the Director of Contingency Operations and Homeland Security (DCO/HS) under USACE. District Commanders shall evaluate the levee sponsors' request for an extension, based on the criteria outlined in this memorandum. If the District recommends approval of an eligibility reinstatement, the District Commander shall forward this recommendation to the Division Commander for concurrence. The Division Commander will review the request and, if in concurrence, will endorse the recommendation and submit the request to the DCO/HS through the Regional Integration Team. The District and MSC Commanders shall coordinate these requests with their Levee Safety Officers for technical input. Eligibility reinstatement will not be implemented until the request is approved by DCO/HS. District Commanders are also responsible for monitoring levee sponsor milestones in implementing SWIFs, conducting reviews for eligibility extensions following initial reinstatement, submitting an accepted SWIF to the local FEMA regional office, and providing approval recommendations through the approval process described herein.

9. Progress Reporting and Continued P.L. 84-99 Eligibility. Once a Letter of Intent has been approved through the process in paragraph 8, a levee sponsor(s) has up to two years of reinstated rehabilitation assistance eligibility under P.L. 84-99 to develop a system-wide improvement framework. The District Commander shall review the levee sponsor's progress for development of the SWIF after the first year and, if deemed not satisfactory, the District Commander may recommend to the DCO/HS that the levee sponsor no longer be eligible for P.L. 84-99 rehabilitation assistance. Eligibility after the two-year period for SWIF development will then be dependent on the levee sponsor's progress in achieving the milestones defined in the SWIF. Continued P.L. 84-99 rehabilitation assistance eligibility during the implementation of the SWIF

CECW-HS
SUBJECT: Policy for Development and Implementation of System-Wide Improvement Frameworks (SWIFs)

must be approved by the DCO/HS on a two-year basis following the process in paragraph 8. During implementation of the SWIF, continued P.L. 84-99 rehabilitation assistance eligibility requests should include a copy of the SWIF; description of accomplishments and milestones met; and description of any changes since the last SWIF was submitted. At a minimum, levee sponsors shall submit a memorandum to the District Commander annually that demonstrates system-wide improvement accomplishments.

10. Overlap with Vegetation Variance Policy. The SWIF process may complement the vegetation variance request process, currently set forth in ER 500-1-1 and anticipated to be updated shortly in paragraph 1.d., as a means for a levee sponsor to address levee deficiencies. If required, a vegetation variance request can be part of the SWIF process. The SWIF offers an interagency approach to identify regional solutions and tools that may be useful in development of a vegetation variance request. The end result of the SWIF process will be levees that meet the USACE inspection standards, which may also include an approved vegetation variance. See enclosure for timelines for both the SWIF process and situations with existing vegetation variances or deviations from the standard.

11. Levee Inspections. During implementation of the SWIF, levee systems with P.L. 84-99 rehabilitation assistance eligibility will continue to be inspected, rated, and the results communicated in accordance with USACE inspection processes, paragraph 1.b. and other applicable guidance. Any "Unacceptable" inspection items identified during inspections will be recorded as such and be corrected in accordance with existing policy or noted to be corrected under a system-wide improvement framework. At any time that the condition of a deferred "Unacceptable" inspection item worsens to a point of creating an emergency condition, immediate corrective actions must be taken by the levee sponsor in order to retain eligibility for P.L. 84-99 rehabilitation assistance.

12. Post-Flood Repair Responsibilities Associated with "Unacceptable" Inspection Items. Upon approval of the Letter of Intent by the USACE, levee sponsors who meet milestones will remain eligible for post-flood repair throughout the SWIF development and implementation process; however, levee sponsors will continue to be responsible for the portion of that repair cost associated with "Unacceptable" inspection items in accordance with ER 500-1-1, paragraph 5-2, g.

13. Funding for USACE Participation in the SWIF Process. USACE review of requests submitted by levee sponsors for SWIF implementation and participation of USACE representatives in collaborative frameworks for developing SWIFs may be funded with Inspection of Completed Works funds for federally-authorized levee systems and Flood Control and Coastal Emergency funds for non-federal levee systems. USACE participation in

CECW-HS
SUBJECT: Policy for Development and Implementation of System-Wide Improvement Frameworks (SWIFs)

collaborative frameworks to address SWIF issues implemented through Silver Jackets (SJ) intergovernmental teams may use SJ funding.

14. Point of Contact. The point of contact for this action is Ms. Germaine Hofbauer, (202) 761-4970.

FOR THE COMMANDER

Encl

KAREN DURHAM-AGUILERA, P.E., SES
Director of Contingency Operations
and Homeland Security

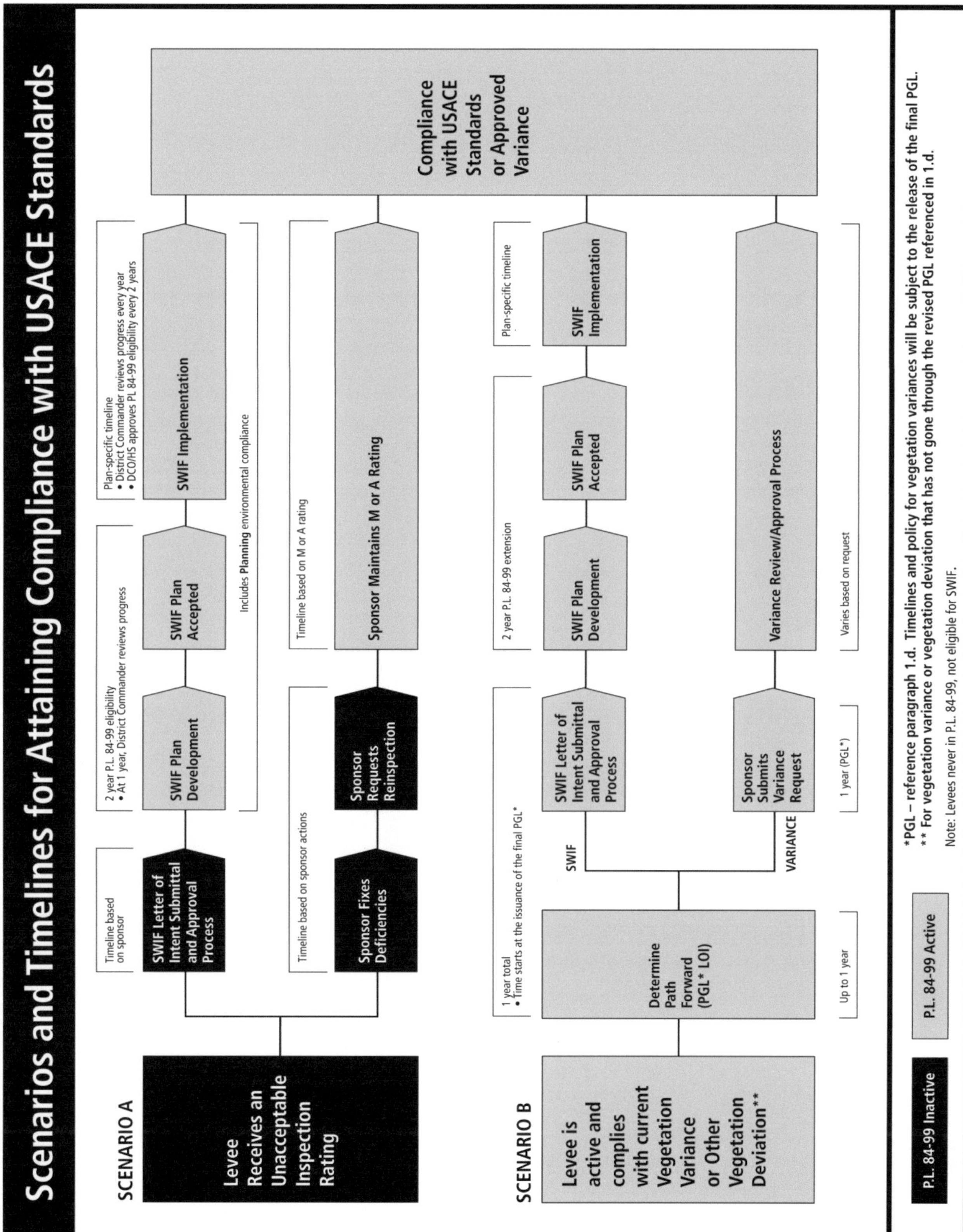
Scenarios and Timelines for Attaining Compliance with USACE Standards
SCENARIO A
Levee Receives an Unacceptable Inspection Rating
Timeline based on sponsor
SWIF Letter of Intent Submittal and Approval Process
2 year P.L. 84-99 eligibility
• At 1 year, District Commander reviews progress
SWIF Plan Development
SWIF Plan Accepted
Plan-specific timeline
• District Commander reviews progress every year
• DCO/HS approves PL 84-99 eligibility every 2 years
SWIF Implementation
Includes Planning environmental compliance
Timeline based on sponsor actions
Sponsor Fixes Deficiencies
Sponsor Requests Reinspection
Timeline based on M or A rating
Sponsor Maintains M or A Rating
Compliance with USACE Standards or Approved Variance
SCENARIO B
Levee is active and complies with current Vegetation Variance or Other Vegetation Deviation**
1 year total
• Time starts at the issuance of the final PGL*
Determine Path Forward (PGL* LOI)
Up to 1 year
SWIF
SWIF Letter of Intent Submittal and Approval Process
2 year P.L. 84-99 extension
SWIF Plan Development
SWIF Plan Accepted
Plan-specific timeline
SWIF Implementation
VARIANCE
Sponsor Submits Variance Request
1 year (PGL*)
Variance Review/Approval Process
Varies based on request
P.L. 84-99 Inactive
P.L. 84-99 Active
*PGL – reference paragraph 1.d. Timelines and policy for vegetation variances will be subject to the release of the final PGL.
** For vegetation variance or vegetation deviation that has not gone through the revised PGL referenced in 1.d.
Note: Levees never in P.L. 84-99, not eligible for SWIF.

Appendix I

Risk Analysis Background

The committee has recommended that the National Flood Insurance Program (NFIP) adopt and implement a modern approach to the analysis of flood risks that will serve as input to the assessment of flood insurance and provide a sound foundation for transitioning to risk-informed floodplain management. The approach recommended is founded on decades of risk analysis experience in the natural hazards and civil infrastructure arena and environmental health and life safety. This experience has evolved in the private and public sectors (USACE, 1996, 2006; USNRC, 2011; AIR, 2012; RMS, 2012; EPA, n.d.) and specifically in the levee–flood risk areas as evidenced by recent studies in Dallas, Texas, New Orleans, Louisiana, Sacramento, California, and in California's Sacramento–San Joaquin Bay Delta (URS/JBA, 2008; IPET, 2009; CA DWR, 2012; USACE, 2012). This appendix provides further background and support for the modern risk analysis recommended in Chapter 3, including continued discussion of elements of a modern flood risk analysis.

NEW CAPACITY ENABLING A RISK-BASED ANALYSIS

At the initiation of the NFIP in the 1970s, simplified analysis was necessary and appropriate. Hydraulic modeling capabilities were limited. Much of the hydraulic analysis for floodplain delineation, for example, was completed with computer program HEC-2 (USACE, 1991), a one-dimensional (1-D) steady-flow open-channel flow analysis application that was executed on large mainframe computers and for which the channel system was described on IBM punch cards, and the channel system description, which includes the channel cross-section geometry, often was derived from U.S. Geological Survey 7.5-minute quadrangle topographic maps; maps with contour intervals of 10 feet that are accurate to ±5 feet (1.5 meters) of the actual elevation. Furthermore, at the onset of the NFIP, knowledge of performance of levees was limited, and the notion of evaluating the potential for levee failure and describing performance in terms of the probability of failure with a fragility curve had not found its way into flood and floodplain analysis. Although a fundamental description of economic consequences and an associated probability of occurrence did play a limited role in flood insurance rate setting and management planning (e.g., in the computation of expected inundation reduction benefit attributable to a flood management project), a detailed understanding and description of risk were not widespread in the floodplain management community, and procedures and tools for risk analysis were not readily available.

Technological advances in the past 40 years now make risk analyses of large insurance portfolios possible within the context of the NFIP. For example, the NRC (2007) notes that Lidar (light detection and ranging) technology has matured to the point that it "is capable of producing a bare-earth elevation model with 2-foot equivalent contour accuracy in model terrain and land-cover types." With geographic information system tools, overland flow on floodplains can be more accurately modeled from these data in a systematic, automated manner, with minimum human processing and few errors. Further, the state of practice of hydraulic modeling has advanced greatly since the early days of the NFIP, with two-dimensional (2-D) unsteady open channel and overland flow models now widely used. In fact, FEMA's list of acceptable models for hydraulic analysis includes five such models.[1] Data for estimation of consequence of inundation, particularly estimation of inundation damage, are readily available. These include, for example, property values, property type categorization, and property elevation data in FEMA's Hazards–United States (Hazus) databases.[2]

The technical capability now exists for implementing and applying risk analysis as proposed in this report. In fact, other state and federal agencies with responsibilities for flood risk management have recognized the need for complete risk analyses to support their decision making, and have incorporated those analyses in their programs. To varying degrees, those analyses:

- Account for the full range of hazard, going beyond the one percent chance event, considering events as common as the 50 percent chance event and as rare as the 0.1 percent or 0.01 percent chance event.
- Include an explicit assessment of the consequences of inundation. Most analyses consider economic consequences, some life safety, and few environmental consequences.
- Evaluate all potential mechanisms for flooding areas that are protected by a flood protection system, including overtopping (without levee or structural failure), misoperation, breaching, and ponding. In addition, damages that may be caused by breaching itself, such as might be caused by the scour that occurs during the breach process, can be evaluated.
- Account for the performance of flood protection systems throughout the range of possible hazards. The analyses consider, for example, that a levee designed for the 10 percent chance event may reduce adverse economic consequence due to the one percent chance event, even though it will not eliminate all damage due to that event. They also represent the possibility, though unlikely, of failure of a FEMA "certified" levee during events that are smaller or larger than the one percent chance event.
- Acknowledge and evaluate the uncertainties about the various inputs to the analysis, including natural variability and lack of perfect understanding and ability to model the natural processes.

For example, all flood damage reduction studies conducted by the U.S. Army Corps of Engineers are required to analyze risk (USACE, 2006). The requirement applies to USACE studies that lead to feasibility reports, general design memorandums, and general re-evaluation reports. The regulation specifically prohibits traditional deterministic approaches—such as inclusion of freeboard to account for what are described as "hydrologic, hydraulic, and geotechnical uncertainties." Instead, USACE analyses must use technical procedures presented in Engineering Manual 1110-2-1619 (USACE, 1996), Engineering Technical Letter 110-2-556, and companion documents.

In addition to the foregoing public-sector applications, private companies have developed flood risk analysis tools for evaluating insurance portfolios (AIR, 2012; RMS, 2012).

EXAMPLES OF CURRENT USE OF A RISK ANALYSIS TO EVALUATE FLOOD RISKS

As discussed in Chapter 3, the use of risk analysis methods to evaluate flood risks and specifically to consider the performance of levees has been ongoing and improving for more than a decade. This includes the development of a probabilistic approach by USACE in the 1990s (USACE, 1996). More recently, after the events of Hurricane Katrina, there has been considerable development and application of risk analysis methods to evaluate the flood

[1] See http://www.fema.gov/national-flood-insurance-program-flood-hazard-mapping/numerical-models-meeting-minimum-requirement-0.
[2] See http://www.fema.gov/protecting-our-communities/hazus.

risks and the performance of levees and flood protection systems in general. As a response to the damage caused by Katrina and the levee failures, the USACE formed a team to conduct a detailed risk and reliability analysis of the New Orleans hurricane protection system, including an assessment of economic consequences as part of the Interagency Performance Evaluation Task Force study (IPET, 2009). During the same period and partly in response to Hurricane Katrina, the California Department of Water Resources carried out a comprehensive risk analysis on the more than 1,200 miles of levees in the Sacramento–San Joaquin Delta (URS/JBA, 2008). This analysis evaluated flood and seismic risks to Delta levees (which are more like dams because they have water against them year round) and the economic risks to the state. Following the Delta risk study, California DWR is conducting flood risk studies in the Central Valley, including the extensive development of levee fragility curves. As part of the California Central Valley Flood Protection Plan, engineers have developed levee fragility curves at approximately 300 locations in approximately 1 year, following an extensive data collection effort (CA DWR, 2012). Similarly, USACE has developed levee fragility curves for use in ongoing planning studies, all of which now employ complete risk analysis (USACE, 2006).

The events of Hurricane Katrina led to significant changes at USACE in terms of their use of risk analysis methods in its levee and dam safety programs. In fact, the IPET evaluation led to USACE making a major paradigm shift, transforming USACE to a risk-informed agency (USACE, 2008). Whereas USACE had been using risk analysis in limited ways prior to Katrina, the impact of failure of USACE-designed structures led to a wholesale move to implementation of risk analysis for the management of the nation's flood risk. As part of these changes, USACE is advancing its risk analysis methods for levees and developing new tools to improve the assessment of flood protection systems (USACE, 2008).

DETAILED DICUSSION OF A MODERN RISK-BASED ANALYSIS

As described in the report, there are a number of key elements to conducting a modern risk analysis. These include:

- the requirement that uncertainties, specifically epistemic uncertainties, be identified and evaluated as an integral part of the analysis;
- modeling and considering the performance of flood protection systems, including the potential for failure and breaching, misoperation, overtopping, and flooding of areas protected by these systems, in the assessment of the potential for flooding and damage; and
- a systems-based approach for the modeling of the hydrologic and hydraulic systems and the flood protection systems that provide protection to a community.

As it pertains to the NFIP, the purpose of a risk-based approach to the analysis of flood risks, including the performance of flood protection systems, is multifold. It is intended to provide a quantitative measure of the flood damages associated with flood hazards that can be used to support the NFIP determination of flood insurance rates. This assessment needs to consider the performance of levee systems and, given the language in 44 CFR §65.10, all elements of flood protection systems and their role in the potential for flooding. In addition to insurance-motivated needs of the NFIP, the risk analysis and, more specifically, the products of the risk analysis will support other floodplain management aspects of the program (risk communication, planning, etc.).

The analysis of flood risks is a multidisciplinary evaluation that is made on the basis of available information. As such, the recognition and evaluation of these epistemic uncertainties is a key concept in modern risk analysis. As a starting point for the development of an NFIP risk analysis methodology, it is important that a clear definition of uncertainty and risk be established. Furthermore, as procedures for the implementation of an NFIP risk analysis method are developed, clear definitions of the responsibilities of the professionals involved in the study, the evaluations they are expected to perform, and the results they generate are critical. Experience suggests that in complex, multidisciplinary evaluations it is important that responsibilities are clearly defined and the technical interface between different parts of the analysis are laid out.

Characterizing Aleatory and Epistemic Uncertainty

As noted above and in Chapter 3, uncertainties are an integral part of the modern risk-based analysis. To evaluate and model aleatory and epistemic uncertainties, a characterization can be made in terms of their effect on models and estimates of model parameters (Table I-1). Modeling epistemic uncertainty represents differences between a physical process (hurricane surge, embankment failure) and prediction models. Modeling epistemic uncertainty can be estimated by comparing model predictions to observed events/performance. Parameter uncertainty is the epistemic uncertainty associated with the estimates of model parameters as derived from available data. Parametric uncertainty is quantified by observing the variation in parameters inferred (either in a direct or an indirect manner).

The distinction between aleatory and epistemic uncertainty is not intuitive and can be difficult to ascertain. Furthermore, the assessment of these uncertainties is model dependent. For example, a simple engineering model of an event (levee performance during a flood) may have higher model aleatory variability than a more complex model that addresses more details of the physical process of a levee dealing with the loads it is exposed to during the flood (i.e., increased hydrostatic loading, wave action, seepage forces, etc.). At the same time, the more complex model may have larger parametric epistemic uncertainty; there are more parameters to estimate and there may be limited data to estimate them. Thus, the characterization of uncertainties is model dependent, making the distinction between different types of uncertainty difficult. Nonetheless, making a distinction between the sources of uncertainty in a logical manner helps ensure that all uncertainties are identified and quantified. In principle, epistemic uncertainties are reducible with the collection of additional data or the use/development of improved models.

A Conceptual Framework for Flood Risk Analysis

As described in the main report, a starting point for the development of an NFIP flood risk analysis is a conceptual framework for flood risk analysis that can be defined as:

$$R = f(H,V,C) \tag{I-1}$$

where risk (R) is a function of flood hazard (H), vulnerability (V), and consequence (C). In this context, risk is a function of the flood hazard a community is exposed to, the vulnerability of flood protection systems and the potential that their failure will contribute to flooding, and the consequences associated with system failures and the damage to a community exposed to flooding, including economic impact and life safety.

As noted in Chapter 3, this framework can be extended to include the concepts of uncertainty. This extension

TABLE I-1 Characterization of Aleatory and Epistemic Uncertainty on Models and Model Parameters

	Type of Uncertainty	
Element	Epistemic	Aleatory
Modeling	Uncertainty about a model and the degree to which it can predict events or outcomes (e.g., levee performance), that is, to what extent a model has a tendency to over- or underpredict observations	Aleatory modeling uncertainty is the variability that is not explained by a model. For instance, this variability is attributed to elements of the physical process that are not modeled and therefore represents a variability (random differences) between model predictions and observations.
Parametric	Uncertainty associated with the estimates of model parameters, given available data, indirect measurements, etc.	This uncertainty is similar to aleatory modeling uncertainty. This is a variability that may be due to systematic, but random, variations associated with parameters of a model.

SOURCES: Abrahamson et al. (1990), USR/JBA (2008), IPET (2009).

provides a quantitative definition of risk (Kaplan and Garrick, 1981; ASME/ANS, 2009; IPET 2009) in probabilistic terms:

$$R = f(H,V,C,\nu,\rho) \qquad \text{(I-2)}$$

where ν = frequency of occurrence or exceedance, and is a measure of the aleatory uncertainty (the randomness of events); ρ = probability as a measure of the confidence to which an estimate of ν is the true value, or the epistemic uncertainty in the estimate of ν. For purposes of a flood risk analysis, the consequences of flooding are measured in terms of the economic damages. Typically, results of risk analysis would be expressed in terms of a frequency distribution on economic damages (dollars). This is denoted

$$R = \nu(\mathrm{c}) \qquad \text{(I-3)}$$

where $\nu()$ = frequency of occurrence per year; and c = consequences measured in dollars.

Another risk metric is the expected annual losses, which can be estimated by

$$E(c) = \int c^*\nu(c)dc \qquad \text{(I-4)}$$

The estimate of flood consequences and their frequency of occurrence are uncertain, that is $\nu(c)$ cannot be determined with certainty; there is epistemic in estimating the frequency and magnitude of floods, uncertainty in the performance of flood protection systems, and the magnitude of consequences. There is therefore a distribution on the estimate of $\nu(c)$. A formal analysis of the uncertainties in the components of a risk analysis (H, V, and C) produces a probability distribution on the estimate of $\nu(c)$, the estimate of risk. This probability distribution that quantifies the uncertainty in the estimate of $\nu(c)$ can be thought of as a distribution from which confidence intervals on the estimate of risk can be derived.[3] Kaplan and Garrick (1981) provide a quantitative definition of risk that includes the quantification of aleatory and epistemic uncertainties and the estimate of consequences:

$$\{\nu,\mathrm{C},\rho\} \qquad \text{(I-5)}$$

or, in terms of the risk example above, equation (I-5) can be written,

$$\{\nu(c)_i,\rho_i\} \qquad \text{(I-6)}$$

where, ν = frequency of occurrence or exceedence, as is a measure of the aleatory uncertainty; C = a consequence metric (e.g., economic impact, public safety); ρ = probability as a measure of epistemic uncertainty in the estimate; and where,

$$\sum_{i=1}^{n} \rho_i = 1.0 \qquad \text{(I-7)}$$

The frequency of occurrence is a measure of the aleatory uncertainty of events or outcomes (consequences), their randomness or stochastic character. An example of aleatory uncertainty is the occurrence of flood events in a watershed, or the performance of a levee for a given flood level. Equation (I-6) defines a discrete probability mass function on the frequency distribution of consequences, where probability is a measure of epistemic uncertainty in the estimate of ν.

[3] The notion of confidence intervals is a useful concept for purposes of illustrating the measure of uncertainty in the risk analysis results. In fact, however the assessment of uncertainties is not same as a statistical analysis of a dataset from which confidence intervals on parameter estimates are derived.

Risk Analysis Methodology

A primary product of a risk analysis for the NFIP is an assessment of the frequency of economic consequences that can occur as a result of flood-related events. For a unit of exposure (a particular property) or for an entire portfolio, the risk analysis will evaluate the potential that a unit may experience flooding and the damage (loss) that will occur as a result of the flooding. For a property behind a levee, the mechanisms that may lead to flooding are

- levee breach, prior to a levee being overtopped;
- inflow to the protected area (behind the levee) due to overtopping that does not result in breaching of the levee;
- flooding that occurs due to failure to close gates or other closures;
- levee breaching as a result of overtopping;
- flooding behind the levee that occurs due to intense rainfall, storm drains getting backed up, etc. that result in flooding/ponding in the protected area behind the levee.

The loss that occurs will vary, depending on

- depth and velocity of flooding associated with the different flooding mechanisms (see above); and
- proximity to the location of the levee breach. For instance, when a levee breaches, the depth of flooding behind the levee might end up at 2 feet above the first-floor elevation. However, a property that is located at or near a breach may be completely destroyed as a result of the depth of water, the high velocities of flow that will occur, and/or the fact the unit is located in the scour zone of the breach which undermines the foundation of the unit, etc.

A primary product of the risk analysis for an individual insured property or for an entire portfolio is an assessment of the frequency of economic consequences, which will vary in scale depending on the size of the flood events and the flooding mechanisms that can occur. This result, repeated from above, can be denoted,

$$\{\lambda(C > c)_i, \rho_i\} \tag{I-8}$$

where $\lambda()$ is the frequency of exceedance of the economic consequences and ρ_i is the probability associated with an estimate of the flood risk, $\lambda(C > c)_i$.

To estimate the result in equation (I-8), there are two fundamental components to the risk model. The first can be referred to as the aleatory flood risk model, which estimates the frequency of occurrence (or exceedance as shown in equation (I-8)) of flood damages. For an individual property, this can generally be denoted

$$\lambda(C > c) = \sum_{i=1}^{n_F} \nu(F_i) \sum_{j=1}^{m_{\text{states}}} \sum_{k=e_o}^{e_{\max}} P(LPS_j \mid F_i) P(e_{ij} \mid F_i, LPS_j) P(C > c \mid e_j) \tag{I-9}$$

where

$\nu(e_j)$ = frequency of occurrence of flood event i;

$P(LPS_j > F_i)$ = probability of levee system performance state j, given flood event, F_i;

$P(e_{ijk} \mid LPS_j)$ = probability of flood elevation k due to flood event, F_i and levee system performance state, LPS_j; and

$P(C > c \mid e_j)$ = probability that consequences of flooding will exceed a level c, given the occurrence of a flood elevation e_j.

The summations are carried out over (1) the range and type of flood events, (2) the number of different levee system performance states, and (3) the range of flood elevations that may occur given the flood events that may occur and the different levee performance states. The summation over flood events applies to all events that could lead to flooding. These could include local precipitation that might lead to ponding/flooding behind levees in the protected area, larger precipitation events that result in high river flows, ice jamming events, etc. The levee performance states can vary considerably depending on the number or type of structures and components that form the levee system. These may include levee breaching at different locations and at different flood elevations (pre- and post-overtopping), failure to close flood gates, structural or stability failure of floodwalls, etc. The performance of the levee system determines the different flooding mechanisms that may occur.

Regarding equation (I-9), first, the use of flood elevation as a metric for the characterization of the flood hazard can be considered a surrogate for the flood threat which may be multidimensional. As noted above, the level of damage that occurs may be a function of the depth and velocity of flooding. It also may be a function of debris content, the scour that may occur near a breach, etc. Second, the flooding that occurs is a function of the occurrence of different flooding mechanisms, some of which may involve levee failures, others of which do not. For purposes of this discussion, the characterization of the hazard is stated in terms of elevation as a surrogate. Equation (I-9) represents the aleatory model for estimating the random chance that flooding will occur and cause property damage.

The second element of the flood risk model is the "model of epistemic uncertainties." If all of the models and parameters that are required to implement equation (I-9) were known with certainty, a single estimate of $(C > c)$ could be determined. This of course is not the case, because there are numerous sources of uncertainty in the assessment of flood hazards, in the estimate of the fragility of levee systems, and in the estimate of the flood damages. As a result, there is uncertainty in the estimate of $(C > c)$. Therefore, as part of the flood risk analysis, an epistemic uncertainty model needs to be developed for the flood hazard analysis, the fragility analysis, and the flood damage model. In addition, a method to propagate these uncertainties to derive the uncertainty in the flood risk needs to be implemented.

Typically, simulation methods are used to carry out this analysis (e.g., Monte Carlo or Latin Hypercube simulation). Other approaches include the use of logic trees to enumerate all of the possible combinations of uncertain models and parameters. Conceptually, the estimate of the uncertainty in the flood risk analysis can be illustrated in a logic tree, as shown in Figure I-1. Across the top of the logic tree are the uncertainty elements of the flood risk model (again, shown here in conceptual form). For elements in the logic tree, there are a series of branches that represent alternative, discrete models are parameter estimates. Collectively, these discrete estimates model the uncertainty in that component of the risk model (e.g., flood hazard model). Associated with each branch is a probability weight, derived from the uncertainty analysis for that part of the flood risk model. The weight corresponds to the degree-of-belief that a particular model or parameter estimate is the true state of nature. Each path through the logic tree represents one, alternative (but uncertain) estimate of the flood risk, based on the uncertain inputs defined along the path and the basis for the risk calculation (equation (I-9).

COMPONENTS OF THE MODERN RISK-BASED ANALYSIS

The modern risk analysis recommended for NFIP insurance, mitigation, and communication programs requires a complete description of the flood hazard. In Chapter 3, a summary description of the modern risk-based analysis is provided which includes a flood hazard analysis, levee and other hydraulic structure and component and fragility analysis, systems analysis, inundation analysis, consequence assessment, and risk and uncertainty quantification. The discussion of these components is expanded, when relevant, here. Also, new components are discussed.

Flood Hazard Analysis

In the context of the committee's recommendation, a description of the flood hazard includes a specification of the magnitude (e.g., peak flood elevation) and its frequency of occurrence (or, more commonly, frequency of exceedance). Hazard at a given location in a floodplain is described with a discharge (flow)-frequency or elevation-frequency distribution. For analysis of riverine flood risk, the latter commonly is derived from the former. For coastal risk analysis, the elevation-frequency function is developed directly.

Methods of development of discharge-frequency functions for rivers are well known and well documented. If an adequate record of unregulated discharge at a site is available, an analytical frequency distribution can be

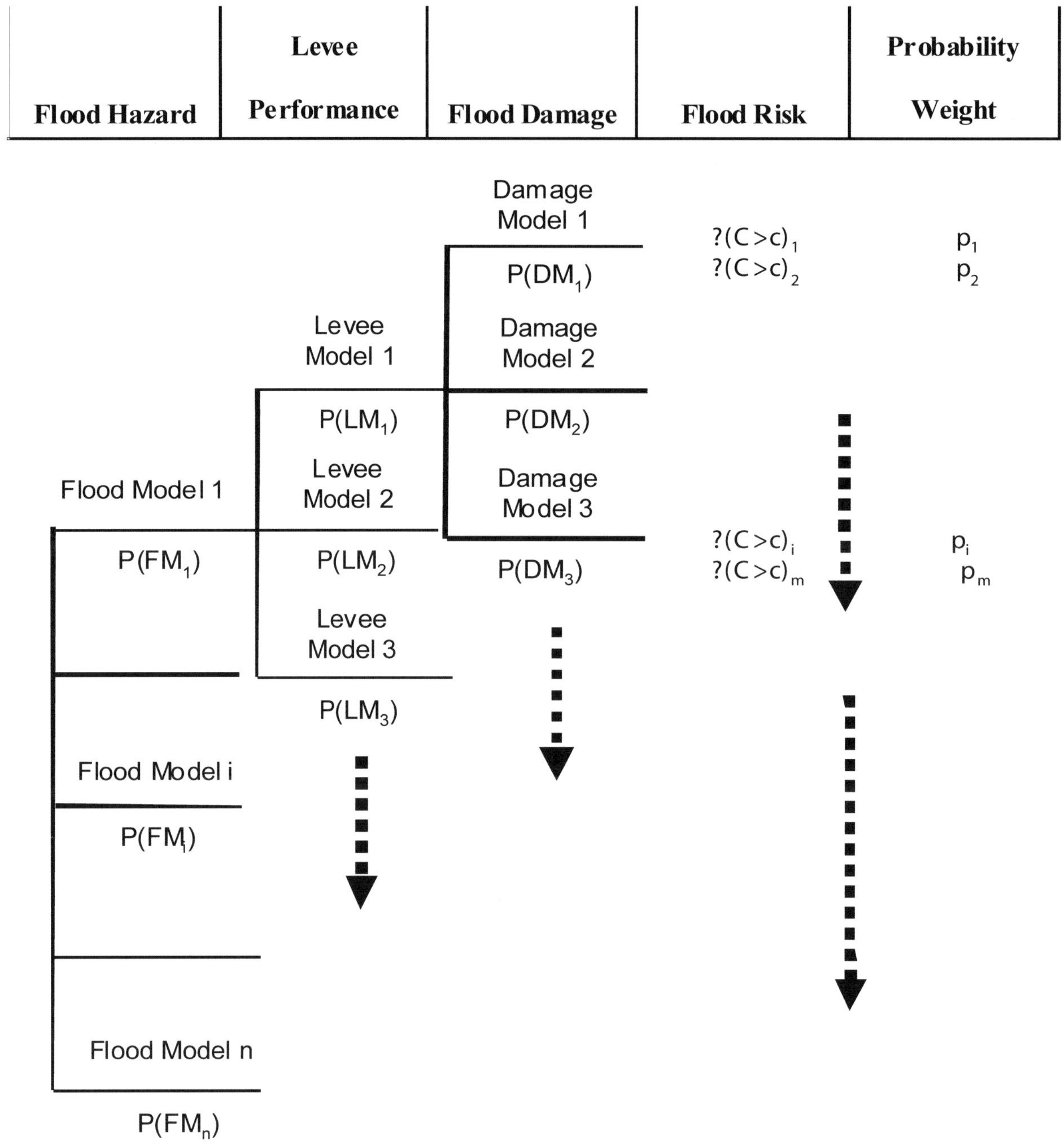

FIGURE I-1 Illustration of a logic tree, showing the uncertainty in different components of risk analysis.

fitted to the data. FEMA generally follows the standard of practice for this distribution fitting with historical flow observations. In cases for which data available are inadequate for distribution fitting, FEMA guidance suggests regional methods, transfer methods, empirical equations, and watershed modeling methods for estimating discharge of specified probability.[4] While certain refinements to both distribution fitting and alternative methods

[4] See http://training.fema.gov/EMIWeb/edu/docs/fmc/Chapter%204%20-%20Flood%20Risk%20Assessment.pdf.

will be necessary for the comprehensive risk analysis proposed herein—notably enhancements to evaluating the epistemic uncertainty about the frequency curves—the hazard description required for risk analysis proposed herein is fundamentally the same as that currently required for FEMA studies. The assessment of epistemic uncertainties in hydrologic modeling, frequency analysis, and estimation of flood elevations is an important element of a comprehensive risk analysis and in the stability of the results (climate change and other physical changes aside). The assessment of epistemic uncertainties must consider the parametric uncertainties associated with limited data, correlation of model parameters, and modeling uncertainties which may include alternative modeling interpretations of the analysts (Yankee Atomic, 1984; IPET, 2009). Thus the same hydrologic engineering methods can be used, with enhancements, and the existing technical knowledge base and experience of analysts will be useful and usable for comprehensive risk analysis.

A key to success of the risk analysis proposed herein is the capability to model the behavior of the riverine, coastal, or floodplain system with sufficient detail and sufficient accuracy to predict water surface elevations and velocities throughout the area of interest, given the discharge values estimated with frequency fitting or alternative methods. Capabilities for such modeling are readily available and well understood in the engineering and floodplain management community.

For flood studies such as those conducted by FEMA, channel flow (or flow contained by the levees) is described well in many cases with a 1-D hydraulic model. Such a model predicts the water surface elevation and velocity in the channel for specified flow rates by solving equations that represent the conservation of mass and momentum in the channel as the water moves downstream. In some cases, a 1-D model is not adequate for representing channel flow, because the inherent assumption that flow is predominately in a single direction is violated. In that case, a 2-D model would be used. The products of application are similar, but with variations both longitudinally and laterally in the channel represented.

Software applications with 1-D and 2-D models currently accepted by FEMA for NFIP studies are usable for the risk analysis proposed herein.[5] Again, certain enhancements are necessary to adequately describe the uncertainty about predicted elevations, but the analysis applications and the knowledge and experience of users are directly usable.

Flood Protection System Performance

The subject of this study and the issues the NFIP is facing are centered around levees, because they are a major component of flood protection systems. However, 44 CFR §65.10 discusses more broadly the fact that flood protection systems that a community relies on are often a composite of many different elements. These include:

- structures of various types, including levees, floodwalls, gates, buildings, etc.;
- natural structures (levee foundations, old streambeds);
- interior drainage systems; pumping systems (pumps, pumphouse), drainage ditches and canals, power supplies, control systems, fuel supplies, etc.; and
- operational requirements for pump system operation, gate closure, etc.

As such, the flood protection that is provided to a community relies on the ability of a "system" to perform as intended during flood events. One of the significant lessons from Hurricane Katrina was the recognition and admission by the Corps of Engineers that the New Orleans Hurricane Protection System (HPS) was a system in name only (IPET, 2009). The post-event evaluations made it clear the HPS was not designed, operated, or maintained as a system and as such did not perform as one during Katrina (IPET, 2009).

Section 65.10 dictates the requirements for elements of the flood protection system, such as an operational plan for the levee system, the installation of gates or other closures, etc. The residual risk in protected areas is a function of the reliability of the flood protection system as a whole, not just of the individual system elements. The

[5] See http://www.fema.gov/national-flood-insurance-program-flood-hazard-mapping/numerical-models-meeting-minimum-requirement-0 and http://www.fema.gov/national-flood-insurance-program-flood-hazard-mapping/numerical-models-meeting-minimum-requirement-3.

assessment of flood risks requires a system-level analysis where the system includes the watershed (the hydrologic system), the hydraulic system (river channels, floodplain), and the flood protection system.

As part of a risk analysis, a systems modeling approach that considers the physical and operational elements of a flood protection system that are required to prevent flooding needs to be used. The system performance (system reliability) over a wide range of flood events is considered in order to estimate the risk in the floodplain, in the protected as well nonprotected, waterside areas of the system. The performance of the system of structures and other components that make up the system is integral to the hydraulic response – whether the protection system is breached or not.

Structure, Component, and Operational Reliability

A key element of the risk analysis is the assessment of the reliability of structures as well as other elements of the flood protection system to perform their intended function. This includes the integrity of the structures (levees, floodwalls, interface between structures, etc.) and operational reliability (actions of personnel) to close gates or other closures, operation drainage systems, etc. For purposes of the risk analysis, the reliability of structures is characterized in terms of the conditional probability of failure expressed as a function of an appropriate metric of the flood hazard (i.e., peak flood elevation) (USACE, 1996; URS/JBA, 2008; IPET, 2009; Schultz et al., 2010) (see Figure I-2). In addition to the assessment of fragility for structures in which the reliability depends on the loading level, there are other elements of the flood protection system whose reliability is independent of the level of flooding. Examples include the reliability of mechanical and electrical equipment (e.g., pump systems), actions of personnel (installation of closures, pump system operations, etc.).

A critical part of the reliability assessment is the collection of information about the various system components so that an understanding of system interactions, potential modes of failure (see Figure I-3 for a view of different levee failure modes) and performance can be developed. With respect to the assessment of fragility curves for structures (levees, floodwalls, etc.), various methods are available (URS/JBA, 2008; IPET, 2009; Schultz et al.,

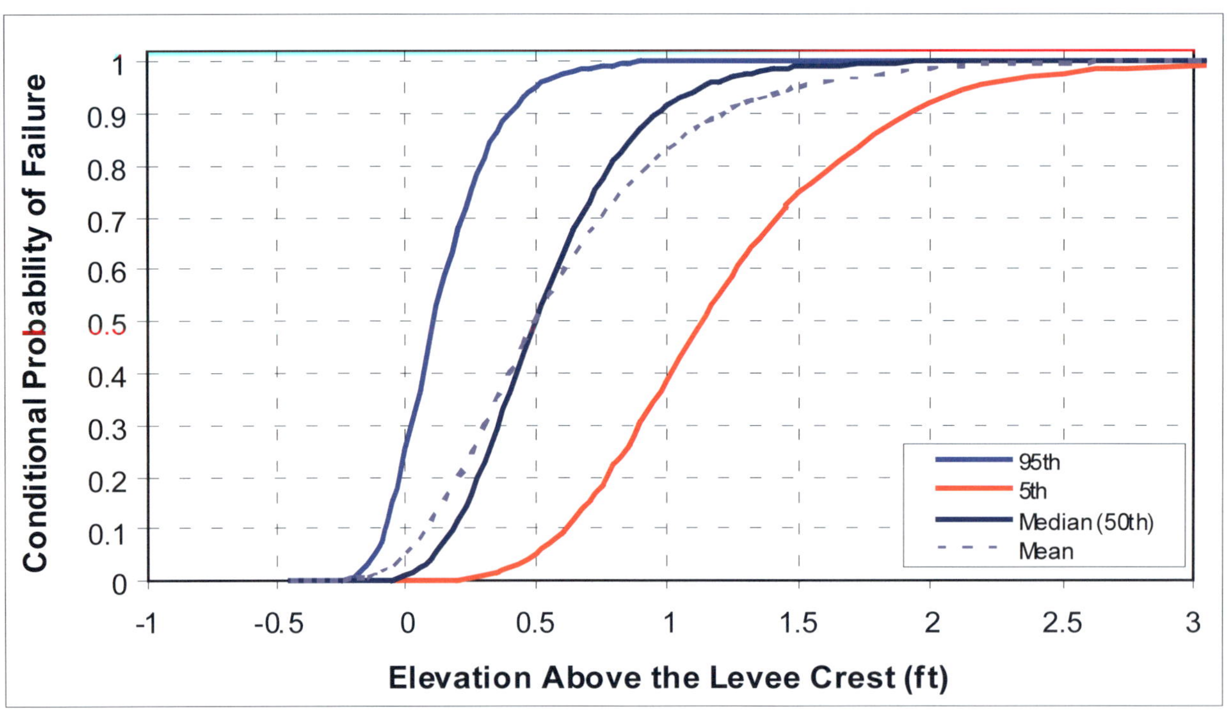

FIGURE I-2 Levee fragility curve.

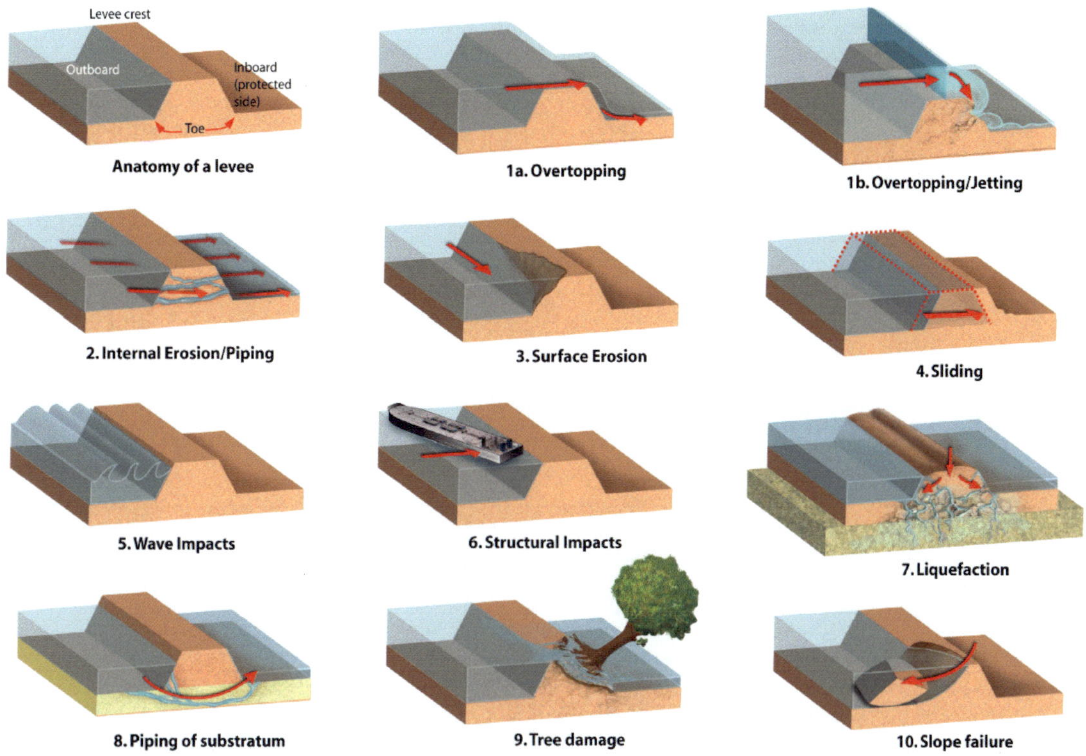

FIGURE I-3 Earthen levees are subject to a suite of potential failure modes. Not all failure modes are correlated to river stage. and this are excluded from consideration of levee fragility curves based on a function between river stage and system probability of failure. For example, dredging immediately adjacent to an earthen levee changes the capacity of that system to resist underseepage and piping (failure mode number 8). This activity, however, would not be captured and communicated to the system managers/operators through levee fragility curves defined by river stage.
SOURCE: Courtesy of Zina Deretsky, National Science Foundation.

2010). These range from judgmental approaches that rely on expert evaluations, often in cases where there is limited data, to empirical, and analytical methods (URS/JBA, 2008; IPET, 2009; Schultz et al., 2010). Empirical methods take advantage of observational data documenting the performance of structures under a variety of loads. Observations may be obtained systematically through controlled experiments or may be collected in an opportunistic fashion from flood events. Analytical methods that can be used to develop fragility curves are based on structural models that evaluate the likelihood that performance limits of the structure are exceeded. These methods have the advantage that they are based on physical models, or at least explicit physical relationships between capacity and demand. A limitation of analytical and empirical methods is that there are potential failure modes that may not be readily addressed by either approach. As a consequence, fragility assessments are often a hybrid of different methods, depending on the information that is available to perform the analysis.

There can be considerable epistemic uncertainty in the estimate of the fragility for structures (Figure I-4). A particularly vexing aspect of levees is the uncertainty associated with foundation conditions. The conditions along a levee can be complex and dynamic (Saucier, 1994). These changing subsurface conditions tend to promote preferential seepage beneath levees, often along buried or abandoned channels, as shown in Figure I-5. Underseepage tends to be "spotty," or inconsistent in some areas, and fairly predictable in other areas, depending on the underlying geology (Kolb, 1976). Underseepage can also be exacerbated by infiltration in riverside borrow pits, excavated to construct the adjacent levees (Mansur and Kaufman, 1957). The quantification of uncertainties and

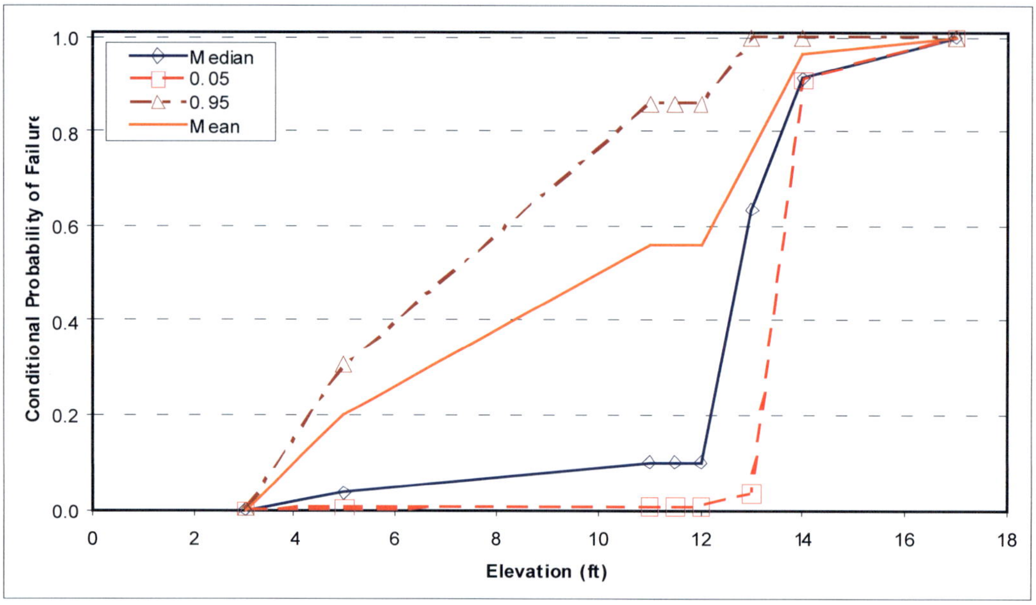

FIGURE I-4 Uncertainty in the estimate of levee fragility for a levee in New Orleans. Dashed lines correspond to the 0.05 and 0.95 uncertainty levels in the estimate of the conditional probability of failure.
SOURCE: IPET (2009).

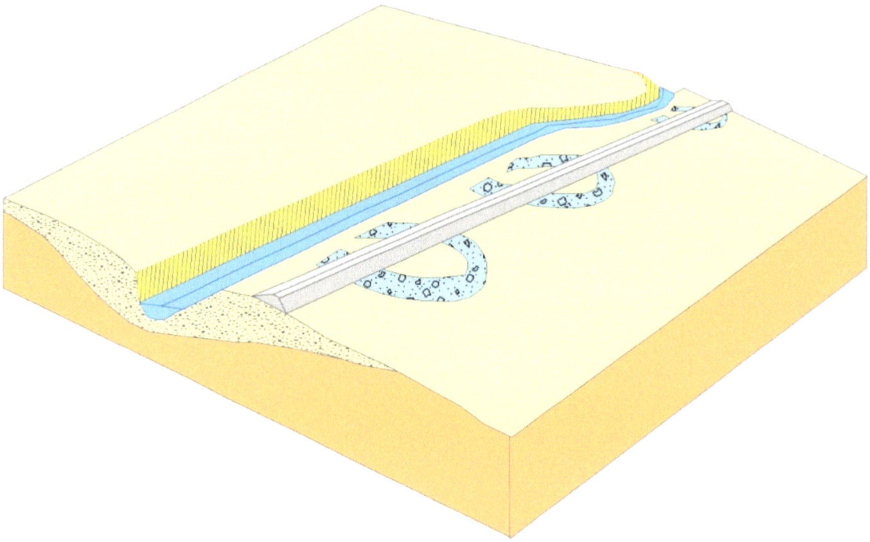

FIGURE I-5 Levees are commonly placed on floodplains adjacent to the present course of the river channels. These embankments are usually underlain by older channel deposits, such as the oxbows sketched above. These dynamic conditions often promote the paths or preferential seepage and hasten differential settlement of the levee embankment.
SOURCE: Rogers (2012).

their potential impact on levee performance will be one of the most challenging aspects of developing fragility curves over the coming decade.

Inundation Analysis Including Levee Breach and Inundation

If the flow rate in a leveed channel increases, the water surface elevation will increase. At a certain flow rate, the water-surface elevation will exceed the top of levee elevation, and water will flow onto the "protected" floodplain—the interior area on the landside of the levee. This floodplain flow may be simulated with a 1-D or 2-D model. Figure I-6 illustrates conceptually such an integrated model of channel and floodplain flow, as would be required for a complete risk analysis as proposed herein. A 1-D model represents the channel flow regime, and a water-surface elevation is computed for a specified exceedence probability. If that water-surface elevation is greater than the elevation of the top of the levee (which is represented in the figure with the heavy line), water spills from the channel. That process is simulated with an appropriate 1-D or 2-D model for the risk analysis proposed. In the illustration, a 2-D model is illustrated. This uses a fixed grid, with elevation and flow resistance parameters described for each cell of the grid. The equations of fluid motion are solved to account for the flow from each cell to all adjacent cells, thus estimating the depth of inundation for each cell. The 1-D and 2-D models are "coupled" at locations where water spills from the channel onto the floodplain and vice versa, as illustrated by the arrows in the figure.

A levee between the channel and floodplain limits the exchange of water between the channel and floodplain, with flow onto the floodplain occurring only when the channel capacity is exceeded if the levee performs as designed. However, the levee may breach prior to (or after) overtopping. In that case, water flows from the channel through the levee breach and onto the floodplain. Shown in Figure I-7, flow onto the protected floodplain is simulated, again, with an appropriate model. As before, model results include water surface elevations and velocity

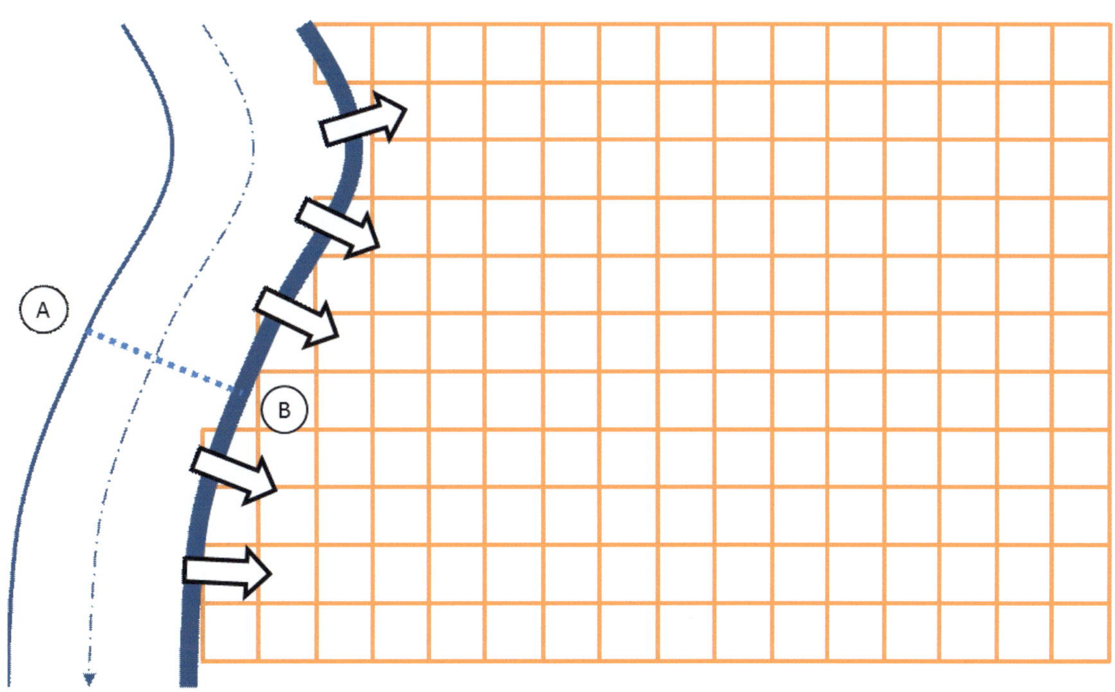

FIGURE I-6 A one-dimensional model commonly is used to represent channel flow (*Left*). The A-B transect illustrates a cross section of the channel in the 1-D model of flow that typically would be used to simulate behavior of the river. A two-dimensional model may be used to represent more complex flow patterns on the adjacent floodplain (*Right*).

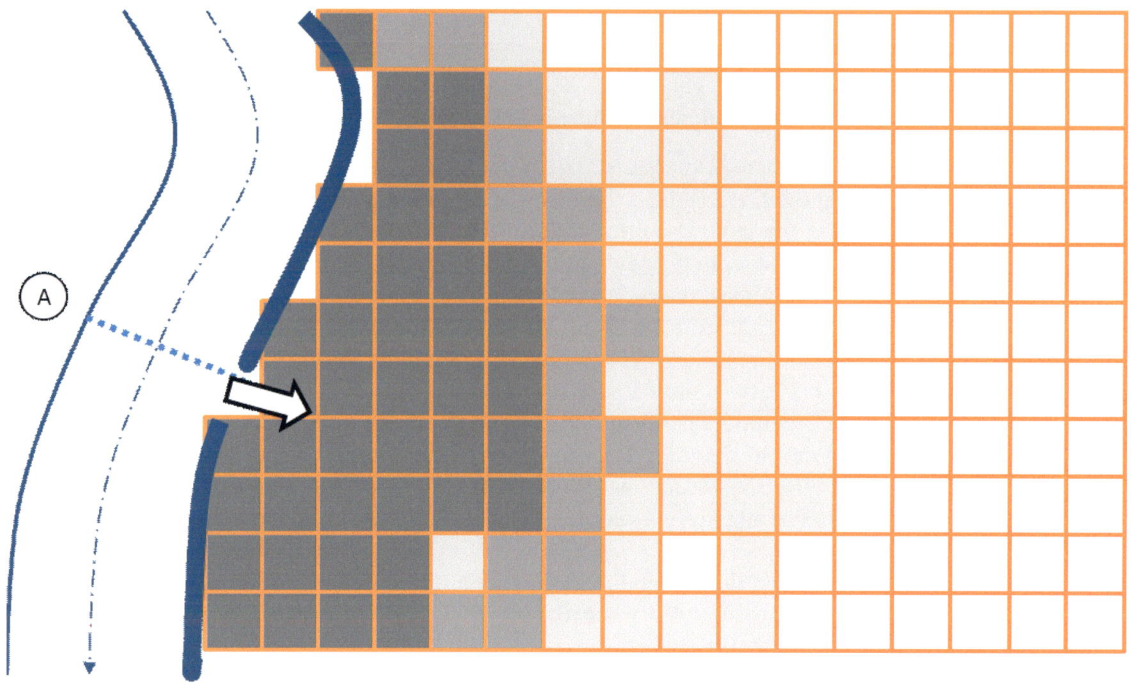

FIGURE I-7 A one-dimensional model of flow in the channel (*Left*) can be coupled with a two-dimensional model of flow on the floodplain (*Right*), with flow moving from the channel onto the floodplain through the levee at the breach. The heavy line represents the levee, and the arrow shows the location of the breach or overflow section through which water flows.

for each cell, in this case with water delivered to the floodplain from the channel through the levee. The floodplain elevations may be mapped, as illustrated, with shading to represent depth. Such analysis can be completed for any flow condition in the channel, including the one percent chance event and larger or smaller events.

For risk analysis, properties of the levee breach are described for modeling the breach. The properties are determined from empirical data, or application of geotechnical engineering analyses and standards of practice, as described earlier in this report.

Consequence Assessment

A modern risk analysis requires the estimation of consequence of inundation for the entire range of hydrologic events and the various states of and levee system performance. This consequence evaluation is commonly accomplished with a model that describes in a quantitative manner the consequences of flooding, given the consequences of the flooding (i.e., depth and velocity of flow), and properties of the exposed floodplain assets (i.e., construction type and material, foundation type and elevation, and content type and value).

A simple elevation-damage function, such as that illustrated by Figure I-8, is typical of models used to characterize the consequence of flooding of the exposed inventory in a floodplain. Such a function may represent economic loss for a single element of the inventory—such as a single residence and its contents—or the function may represent the aggregated loss for groups of items. The function may represent tangible (direct and/or indirect) loss and/or intangible loss. The tangible consequence of flooding may include a broad range of deleterious effects and associated costs, including (a) economic damage to buildings and contents; (b) damage to outbuildings, landscaping, vehicles, boats, and other inventory not within buildings; (c) damage to infrastructure, including transportation facilities, utilities, communications facilities, and so on; (d) damage to vegetation, including crops; (e) emergency response, management, and recovery costs, including flood-fighting costs; security, evacuation,

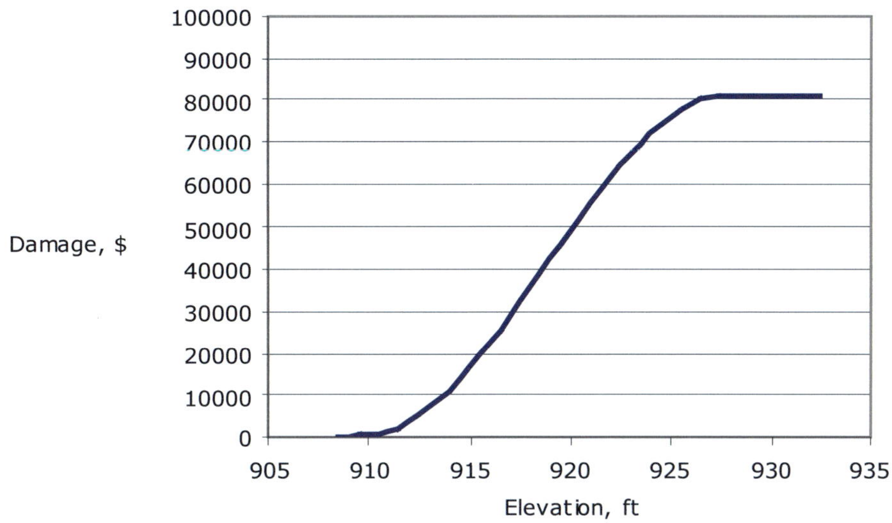

FIGURE I-8 Example of an elevation-damage model prediction of consequence (damage), given flood properties (water-surface elevation).

search and rescue costs, debris removal, and others; and (f) other economic impacts, such as displacement costs for temporary quarters, disruption time for affected people, economic impacts of loss of service of transportation and utility functions, loss of business income, loss of public services, and others.

Consequences considered in a flood risk analysis need not be restricted to economic loss. Thus, the elevation-damage function may be interpreted broadly to represent any undesired consequences that can be predicted from knowledge of flood properties.

FEMA has developed consequence models for various structure types, content of structures, infrastructure, vehicles, and other relevant categories. These models, derived from information gathered from insurance claims, are included in FEMA's Hazus software application. With similar analysis of historical inundation damage data, augmented with analysis of building methods, USACE has developed, published, and used generic elevation-damage models (USACE, 2000, 2003). For flood situations or structures not well represented by these readily available models, site-specific models are required. Those are developed from (a) empirical analysis of post-flood damage surveys; (b) detailed engineering analysis of cost of repair, considering various flood depths at a property and using cost estimates from, for example, Marshall & Swift's *Residential and Commercial Estimator*, *Means Square Foot Costs*, and the BNI *Home Builder's Costbook* (2013); or (c) expert elicitation, using professional judgment offered by knowledgeable economists and engineers.

Risk analyses may be completed independently for each property within the floodplain. This will provide the economic information needed for insurance rate setting for individual properties. In addition, or as an alternative, economic values can be aggregated by political boundary, by structure type, and so on as appropriate for decision making for mitigation or for risk communication.

Risk Quantification

The final step in the risk analysis is the probabilistic combination of the different parts of the risk analysis to estimate the frequency of flooding, performance of the flood protection systems, and flood damages. The quantification also includes the propagation of epistemic uncertainties in the flood hazard, levee system performance and economic damage assessment, to derive the uncertainty in the risk analysis results. With current analytical and computational capabilities the estimation of flood risks can be carried out at the individual property level, the community level, and for the NFIP portfolio. These capabilities make it possible for a modern NFIP risk analysis to provide comprehensive support to all aspects of the program: insurance, flood risk communication, floodplain management, etc.

REFERENCES

Abrahamson, N. A., P. G. Somerville, and C. A. Cornell. 1990. Uncertainty in numerical strong ground motion predictions. Pp 407-416 in Proceedings of the Fourth U.S. National Conference on Earthquake Engineering: May 20-24, 1990, Palm Springs, California, Vol. 1. Oakland, CA: Earthquake Engineering Research Institute.

AIR (American Institutes for Research). 2012. AIR Models Overview. Available online at http://www.air-worldwide.com/Models/Overview/. Accessed November 28, 2012.

ASME/ANS (American Society of Mechanical Engineers and American Nuclear Society). 2009. Standard for Level 1/Large Early Release Frequency Probabilistic Risk Assessment for Nuclear Power Plant Applications, Addenda to ASME/ANS RA-S-2009. LaGrange Park, IL: ANS.

CA DWR (California Department of Water Resources). 2012. 2012 Central Valley Flood Protection Plan: A Path for Improving Public Safety, Environmental Stewardship, and Long-Term Economic Stability. Available online at: http://www.cvfpb.ca.gov/CVFPP/index.cfm.

EPA (U.S. Environmental Protection Agency). n.d. Risk Analysis Guidelines and Tools, Available online at http://www.epa.gov/riskassessment/guidance.htm. Accessed December 10, 2012.

IPET (Interagency Performance Evaluation Task Force). 2009. Performance Evaluation of the New Orleans and Southeast Louisiana Hurricane Protection System: Final Report.. Available online at https://ipet.wes.army.mil/. Accessed December 4, 2012.

Kaplan, S., and B. J. Garrick. 1981. On the quantitative definition of risk. Risk Analysis 1(1): 11-27.

Kolb, C. R. 1976. Geologic control of sand boils along Mississippi River levees. Chapter 6 in Geomorphology and Engineering, D. R. Coates, ed. Stroudsburg, PA: Dowden, Hutchinson and Ross.

Mansur, C. I., and R. I. Kaufman. 1957. Underseepage, Mississippi River Levees, St. Louis District. ASCE Transactions 122: 985-1008.

NRC. 2007. Elevation Data for Floodplain Mapping. Washington, DC. The National Academies Press.

RMS (Risk Management Solutions). 2012. RMS Models. Available online at http://www.rms.com/models/. Accessed November 28, 2012.

Rogers, J. D. 2012. Evolution and Development of Flood Control Engineering. GE 301 Lectures. Available online at http://web.mst.edu/~rogersda/umrcourses/ge301/. Accessed November 15, 2012.

Saucier, R. T. 1994. Geomorphology and Quaternary Geological History of the Lower Mississippi Valley. Vicksburg, MS: USACE.

Schultz, M. T., B. P. Gouldby, J. D. Simm, and J. L. Wibowo. 2010. Beyond the Factor of Safety: Developing Fragility Curves to Characterize System Reliability. Washington, DC: USACE.

USACE (U.S. Army Corps of Engineers) 1991. HEC-2 Water Surface Profiles: User's Manual. Davis, CA: Hydrologic Engineering Center.

USACE. 1996. Risk-Based Analysis for Flood Damage Reduction Studies. EM 1110-2-1619. Washington, DC: USACE.

USACE. 2000. Generic Depth Damage Relationships: Economic Guidance Memorandum EGM 01-03. Washington, DC: USACE.

USACE. 2003. Generic Depth Damage Relationships for Residential Structures with Basements: Economic Guidance Memorandum EGM 04-01. Washington, DC: USACE.

USACE. 2006. Risk Analysis for Flood Damage Reduction Studies. ER 1105-2-101. Washington, DC: USACE.

USACE. 2008. Improving Public Safety—From Federal Protection to Shared Risk Reduction. Memo to Honorable Jim Nussle, Director, Office of Management and Budget. Washington, DC: USACE

USACE. 2012. Risk Assessment: Trinity River Corridor, Dallas Floodway Near Dallas, TX. Washington, DC: USACE.

USNRC (U.S. Nuclear Regulatory Commission). 2011. Practical Implementation Guidelines for SSHAC Level 3 and 4 Hazard Studies. NUREG-2117. Rockville, MD: USNRC.

URS/JBA (URS Corporation/Jack R. Benjamin & Associates, Inc.). 2008. Delta Risk Management Strategy (DRMS), Phase 1, Risk Analysis Report (Draft). Sacramento: California Department of Water Resources.

Yankee Atomic Electric Company. 1984. Probability of Extreme Rainfalls. YAEC-1405. Dallas, TX: Yankee Atomic Energy.